The central thesis of this book is that *Volvox* and its unicellular and colonial relatives provide a wholly unrivaled opportunity to explore the proximate and ultimate causes underlying the evolution, from unicellular ancestors, of multicellular organisms with fully differentiated cell types.

A major portion of the book is devoted to reviewing what is known about the genetic, cellular, and molecular bases of development in the most extensively studied species of *Volvox: V. Carteri*, which exhibits a complete division of labor between mortal somatic cells and immortal germ cells. However, this topic is put in context by first considering the ecological conditions and cytological preconditions that appear to have fostered the evolution of organisms of progressively increasing sizes and with progressively increasing tendencies to produce terminally differentiated somatic cells. The book concludes by addressing the question whether the germ–soma dichotomy evolved by similar or different genetic pathways in different species of *Volvox*.

Biologists interested in development, genetics, and cellular evolution will find this a fascinating work.

DEVELOPMENTAL AND CELL BIOLOGY SERIES
EDITORS
J. B. L. BARD   P. W. BARLOW   P. B. GREEN
D. L. KIRK

VOLVOX

Developmental and cell biology series

The aim of the series is to present relatively short critical accounts of areas of developmental and cell biology where sufficient information has accumulated to allow a considered distillation of the subject. The fine structure of cells, embryology, morphology, physiology, genetics, biochemistry, and biophysics are subjects within the scope of the series. The books are intended to interest and instruct advanced undergraduates and graduate students and to make an important contribution to teaching cell and developmental biology. At the same time, they should be of value to biologists who, while not working directly in the area of a particular volume's subject matter, wish to keep abreast of developments relevant to their particular interests.

VOLVOX

Molecular-Genetic Origins of Multicellularity and Cellular Differentiation

DAVID L. KIRK
Washington University, St. Louis

CAMBRIDGE
UNIVERSITY PRESS

CAMBRIDGE UNIVERSITY PRESS
Cambridge, New York, Melbourne, Madrid, Cape Town, Singapore, São Paulo

Cambridge University Press
The Edinburgh Building, Cambridge CB2 2RU, UK

Published in the United States of America by Cambridge University Press, New York

www.cambridge.org
Information on this title: www.cambridge.org/9780521452076

First published 1998
This digitally printed first paperback version 2005

A catalogue record for this publication is available from the British Library

Library of Congress Cataloguing in Publication data
Kirk, David L., 1934–
Volvox : molecular genetic origins of multicellularity and
cellular differentiation / David L. Kirk
p.   cm. – (Developmental and cell biology series)
Includes bibliographical references (p.   ) and index.
ISBN 0-521-45207-4
1. Volvox.   I. Title.   II. Series.
QK569.V9K57   1997
579.8′32 – dc21   96-40031
                                                      CIP

ISBN-13 978-0-521-45207-6 hardback
ISBN-10 0-521-45207-4 hardback

ISBN-13 978-0-521-01914-9 paperback
ISBN-10 0-521-01914-1 paperback

In some colony like Volvox *there once lay hidden the secret of the body and mind of man.*
Huxley 1912

Contents

Preface

In the late nineteenth century, "development" and "evolution" were considered to be so intimately related that many biologists used the terms interchangeably, applying them rather indiscriminately to both the process that generates a new individual resembling its parents and the process that generates a new species different from its ancestors. For most twentieth-century biologists, continuation of that practice would have been unthinkable, because evolution and development have seemed to us to be such fundamentally different phenomena. However, history has a way of repeating itself. After a century-long estrangement that began with widespread rejection of Haeckel's dogma that "ontogeny recapitulates phylogeny," evolution and development are now undergoing a dramatic rapprochement, with genetics acting as the broker for their remarriage. Increasingly, those investigating the mechanisms by which differentiated cells and organs arise in the course of embryonic development and those seeking to understand how morphological novelties arise in the course of evolution find themselves converging on the study of related sets of genes, and talking to one another again!

Contemporary studies of *Volvox*, the rolling green spheroid that first fascinated Antoni van Leeuwenhoek 300 years ago, illustrate this sort of convergence. When my wife, Marilyn, and I began to study *Volvox* more than 20 years ago, our objective was to capitalize on its simplicity to address a central problem of development that we had found it extremely difficult to address with any clarity by studying vertebrate embryos: How do cells with very different phenotypes arise from the progeny of a single cell? But the longer we have worked with *Volvox*, and implored it to reveal its secrets to us, the more difficult we have found it to think about its development in other than an evolutionary framework. This is because it has become increasingly evident that the genes that *Volvox* now employs to establish a germ–soma division of labor during development exert their effects by modifying particular aspects of a genetic program and a set of cytological specializations that were inherited from ancestors (resembling modern *Chlamydomonas* and various colonial flagellates) that had only a single cell type. Therefore, the central premise of this book is that *Volvox* and its relatives provide us a unique window through which to examine the important roles that a few regulatory genes can play in both development and evolution, and gain fresh insights into a problem that has fascinated both developmental and evolutionary biologists for many generations: How do multicellular organisms with differentiated cell types arise from unicellular antecedents?

An effort has been made to write this book at a level that will be accessible to prospective biologists at the advanced undergraduate level and yet will be capable of holding the interest of more mature practicing biologists with a diversity of research interests. I would be less than forthcoming, however, if I did not admit to being motivated by an evangelistic spirit as I wrote, hoping to capture for an organism that I have come to love the vocational interests of young (or perhaps not so young) biologists who may be looking for a new window through which to view the living world.

In a very real sense, this book had its roots in a piece of advice given to me by my first mentor and scientific father figure, Jerome Gross (a piece of advice that other biologists caught up in the frenetic competitiveness of our current age might do well to consider). On the day that I was saying my farewells, about to leave for graduate school after spending a major portion of my undergraduate years in his laboratory, Jerry said, "If you want a happy and productive career in science, find yourself a quiet backwater where there are lots of big fish to be caught, but not many people fishing, and go after them." It took many years to find such a niche and begin to follow Jerry's sage advice.

It was Richard Starr who eventually led us to this niche, by introducing Marilyn and me to *Volvox*. For that, and for the friendship, help, and encouragement that he has offered over all the intervening years – from the days he first hosted us in his laboratory and shared his cultures with us, up to and including his recent willingness to scrutinize a draft of this manuscript, point out various factual errors, and make many useful suggestions for improvement – I wish to express profound and lasting gratitude.

I also wish to thank two "life-history biologists" who introduced me to a completely new and different way of looking at *Volvox* and its relatives several years ago: Graham Bell and his then-student, Vassiliki Koufopanou.

I was initially astonished at how we could look at the same set of simple organisms, see such different things, use such different lexicons to discuss them, and ask such different questions about their evolutionary origins. While I was consumed with a desire to understand the proximate (i.e., genetic and cytological) causes of the *Volvox* germ–soma dichotomy, they were consumed with a desire to understand its ultimate (i.e., ecological and adaptive) causes. Over the years the gap has been narrowed (and the minds broadened, I trust) by centripetal movement on both sides. Chapters 3 and 4 of this book constitute my attempt to incorporate into this narrative many ideas gained from interactions with Graham and Vasso, from their papers, and from the additional literature to which they directed me. Some of the original shortcomings of these chapters have been mitigated by suggestions they (particularly Vasso) made after reading a draft version. Thus, to the extent that I have succeeded in my effort to integrate into this book a consideration of the ultimate causes underlying the *Volvox* germ–soma dichotomy, credit goes to Vasso and Graham. But to the extent that I have failed, the fault is mine alone.

At the other end of the spectrum, the molecular end, an enormous debt of gratitude is owed to our good friend and close collaborator Rüdiger Schmitt and his colleagues. When Rudy and I met at a conference about 10 years ago and realized the extent to which we were attacking related aspects of *Volvox* biology from different perspectives, we drew up detailed plans for collaborative studies that have dominated the research agendas of both laboratories ever since. Little did we appreciate at the outset what a long-term program that would turn out to be, or how many obstacles would have to be overcome along the way. But as a result of this collaboration, we now have molecular-genetic methods for the study of *Volvox* that clearly would not yet exist if our two research groups had not worked together so closely over the past decade. Each major step along the way has been the product of a synergistic research relationship and has been enlivened and made much more enjoyable by frequent exchange visits between St. Louis and Regensburg by students, postdoctoral fellows, and technical assistants, as well as by Rudy and myself.

Above all, however, my gratitude extends to my wife and closest collaborator, Marilyn, without whom it is doubtful that *Volvox* research would still be alive and well in St. Louis, or in the larger community for that matter. For more than two decades she played the critical role of keeping our laboratory running efficiently during all the dreary hours when I was tied up with interminable committee meetings and other academic responsibilities. More importantly, she overcame the innate distaste of a classic biochemist for molecular biology – where one seldom can see a product, let alone weigh it –

and was responsible for most of the critical breakthroughs required to bring *Volvox* research into the modern era. From the weekend on which (while cruising the northern Mississippi on an ancient stern-wheeler) we decided to visit Richard Starr and "go *Volvox*," to the present day (when she ignores her recent retirement and finds her way back to the bench whenever there is an important experiment that could benefit from her attention), it has been a productive relationship. I only hope that it has been half as rewarding and enjoyable for her as it has been for me. Her painstaking and critical reading of every word of this text, in each of several versions, has engendered both my sympathy and gratitude and has deepened my affection.

Finally, I wish to thank the National Institutes of Health, which supported our *Volvox* research over many years, the National Science Foundation, which has done so in recent years, and NATO, which provided travel grants to Rudy Schmitt and myself to facilitate our transatlantic collaboration.

Prologue

I had got the foresaid water taken out of the ditches and runnels on the 30th of August: and on coming home, while I was busy looking at the multifarious very little animalcules a-swimming in this water, I saw floating in it, and seeming to move of themselves, a great many green round particles, of the bigness of sand grains.

When I brought these little bodies before the microscope, I saw that they were not simply round, but that their outermost membrane was everywhere beset with many little projecting particles . . . all orderly arranged and at equal distances from one another; so that upon so small a body there did stand a full two thousand of the said projecting particles.

This was for me a pleasant sight, because the little bodies aforesaid, how oftsoever I looked upon them, never lay still; and because their progression was brought about by a rolling motion. . . .

Each of these little bodies had enclosed within it 5, 6, 7, nay, some even 12, very little round globules, in structure like to the body itself wherein they were contained.

While I was keeping watch, for a good time, on one of the biggest round bodies . . . I noticed that in its outermost part an opening appeared, out of which one of the inclosed round globules, having a fine green colour, dropt out, and took on the same motion in the water as the body out of which it came. . . . soon after a second globule, and presently a third, dropt out of it; and so one after another till they were all out, and each took on its proper motion. . . .

What also seemed strange to me, was that I could never remark, in all the motions that I had observed in the first round body, that the contained particles shifted their positions; since they never came into contact, but remained lying separate from one another, and orderly arranged withal.

Many people, seeing these bodies a-moving in the water, might well swear that they were little living animals; and more expecially when you saw them going round first one way and then t'other. . . .

1

> Without a break I continued from day to day to watch these little round particles that had issued from the bigger ones; and I always saw that they not only waxed in bigness, but that the particles contained within them got bigger too. . . .
>
> Now as we see that the oft-mentioned round bodies come into being not spontaneously, but by generation, as we know all plants and seeds do (inasmuch as every seed, be it never so small, is as it were endowed with its enclosed plant); so can we now be more assured than we ever before were heretofore concerning the generation of all things. For my part, I am fully persuaded that the little round bodies, which are found in the bigger ones, serve as seeds; and that without them these big round bodies couldn't be produced. . . .
>
> van Leeuwenhoek (1700), quoted by Dobell (1932)

Thus, nearly three centuries ago, did Antoni van Leeuwenhoek first describe the little round, green "animalcules" that Linnaeus (1758) would later name "*Volvox*" (the "fierce roller"). Two aspects of this record of the first human encounter with *Volvox* are noteworthy as harbingers of many such interactions to come.

Although van Leeuwenhoek was approaching his 70th birthday when he first encountered *Volvox*, had already discovered a wider range of bizarre and beautiful life forms than any person before or since, and could scarcely have been faulted if he had grown somewhat jaded, he felt compelled to interrupt his scholarly description to note the pleasure he experienced as he first viewed the restless, rolling progression of the globes beneath his lens. That excitement has been shared by countless thousands of youthful students – and not a few worldly, aging scientists – over the intervening centuries as each has first observed *Volvox* in its stately, ceaseless, forward-rolling motion.[1] It is a sensation that can be as enthralling – and as ineffable – as the first taste of a choice exotic fruit.

The second recurrent theme emerging from that first human encounter with *Volvox* was the use of the organism – particularly its visibly differentiated reproductive cells – as a model system for addressing one of the central issues of biology. Few cells or organisms (if any) have been cited as often in the literature over the past three centuries as potentially useful models for ad-

---

[1] Witness the following account, penned nearly three centuries after van Leeuwenhoek's, but carrying very much the same flavor: ". . . a visit was made to the lake to pond hunt. . . . Samples were taken . . . [and] there were *surprise, surprise* quite a number of Volvox to be seen. Now, Volvox is . . . regarded by pond hunters as a prize. It is a beautiful minute green sphere which just rolls along in the water and inside each sphere can be seen smaller green spheres. Under . . . the microscope they are a pleasure to observe. The spheres are composed of hundreds of single cells each with two flagella and the action of these flagella beating in unison enables the sphere to majestically roll along in the water" (McInness 1994).

dressing one important biological problem or another. Not content to limit his interaction with *Volvox* to his initial pleasurable sensation, van Leeuwenhoek watched the organism "without a break" for "day after day," noting that even as the smaller globes grew into replicas of the larger globes within which they were contained – and eventually emerged to roll about on their own – they contained yet smaller globes that were already reiterating the generative cycle. He viewed this repetitive cycle of growth and reproduction as a powerful paradigm, from which one could "now be more assured than we ever before were . . . concerning the generation of all things," namely, that living beings do not arise spontaneously from nonliving matter, but only from other living beings of like type.

Ironically, however, such pronouncements did not put the question of spontaneous generation to rest. Quite to the contrary, van Leeuwenhoek's discovery of a teeming world of microbes and worms thriving on lifeless substrates reinvigorated it. So about 70 years later, Lazzaro Spallanzani, who was in the forefront of a renewed effort to nail the lid on the coffin of spontaneous generation, and who apparently was unfamiliar with van Leeuwenhoek's writings in this regard, would reconstruct in considerable detail the argument that *Volvox* reproduction constituted a visible refutation of the concept (Spallanzani 1769).

At nearly the same time, Charles Bonnet (1762) used the reproductive cells of *Volvox* as a paradigm to anathematize the twin concepts of epigenesis and spontaneous generation, but with a twist slightly different from that of Spallanzani. For Bonnet, the asexual reproductive cycle of *Volvox* that van Leeuwenhoek first described was a prime exemplar of the principle of *l'emboîtement*: the concept that future generations of an organism are in a sense "preformed" within the germ cells of the present generation – not so much like a set of Russian dolls or Chinese boxes of increasingly smaller sizes, as in the common textbook caricature of his views, but as a set of self-perpetuating regulatory principles (Farley 1974; Gould 1977).

In the first half of the nineteenth century, it was the somatic cells of *Volvox*, rather than the germ cells, that were enlisted to do battle against the concept of spontaneous generation. Blessed with one of the best microscopes and one of the best imaginations of his age, Christian Ehrenberg described, in richly embroidered detail, countless "infusoria," including two species of *Volvox* (Ehrenberg 1831, 1832a,b, 1838). Within each *Volvox* somatic cell, as in so many other infusorians that he examined, Ehrenberg claimed to see an integrated set of miniature organs analogous to those of well-known higher animals: a mouth, a digestive tract, a pair of testes, a sperm bladder, and a collection of eggs, as well as a bright red eye (the one organelle to which he did assign the correct function!). In Ehrenberg's view, *Volvox* was not merely a tiny animalcule, but a collection of cooperating, sentient individuals: a "social animal." Forty years later, Ferdinand Cohn would characterize Ehrenberg's views of *Volvox* (and his own reactions to them) thus:

... in the aggregation of so many individuals he sees the effect of a social drive composed of power and respect for common goals, a state which would require the activity of a mind, which we are not entitled to reject (although we are certainly tempted to). One should also not forget that all the individuals have sensory organs comparable to eyes, and that they are therefore not blindly rolling in the water, but that they share the pleasure of sensitive existence as citizens of a remote world far from the reach of our judgment – a fact we have to admit, no matter how proud we might be of ourselves [Cohn 1875; translation by Heribert Gruber, personal communication].

That hitherto unimagined structural and functional complexity, Ehrenberg argued, made the spontaneous generation of such beings extremely unlikely (Ehrenberg 1832b).

Contrary to widespread belief, in the latter half of the nineteenth century the concept of spontaneous generation was still far from dead, despite Pasteur's widely heralded efforts to inter it (Farley 1974). Indeed, the concept that spontaneous generation had at least occurred in the past was an essential underpinning of Haeckel's biogenetic law, which became *the* dominant force to be reckoned with in the biology of the late 1800s (Farley 1974; Gould 1977). In that context, *Volvox* was cast in a new but no less fantastic role. Many of the zealous recapitulationists following in Haeckel's train would propose *Volvox* as the "blastea" from which his infamous "gastrea," the postulated ancestor of all metazoans (Haeckel 1874), might have been derived. Going a step further, many of these individuals, after noting the combination of animal-like and plant-like qualities that it possessed (e.g., its motility and photoautotrophy), would propose *Volvox* as the "missing link" between the plant and animal kingdoms. A generation later, however, such "Haeckelian philosophers" would be castigated for burdening "our delicate *Volvox*" with such a "ponderous weight of phylogenetic fancy" (Powers 1908).

One intellectual giant of Haeckel's age, August Weismann, rejected the concept that *Volvox* was the primal metazoan, but he nonetheless first articulated the concept that *Volvox*, together with its simpler relatives, might provide a useful model from which we might gain some insights about how metazoans with a germ–soma dichotomy might have arisen from organisms lacking differentiated cell types. In his classic treatise on the continuity of the germ plasm, he wrote as follows:

If we desire to investigate the relationship between the germ-cells and somatic cells we must not only consider the highly developed and strongly differentiated multicellular organisms, but we must also turn our attention to those simpler forms in which

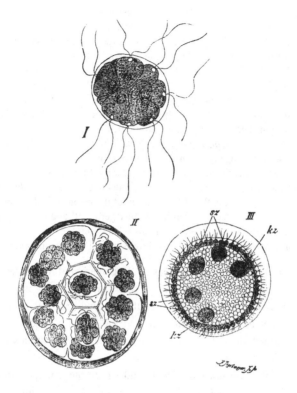

August Weismann's drawings of *Pandorina* and *Volvox* (from Weismann 1892b).
His caption read: "I. *Pandorina morum*—A colony of swarming cells.
II. A colony which has given rise to daughter colonies:—all the cells are similar
to one another. (After Pringsheim.) III. A young individual of *Volvox minor*,
still enclosed within the parent (after Stein): the cells are differentiated into
somatic- and germ-cells."

phyletic transitions are represented. In addition to solitary unicellular organisms, we
know of others living in colonies of which the constituent cells are morphologically
and physiologically identical. Each unit feeds, moves, and . . . is capable of reproducing . . . a new colony by repeated division. The genus *Pandorina* (Figure I), belonging
to the natural order Volvocineae, represents such . . . organisms . . . [and] each cell in
the colony acts as a reproductive cell; in fact it behaves exactly like a unicellular
organism. It is very interesting to find in another genus belonging to the same natural
order, that the transition from the homoplastid to the heteroplastid[2] condition, and the

---

[2] In Weismann's time, the terms "homoplastid" and "heteroplastid" provided an economical
way of making the distinction between organisms composed of multiple cells of like type and
organisms containing two or more differentiated cell types that executed different functions.
Subsequently, however, it appears that these terms first fell into disuse and then were appro-
priated by botanists to mean something entirely different, namely, "containing one *versus*

separation into somatic and reproductive cells, have taken place. In *Volvox* (Figure III) the spherical colony consists of two kinds of cells, viz. of very numerous small ciliated cells and of a much smaller number of large germ-cells without cilia. The latter alone possess the power of producing a new colony.... The germ cells of *Volvox* are differentiated during embryonic development, that is before the escape of the young heteroplastid organism from the egg capsule. We cannot therefore imagine that the phyletic development of the first heteroplastid organism took place in a manner different from that which I have previously advocated on theoretical grounds before this striking instance occurred to me. The germ–plasm . . . of some homoplastid organism (similar to *Pandorina*) must have become modified in molecular structure during the course of phylogeny, so that the colony of cells produced by division was no longer made up of identical units, but of two different kinds. After this separation, the germ-cells alone retained the power of reproduction.... Thus *Volvox* seems to afford distinct evidence that in the phyletic origin of the heteroplastid groups, somatic cells were not . . . intercalated between the mother germ-cell and the daughter germ-cells in each ontogeny.... Thus the continuity of the germ–plasm is established at . . . the beginning of the phyletic series of development [Weismann 1892a].

Many of the most provocative ideas of Weismann and his contemporaries failed to stand the tests of time and further study. Among those was the concept that *Volvox* represented the primal metazoan and/or the missing link between the plant and animal kingdoms, as well as the concept that germ cells alone retained the full complement of nuclear determinants that characterize a species.

Nonetheless, the concept that *Volvox* and its simpler relatives might provide a uniquely useful model for exploring both the ontogeny and phylogeny of a very simple form of multicellularity and cellular differentiation has endured, and it is this concept that animates the present work.

more than one kind of plastid." As a result of that etymological accident, in most contemporary discussions of the origins of cellular differentiation, much more verbose (and often less precise!) expressions are usually used to distinguish multicellular organisms containing one cell type from those possessing two or more cell types. In Chapter 1, I will propose a variant of Weismann's terms that I hope will recapture his verbal economy and precision, while avoiding the ambiguity that would result from an attempt to resurrect his terms in unmodified form.

1

Introduction

... we want to understand the mechanics of ... development. Today this is one
of the central problems in biology. In primitive colonial organisms we see not
only the origin of multicellularity, but also the origin of development.

Bonner (1993)

In all probability the first objects on this planet worthy of the term "living"
were single cells. At present, after nearly 4 billion years of intense competition among millions of kinds of life forms, the vast majority of individuals
inhabiting this planet are still unicellular. This is proof enough of the survival
value of a unicellular body plan. Nevertheless, all of the conspicuous organisms on the land, in the waters, and in the air above our planet are multicellular organisms with an impressive diversity of differentiated cell types
that share the labors of resource accumulation and reproduction in a highly
integrated and effective manner.

These conspicuous and highly successful multicellular organisms, with
their differentiated cell types and division of labor, have provided biologists
some of their most intriguing and persistent puzzles: From what sorts of
unicellular ancestors – and why, and when, and how – did multicellular
organisms with differentiated cells arise? Such questions tend to fall into two
categories: What were the *ultimate* (e.g., ecological) causes that *fostered* the
evolution of multicellular organisms? What were the *proximate* (e.g., genetic
and cytological) causes that *permitted* their evolution?

Shunning the intellectual division of labor that has too often characterized
biology in this century, in this book we will attempt to examine both the
ultimate causes and the proximate causes of the ontogeny and phylogeny of

one elegantly simple and beautiful example of multicellularity and cellular differentiation: *Volvox*, "the fierce roller."

1.1 Multicellularity and cellular differentiation have evolved many times

The generation of multicellular organisms from unicellular ancestors has been a recurrent aspect of evolution. It is now widely accepted, on the basis of structural, biochemical, and molecular evidence, that the five major types of large, complex, multicellular organisms (i.e., red algae, brown algae, land plants, fungi, and animals) arose separately from different types of unicellular ancestors (Devereux, Loeblich & Fox 1990; Sogin 1991; Wainwright et al. 1993). But clearly those were neither the first nor the last groups to develop something other than a unicellular existence. When we consider all of the examples of communal living that now are observed among discrete lineages of microorganisms (a small sample of which will be discussed here), it becomes clear that such transitions must have occurred numerous times in the history of this planet.

In one form, such communal living results in what I propose to call a "homocytic"[1] organism: a string, sheet, or more complex assembly of prokaryotic or eukaryotic cells, all of which are structurally and functionally equivalent, but which nonetheless often are able to construct colonies of astonishing beauty and regularity (Shapiro 1988). In the cases of greater interest here, however, cells within such an assemblage become differentiated from one another in structure and function, divide the communal labors of resource acquisition and reproduction, and thereby establish a "heterocytic" condition.

The multicellular condition is far from being a modern invention. Fossils embedded in the oldest rocks bearing unmistakable evidence of life – certain cherts of the Swaziland region of southern Africa and the outback country of Western Australia – indicate clearly that by about 3.5 billion years ago a diverse array of filamentous prokaryotes, very likely including many kinds of oxygen-producing cyanobacteria not greatly different morphologically

---

[1] As noted in the Prologue, a century ago those organisms containing multiple cells of a single type were economically distinguished from organisms containing cells of two or more different types and functions by use of the adjectives "homoplastid" and "heteroplastid," respectively, but those adjectives have now taken on entirely different meanings, and the language has been left without a simple way to express this important distinction. In an attempt to relieve this deficiency, and thereby avoid the need to employ verbose circumlocutions repeatedly in this book, I propose to replace the terms "homoplastid" and "heteroplastid" (in their original meanings) with "homocytic" and "heterocytic," respectively.

from some species alive today, had already developed the colonial, homocytic lifestyle (Schopf 1992, 1993). Other fossil deposits dating to about 2 billion years ago indicate that by that time some filamentous cyanobacteria had already gone a step further, developing two morphologically distinct cell types closely resembling the photosynthetic cells and nitrogen-fixing cells of contemporary cyanobacteria, such as *Anabaena*, thereby becoming the first heterocytic organisms to be preserved in the fossil record (Schopf 1974). The extraordinary success of those very early experiments in group living is attested by the astonishing morphological similarities between many species that are still abundant in parts of the modern world and species that were buried in the rocks 2 billion years ago or more (Schopf 1992).

Although homocytic colonial organisms precede heterocytic ones in the fossil record, several modern prokaryotes illustrate that a colonial lifestyle is not a necessary precursor to the heterocytic state. For example, certain stalked bacteria, of which *Caulobacter* has been the most thoroughly studied, live as single cells, but regularly divide to produce two cells of different types: a stalk cell that remains attached to the substrate to continue the cycle, and a swarmer cell that swims away to find another location where it can settle down, transform into a stalk cell, and establish the same sort of cycle in a new area (Brun, Marczynski & Shapiro 1994; Wingrove & Gober 1995). Several genera of unicellular bacteria display a form of cellular differentiation that we can, with no stretch of the imagination, think of as an exquisitely simple form of a germ–soma division of labor, analogous to that which characterizes so many higher animals. A particularly well known example is *Bacillus subtilis,* in which starvation causes each cell to divide asymmetrically and then produce a resistant spore cell (the ''germ''), which will carry the genome forward, and a spore mother cell (the ''soma''), which undergoes irreversible genetic rearrangement and eventual death, in the process of assisting in the development of the spore (Errington 1993; Levin & Losick 1995).

Another well-known example of prokaryotic cellular differentiation is provided by the various species of myxobacteria that, when resources dwindle, swarm together to form a mound in which some cells (the ''somata'') perish when trapped in the acellular stalk that raises the remaining cells (the ''germs'') a short distance into the air, where they form a ''fruiting body,'' encyst, and await the physical or biological forces that will disperse them to more favorable environments (Brock, Smith & Madigan 1984; Rosenburg 1984; Reichenbach & Dworkin 1992).

Astonishingly parallel forms of differentiation, leading to the production of a stalk that supports a fruiting body full of dormant spores awaiting dispersal, have arisen at least six separate times among the eukaryotes: in the acellular or syncytial slime molds (the myxomycetes) (Alexopoulos 1963; Frederick 1990), in two discrete groups of cellular slime molds (the acrasids and the dictyostelids) (Bonner 1959), in a ciliate (Olive 1978), in a labyrinthulan-like protist of uncertain affinity (Dykstra & Olive 1975), and, of course, in the fungi (Alexopoulos & Mims 1979; Ross 1979).

Variations on this theme of communal living accompanied by differentiation of specialized reproductive cells are found in a number of other protist groups. For example, among the several genera of peritrichous ciliates that have developed a sedentary, colonial lifestyle there is at least one, *Zoothamnion*, that has established a unique form of germ–soma differentiation. In *Zoothamnion,* branched colonies comprising many feeding cells (or ''microzooids'') are produced, but it is only the cells located at branch points of the colony that are able to differentiate as ''macrozooids,'' detach, and swim to a new location, where they settle down, reproduce asexually, and establish a new colony (Summers 1938; Grell 1973).

This list of homocytic and heterocytic organisms that appear to have arisen independently from unicellular ancestors is far from exhaustive. Other examples could be drawn from the diatoms, the radiolaria, the foraminifera, and so on. But it appears that the record for this sort of accomplishment belongs to the diverse group of protists known collectively as the flagellates. It now appears probable that over the course of the ages the flagellates have given rise to many more than a dozen independent multicellular lineages (Buss 1987); these include, in addition to the *Volvox* lineage that will be discussed at length in this book, the fungi, the land plants, and the higher animals.

In short, successful experiments in communal living by cells of like type have occurred countless times in the long history of life on earth, and in a considerable number of cases that has eventually been accompanied by differentiation of cells within the commune to perform different tasks, quite frequently with the division of labor being between germ and soma (Buss 1987; Denis & Lacroix 1993).

1.2   The evolutionary origins of most forms of multicellularity are obscure

The genesis of the major groups of complex modern organisms remains largely (to pilfer a phrase from Winston Churchill) a riddle wrapped in a mystery inside an enigma. According to a widespread contemporary view, the history of life is best read as a series of ''punctuated equilibria'' in which long periods of relative evolutionary stasis were separated by periods of sudden and rapid change that moved the biosphere to new equilibrium levels (Gould & Eldredge 1977). If so, then surely one of the most dramatic and fascinating punctuation marks in this historical account was the explosive diversification of animals that was recorded with exceptional clarity in rocks 544–530 million years old (Bowring et al. 1993; Butterfield 1994) and that serves as a giant exclamation mark to terminate a languid chapter telling of a world dominated for billions of years by rather simple, slowly changing life forms. Some vague traces in older rocks, some well-preserved, often bizarre Ediacaran fossils of uncertain affinities, and various types of indirect evidence indicate that the revolution that took place at the Precambrian–Cambrian boundary may have had its beginnings as much as 400 million years earlier (Runnegar 1992; Conway Morris 1993, 1994). Nevertheless, it now seems unlikely that anything approaching a detailed record of the origins of Precambrian protometazoans and the important transitions occurring among them will ever be reconstructed from the traces left in the rocks.

Fragmentary and disconnected as they are, however, the pictographs that primitive metazoans left in the rocks to record their passage constitute a rich source of information when compared with the meager traces left by the ancestors of the other major eukaryotic groups. On the basis of the simplest of ecological assumptions, we must assume that in order to support the diversity and abundance of heterotrophic animals that it did, the late Precambrian and early Cambrian community must also have had a considerable abundance and diversity of autotrophs; however, apart from the prokaryotes referred to earlier, such autotrophs have left precious few clues to their nature.

The molecular clock derived from ribosomal RNA (rRNA) sequences (which, however imperfect it may be, appears to be the best chronometer we now have for estimating the time course of early eukaryotic evolution) indicates that although the red and brown algae had set off on their separate ways somewhat earlier, the ancestors of the three major kingdoms of multicellular organisms (plants, fungi, and animals) diverged from a pool of related eukaryotic ancestors at very nearly the same time, deep in the geological gloom of the Precambrian, more than a billion years ago (Sogin 1991). If there were sufficient residues of the community of life preserved in rocks formed 1–1.5 billion years ago, undoubtedly we would be able to reconstruct a fascinating panorama of relatively rapid eukaryotic experimentation and diversification. But as it is (undoubtedly in large part because of the soft, poorly preservable nature of most organisms extant at that time), we are left largely to our own imaginations as we read a record of several hundred million years in which the primary fossilized eukaryotes were "acritarchs" (literally, "undecided," or "confused" forms) of uncertain affinities (Taylor 1978). Certain Precambrian fossils suggest to some investigators that both multicellular red algae (which the rRNA clock indicates were the first of the five major groups of modern multicellular eukaryotes to have diverged from the main line) and unicellular green algae were present by about 1.25 billion years ago (Cloud et al. 1969; Butterfield, Knoll & Swett 1990). A few fossils in Cambrian rocks clearly indicate that by about 540 million years ago these two groups – red and green algae – had diversified sufficiently to produce individuals of substantial size, but the record of how such diversification occurred is essentially nonexistent. Although molecular evidence indicates that the fungi originated at about the same time as the most primitive animals and are more closely related to animals than to any of the other major modern groups (Wainwright et al. 1993), they left precious few traces during the first two-thirds of their long history. The oldest fossils easily related to modern fungi are first seen in association with roots of primitive vascular plants, in Silurian fossils that date to about 400 million years ago, and the fossil and molecular data bearing on the origins of the "true" fungi are in much less than perfect accord (Berbee & Taylor 1993). The brown algae have left a record that is even more miserly. Moreover, the origins of most multicellular protists, such as the dictyostelid slime molds, appear to be buried even deeper in prehistory than the origins of the major groups just discussed (Sogin 1991; Wainwright et al. 1993) and to be even more inscrutable.

Even if the fossil records of one or more of these groups were close to being complete, however, such records would not provide answers to all the questions we might wish to ask about the origins of their particular forms of multicellularity. A more extensive fossil record might permit us to speculate with greater confidence about the sequence in which related forms with different levels of morphological complexity appeared and became abundant, but it would not permit us to examine experimentally what sorts of ecological conditions fostered such a progression, let alone examine the detailed changes within the genome that permitted it.

As mentioned earlier, modern studies (e.g., Devereux et al. 1990; Wainwright et al. 1993) have reinforced suggestions repeatedly made in the past (e.g., Lankester 1877; Bütschli 1883; Pascher 1918; Hyman 1942) that fungi, plants, and animals were all derived from flagellated protists. But even if we could identify the modern flagellate most similar to the immediate predecessor of one of these groups, and the most primitive member of that particular group, it seems unlikely that either would still retain within its genome – after a billion years of genetic drift – a legible record of the genetic pathways that led to that version of multicellularity and cytodifferentiation.

In short, the great antiquity of the major groups of modern multicellular eukaryotes makes it seem extremely unlikely that at any time in the foreseeable future will we be able to develop a detailed understanding of the origins of multicellularity in any one of them that will go significantly beyond the theoretical musings of the past.

But might there be an alternative group of organisms, in which the pathway to multicellularity was traversed much more recently, that might serve as a model system from which we could obtain detailed, meaningful insights into both the proximate and ultimate causes that account for the evolution of at least one form of multicellularity?

1.3   Desirable features of a model system for studying the origins of multicellularity

If, in a flight of fantasy, we were to imagine a group of organisms that could be used as a model to explore the ecological, cytological, and molecular-genetic foundations for the evolution of a group of heterocytic multicellular organisms from simple unicellular ancestors, what qualities might we envision that such a group should possess? Clearly, any sufficiently interested individual could probably append several subtle niceties to any such wish list assembled by another. Nonetheless, most who were interested enough to give this question a second thought might well agree that, at least as a starting point, it would be quite satisfying if one could identify a group that (1) comprises an extant collection of closely related organisms that range in complexity from unicellular forms through homocytic colonial forms to heterocytic multicellular forms, with different cell types and a complete division of labor, (2) includes predominantly organisms that can readily be obtained from nature, maintained over long periods in the laboratory, and studied in detail under a wide range of controlled environmental conditions, (3) has already been studied in sufficient detail by cytologists, biochemists, developmental biologists, geneticists, and/or molecular biologists that some generalizations can be made concerning both the range of ultrastructural, developmental, molecular, and genetic features that unify the group and the variants of such features that distinguish members at one level of complexity from those at another, (4) has also been sufficiently well studied with respect to distribution, natural history, and response to environmental variables that generalizations can be made about features of the natural world that appear to influence the relative abundance of members that differ from one another in their levels of organizational complexity, (5) is of such recent origin (in a geological frame of reference) that there is some hope that its various members may still retain within their genomes – unblurred by long eons of genetic drift – traces of the genetic changes that permitted transitions from one level of organizational complexity to the next, (6) bears significant evidence that within this single group the transition from the unicellular to the homocytic condition and from the homocytic to the heterocytic condition may have occurred more than once, thus raising the hope that the genetic changes required to cross one or another of those boundaries may have been so modest that there is some hope of analyzing them within a human lifetime, and (7) is capable of being manipulated by modern molecular-genetic methods – such as DNA transformation – with sufficient ease that it is not wholly unrealistic to hope that one day it might become possible to test hypotheses about the molecular-genetic basis underlying different levels of organizational complexity by transferring genes among selected species.

In general terms at least, it was already obvious to August Weismann more than 100 years ago (see the Prologue) that green flagellates within the group he called the "natural order Volvocineae"[2] met the first criterion listed in the preceding paragraph. For the first 250 years after van Leeuwenhoek first described *Volvox*, however, reports on the biology of *Volvox* and its homocytic, colonial relatives were somewhat sporadic and (with very few exceptions) were based on examination of specimens fixed or examined live shortly after isolation from nature.[3]

In the late 1950s, however, a new era of volvocalean research opened with the establishment of new methods for collection of specimens from nature and the development of defined media that first made it possible to maintain many of them in the laboratory over prolonged periods in axenic cultures and thus to study their complete asexual and sexual life cycles in detail. Thus, the 20 years between 1955 and 1975 brought a flood of detailed laboratory studies of *Volvox* and its relatives that were stimulated by the studies of Mary Agard Pocock and others, but were then spearheaded by Richard Starr and his students and collaborators.[4] By the end of this period it was clear

---

[2] Now more commonly called the family Volvocaceae, within the order Volvocales (Mattox & Stewart 1984; Melkonian 1990), as will be discussed in more detail later.

[3] For example, see Baker (1753), Linnaeus (1758), Bonnet (1762), Ehrenberg (1831, 1832a,b, 1838), Williamson (1851, 1853), Busk (1853), Carter (1859), Pringsheim (1870), Cohn (1875), Henneguy (1879a,b), Cooke (1882), Kirschner (1883), Wolle (1887), Pfeffer (1888), Klein (1889a,b, 1890), Overton (1889), Ryder (1889), Oltmanns (1892, 1917), Shaw (1894, 1916, 1919, 1922a-c, 1923), Meyer (1895, 1896), Carlgren (1899), Terry (1906), Bancroft (1907), Mast (1907, 1917, 1919, 1926, 1927), Powers (1907, 1908), Smith (1907), Merton (1908), Collins (1909, 1918), West (1910, 1918), Harper (1912, 1918), Janet (1912, 1914, 1922, 1923a,b), Fritsch (1914, 1935, 1945), Rousselet (1914), Playfair (1915, 1918), Crow (1918), Laurens & Hooker (1918), Smith (1918, 1920, 1933, 1944), Iyengar (1920, 1933), Zimmermann (1921, 1923, 1925), Kuschakewitsch (1923, 1931), Hartmann (1924), Korschikoff (1924, 1938, 1939), Schreiber (1925), Bock (1926), Pascher (1927), Printz (1927), Lander (1929), Mainx (1929a, b), Pringsheim (1930), Pocock (1933a,b, 1938), Rich & Pocock (1933), Apte (1936), Hyman (1940, 1942), Metzner (1945a,b), Fott (1949), Bonner (1950), Cave & Pocock (1951a,b).

[4] See, e.g., Pocock (1955, 1960, 1962), Starr (1955, 1962, 1968, 1969, 1970a,b, 1971, 1972a,b 1973a,b, 1975), Cave & Pocock (1956), Skvortzow (1957), Pringsheim (1958, 1970), Stein

that the volvocine algae satisfied the foregoing criteria (1) and (2), and were well on their way to satisfying criterion (3). Studies over the intervening years by biologists with a wide range of interests and backgrounds have resulted in a combination of theoretical, field, and laboratory data that, when considered in toto, foster a growing realization that this group of modest beings meets all of the other criteria in the foregoing wish list to an extent that is wholly unrivaled by any other group of organisms yet studied.

(1958a,b, 1959, 1965 a,b, 1966a,b), Coleman (1959, 1962, 1975), Gerisch (1959), Pringsheim & Pringsheim (1959), Provasoli & Pintner (1959), Lang (1963), Goldstein (1964, 1967), Rayns & Godward (1965), Bisalputra & Stein (1966), Bourrelly (1966), Brooks (1966), Carefoot (1966), Darden (1966, 1968, 1970, 1971, 1973a,b), Ikushima & Maruyama (1968), Kochert (1968, 1971, 1973, 1975), Mishra & Threlkeld (1968), Darden & Sayers (1969, 1971), Deason, Darden & Ely (1969), Harris & Starr (1969), Kochert & Olson (1970a,b), Kochert & Yates (1970, 1974), McCracken (1970), McCracken & Starr (1970), Olson & Kochert (1970), Pickett-Heaps (1970, 1972, 1975), Deason & Darden (1971), Hand & Haupt (1971), Harris (1971a,b), Kemp & Wentworth (1971), Kochert & Sansing (1971), Palmer & Starr (1971), Vande Berg & Starr (1972), Kemp, Tsao & Thirsen (1972), Tucker & Darden (1972), Burr & McCracken (1973), Pall (1973, 1974, 1975), Sessoms & Huskey (1973), Bradley, Goldin & Claybrook (1974), Harris & James (1974), Karn, Starr & Hudcock (1974), Roberts (1974), Starr & Jaenicke (1974), Wirt (1974a,b), Lee & Kemp (1975), Lembi (1975), Meredith & Starr (1975), Stewart & Mattox (1975), Toby & Kemp (1975), Yates, Darley & Kochert (1975).

$$2$$

The Volvocales: Many
Multicellular Innovations

Few groups of organisms hold such a fascination for evolutionary biologists
as the Volvocales. It is almost as if these algae were designed to exemplify the
process of evolution. They.... display, within a narrow taxonomic compass,
extremes of somatic and sexual organization which could normally be found
only in entirely unrelated groups: their somatic organization may be unicellular
at one extreme, or multicellular with a macroscopic, functionally differentiated
... body at the other; their sexual reproduction may be isogametic, with no
distinction of male and female, ... [or] oogametic, with a massive immotile
ovum and a tiny motile sperm. These algae, therefore offer an unparalleled
opportunity to describe and interpret the evolution of fundamental features of
biological organization.

Bell (1985)

The Volvocales discussed in that quotation are a subset of the green algae,
or Chlorophyta,[1] the most diverse and ubiquitous group of modern eukaryotic

[1] The taxonomy of the green algae is very much in a state of flux. As ultrastructural and
molecular data accumulate, ideas regarding probable phylogenetic relationships among various
chlorophyte groups shift accordingly, and various authors have proposed different ways of
subdividing these organisms at the levels of class, order, and family to take such information
into account and establish (it is hoped) more natural, phylogenetically informative subdivisions
(e.g., Pickett-Heaps 1972, 1975; Irvine & John 1984; Bold & Wynne 1985; Hoek, Stam &
Olsen 1988; Kantz et al. 1990; Melkonian 1990; Zechman et al. 1990; Buchheim et al. 1996).
There is even active disagreement on the proper boundaries of the division (= phylum) Chlorophyta and its taxonomic relationship to the higher plants (Cavalier-Smith 1981; Bremer
1985; Chapman 1985; Hori, Lim & Osawa 1985; Sluiman 1985; Bremer et al. 1987; Devereux
et al. 1990). Although this is undoubtedly a healthy state of affairs in one sense, it has the

algae. Chlorophytes are characterized by a number of shared features, such as chloroplasts that contain chlorophylls *a* and *b* and are enclosed in a double membrane, the presence of certain characteristic carotenoids and xanthophylls, the ability to store excess photosynthate as starch, and so forth (Pickett-Heaps 1975; Mattox & Stewart 1984; Bold & Wynne 1985; Melkonian 1990). Many of those features are shared, of course, by a large and conspicuous set of descendants of the green algae: the vascular plants. To put the origins of *Volvox* in context, it will be useful to review briefly some of the major evolutionary trends that are evident when the green algae are viewed in a broad framework.

2.1 The green algae: an ancient and inventive group

The green algae apparently arose deep in the Precambrian (Cloud, Licori et al. 1969; Sogin 1991), probably as a result of a stable relationship established between a colorless flagellate and an oxygen-producing endosymbiotic cyanobacterium containing chlorophylls *a* and *b* (Taylor 1978; Margulis 1981). Initially the chlorophytes probably were a small group of unicellular green

unfortunate effect of causing different authors to use a given taxonomic term to cluster different assemblages of organisms, thereby leading to potential confusion.

An example of this that must be considered in any discussion of *Volvox* and its presumed evolutionary history is the diversity of views that have arisen regarding the boundaries of the order Volvocales, within which *Volvox* is placed by nearly all taxonomists. Until rather recently it was standard taxonomic practice to group all motile unicellular and colonial green algae within the order Volvocales (e.g., Collins 1918; Iyengar 1920, 1933; Pascher 1927; Printz 1927; Smith 1933; Fritsch 1945; Huber-Pestalozzi 1961; Bourrelly 1966, 1972; Round 1971; Iyengar & Desikachary 1981; Bold & Wynne 1985). However, more recent comparative studies have led to the clear conclusion that the green flagellates grouped in this way are not monophyletic. Thus, some have suggested that certain green flagellate genera should be transferred into other classes that formerly contained only nonmotile members, while the rest of the green flagellates should be subdivided into two orders: the unicellular order Chlamydomonadales and the colonial/multicellular order Volvocales (Mattox & Stewart 1984; Melkonian 1990). Despite recognizing certain virtues of such a split, however, many contemporary authors continue to use the term "Volvocales" in something closer to its older sense (e.g., Bell 1985; Nozaki 1986a; Matsuda et al. 1987; Koufopanou 1994), either as a matter of simple convenience or because they believe that when the unicellular flagellates are split off the new, smaller order Volvocales deviates from monophyly in a way that the earlier, larger order did not – by grouping colonial forms that are derived from quite different members of the new order Chlamydomonadales.

Given this state of affairs, and a fear of having concepts buried under a plethora of taxonomic details and controversies, I have made an effort to keep the use of formal (capitalized) names for taxonomic categories above the generic level to a minimum in this book. When such formal names are used, the author whose precedent is being followed will usually be noted. In many cases, however, less precise, informal (uncapitalized) terms are used. Specifically, the terms "the volvocales" (uncapitalized), "volvocaleans," or "volvocine algae" will be used (more or less synonymously) to refer to a group that includes the genus *Volvox* and the varous unicellular and colonial forms to which it appears to be related by descent.

flagellates whose cell surfaces and flagella were covered with a loose assemblage of nonmineralized, glycoprotein-rich scales, resembling, in that regard, both a number of primitive colorless flagellates and one small group of modern green algae known as prasinophytes (Manton 1966; Mattox & Stewart 1984; Melkonian 1984a, 1990).

The basic cellular unit of the first green algae appears to have been exceptionally adaptable and malleable, permitting their descendants to undergo substantial modifications in morphology and life history as they moved out to exploit a variety of new environments. Both the fossil record (limited as it is) and the molecular clock, based on rRNA sequences, suggest that such diversification began long before the beginning of the Cambrian period. Today, green algae are regularly found in an extraordinarily diverse set of habitats. Although most species live in fresh water, some inhabit brackish or ocean waters, some are found in the soil (from the moist edges of lakes, ponds, and swamps to hot deserts and volcanic ashes), some live on or under permanent ice in the polar zones, and yet others make their homes on the surfaces (and, in many cases, within the cells) of a wide range of protists, plants, and animals (Bold & Wynne 1985; Raven, Evert & Eichorn 1986).

This ecological diversity is accompanied by equally impressive diversity in morphology and life cycles: Some chlorophytes appear to reproduce exclusively by asexual means, but others have an active sex life, which, as the foregoing quotation indicates, may involve differentiated eggs and sperm, reminiscent of those of animals. Large numbers of modern green algae remain flagellated and motile throughout most or all of the life cycle. However, in many separate and distinct chlorophyte lineages the organisms have abandoned flagellar motility in the vegetative phase; some of these remain unicellular, but a greater number form multicellular clusters, spheres, filaments, sheets, or much more complex bodies, sometimes with a variety of differentiated cell types. Many sedentary multicellular green algae reveal their flagellate ancestry, however, by retaining in their nonmotile cells such flagellate features as eyespots, basal bodies, or "pseudocilia." Others betray their ancestry when they produce flagellated gametes. Still others have lost their flagellated stages entirely, but reveal their flagellate ancestry through other aspects of their biochemistry and cytology.

As the dependence of various chlorophyte groups on flagellar motility underwent modification, so did the nature of the cell coat. It now appears that the scales covering primitive green flagellates became transformed into a tough, coherent cell wall in at least three separate chlorophyte lineages (Domozych, Stewart & Mattox 1980; Mattox & Stewart 1984), of which two are worthy of brief discussion here.

In charophytes (members of the class Charophyceae, in the sense of Mat-

Figure 2.1 A multicellular charophyte, such as *Chara* or one of its close relatives. These plant-like algae are anchored in stream beds by a well-developed root-like rhizoid and can grow to a height of 30 cm or more. The ability of these algae to remain erect in a flowing stream can be attributed to the fact that, as in higher plants, each cell is encased in a stiff, cellulosic cell wall (though in many cases the wall is reinforced with calcium or magnesium carbonates, giving rise to the common name of "stoneworts" or "brittleworts"). Similarities in growth form, cell-wall composition, and other features cause many to believe that the higher plants were derived from a charophyte-like ancestor. (Adapted from Bold & Wynne 1985.)

tox & Stewart 1984), of which *Chara*, a common fresh-water, "plant-like" multicellular alga (Figure 2.1), is one of the best-known examples, cells are typically enclosed in stiff, cellulose-containing cell walls. Such stiff walls appear to be incompatible with flagellar motility, because charophytes are immotile during most of the life cycle, and when they do produce flagellated reproductive cells those cells lack a characteristic charophyte wall and either are naked or are covered with scales resembling those of more primitive green algae. It is now generally believed that a primitive charophyte was ancestral to the land plants (Raven et al. 1986; Devereux et al. 1990), and hence the charophyte cell wall was ancestral to the cellulose-reinforced cell walls that permit land plants to achieve great stature, despite the battering of wind and rain (Domozych et al. 1980).

In marked contrast, another major lineage of green algae (the class Chlorophyceae, in the sense of Mattox & Stewart 1984), of which *Chlamydomonas* (Figure 2.2) is the best-known unicellular example, evolved a very different sort of wall that clearly is compatible with flagellar motility: a tough, coherent, but flexible cell wall composed primarily of fibrous glycoproteins and lacking any detectable cellulose (Lamport 1974; Roberts 1974; Domo-

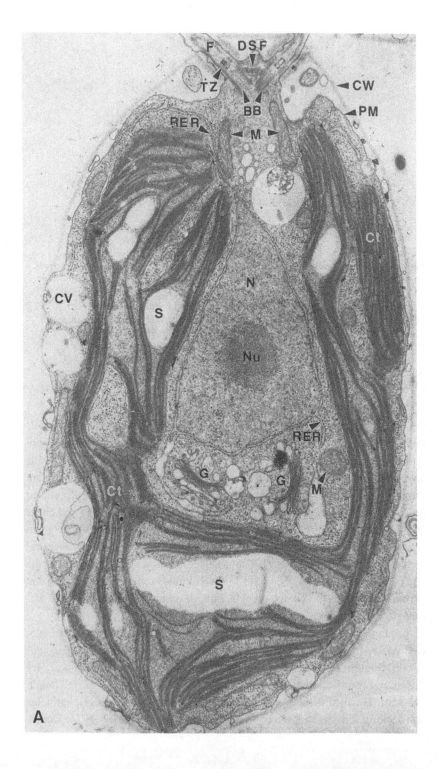

A

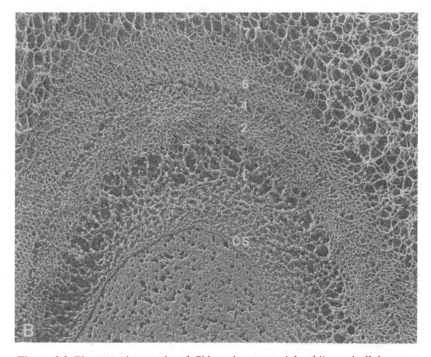

Figure 2.2 Electron micrographs of *Chlamydomonas reinhardtii*, a unicellular chlorophycean alga, labeled to illustrate features shared by members of this group, many of which will be referred to repeatedly in this book. Features seen in a typical stained, sectioned specimen (**A**) include BB, basal bodies; Ct, chloroplast; CV, contractile vacuole; CW, cell wall; DSF, distal striated fiber linking BBs; F, flagellum; G, Golgi complex; M, mitochondrion; N, nucleus; Nu, nucleolus; PM, plasma membrane; RER, rough endoplasmic reticulum (continuous with nuclear envelope); S, starch granules (within the chloroplast); TZ, transition zone between BB and axoneme of the flagellum. In a sectioned specimen, the cell wall appears as a rather inconspicuous structure composed of alternating electron-dense and electron-lucid layers, but in a freeze-fractured, deeply etched specimen (**B**) it is seen to be a highly coherent structure composed of an assortment of fibrils that are woven into an intricate meshwork. CS, cell surface; numbers refer to regions equivalent to layers of the *Chlamydomonas* cell wall previously identified by Roberts (1974) in high-resolution images of sectioned specimens (the missing numbers correspond to looser areas in the meshwork rather than discrete structures). (**A** reproduced from Johnson & Porter, *The Journal of Cell Biology*, 1968, **38**, 403–25, and **B** from Goodenough & Heuser, *The Journal of Cell Biology*, 1985, **101**, 1550–68, by copyright permission of The Rockefeller University Press.)

zych et al. 1980; Adair et al. 1987; Goodenough & Heuser 1988; Harris 1989). It has been postulated that such a wall arose either as a fusion of glycoprotein scales, like those covering more primitive green flagellates (Mattox & Stewart 1984), or in a more complex two-step process (Melkonian 1990). In any case, a one-piece, fibrous-glycoprotein cell wall lacking detectable cellulose now surrounds most chlorophyceans, and although there are numerous non-motile members of the group, the majority of chlorophycean species retain flagella and remain motile through most of the life cycle, despite being enclosed in such a coherent cell coat. As we will see later, the establishment of a cell wall that was compatible with flagellar motility was accompanied by the appearance of a unique form of cell division, which appears to have been an important precondition leading to the development of multicellular organisms such as *Volvox* from unicells like *Chlamydomonas*.

2.2 *Chlamydomonas* and its relatives: master colony-formers

"*Chlamydomonas* . . . is representative of the primitive stock from which all other groups of green algae are thought to have evolved" (Scagel et al. 1965). A generation ago, statements like the preceding were commonplace in widely respected textbooks. More detailed comparative studies, however, have led to the conclusion that – far from being the most primitive group of green algae – the Chlorophyceae[2] (of which *Chlamydomonas* is a prototype) may well be the most highly differentiated class of green algae (Mattox & Stewart 1984; Melkonian 1984a). That evaluation has been reinforced by molecular-phylogenetic studies, which have led to the conclusion that *Chlamydomonas* and its close relatives are not a great deal more closely related to higher plants than they are to animals or fungi (Rausch, Larsen & Schmitt 1989; Devereux et al. 1990; Sogin 1991). Considering the modern cytological and molecular evidence together, it now appears that the lineages leading to various modern green-algal groups diverged from the lineage leading to the higher plants (and then diverged from one another) not long after the first green algae had diverged from the flagellate lineages leading to the fungi and animals (Figure 2.3).

Modern chlorophyceans are exceptionally diverse, with 11 orders containing well over 100 genera and 1,000 species (Melkonian 1990). Although members of most chlorophycean genera and species are unicellular flagellates, multicellular forms are present in 9 of the 11 chlorophycean orders (Melkonian 1990). Multicellularity is believed to have arisen independently in each of these orders, and in some orders more than once. Clearly, the basic chlorophycean cell, typified by *Chlamydomonas*, appears to have features that regularly lent it to the establishment of cellular communes, and as a result the chlorophyceans undoubtedly hold the world record for successful multicellular innovations.

What is the source of this chlorophycean proclivity for colony formation?

---

[2] Terms for higher taxonomic categories used in this section follow Melkonian (1990).

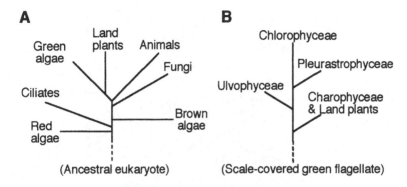

Figure 2.3 Phylogenetic branching order inferred for the major groups of modern eukaryotes (**A**) and various groups of green algae plus higher plants (**B**). Lengths of lines are arbitrary and do not imply any time scale. A is derived from a combination of sources; B is based on Mattox & Stewart (1984). The Chlorophyceae include the volvocine algae that are the focus of this book.

There can be little doubt that it resides in their propensity to form two particular types of extracellular materials. In contrast to the multicellular animals, in which adjacent cells are commonly attached to one another by various specializations of the plasmalemma (such as desmosomes, tight junctions, etc.), chlorophycean colonies are held together predominantly by a shared extracellular matrix. In addition to the fibrous glycoprotein cell walls that characterize all but an exceptional few "naked" genera, most chlorophyceans (whether naked or walled, unicellular or multicellular) produce a second type of extracellular material that is commonly called "mucilage" (a relatively amorphous, sticky mixture of glycoproteins, acidic mucopolysaccharides, etc.). Both types of extracellular material (walls and mucilage) are involved, albeit in different species-specific proportions, in stabilizing various types of chlorophycean multicellular assemblages.

Even chlorophyceans commonly characterized as unicellular have a strong tendency to form multicellular arrays. For example, when common laboratory strains of *Chlamydomonas reinhardtii* are grown in liquid medium under optimum conditions, they normally are actively motile and relatively free of adherent mucilage; but under other conditions they will attach to the bottom or sides of the culture vessel, resorb their flagella, and secrete copious amounts of mucilage that binds them to the substrate and to one another in an amorphous mass known as a "palmella." In many other species of *Chlamydomonas*, such loss of flagellar motility and conversion to the palmelloid state occur regularly, even under optimum growth conditions in liquid medium. Such species of *Chlamydomonas* are motile for only a brief period after each round of cell division, whereupon they resorb their flagella and

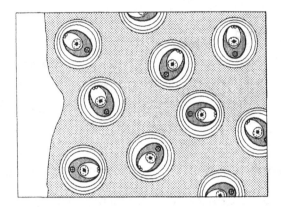

Figure 2.4 Diagrammatic representation of the organization of a typical *Palmella* colony. Except for the lack of flagella and eyespots, the cells resemble *Chlamydomonas*, but they are surrounded by concentric layers of fibrous extracellular matrix and are embedded in an amorphous mucilaginous mass. Under less than optimum conditions, many species of *Chlamydomonas* can adopt a similar "palmelloid" lifestyle. (Adapted from Bold & Wynne 1985.)

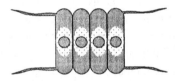

Figure 2.5 A *Scenedesmus quadricauda* colony, or coenobium, consists of four cells that are tightly associated because of their fused cell walls. Two of the cells bear the four projections (or "tails") that are the basis for the organism's specific epithet. In the asexual life cycle each cell enlarges about fourfold and then divides twice to produce a new four-cell, four-tail colony that is subsequently released from the parental cell wall. (Adapted from Bold & Wynne 1985.)

convert to the palmelloid form (Schlösser 1976, 1984). In its palmelloid phase, *Chlamydomonas* mimics the usual form of certain colonial relatives, such as *Palmella* (a member of the order Tetrasporales), from which the term for this lifestyle is derived (Figure 2. 4).

Although palmelloid colonies often are highly irregular in size, shape, and cell number, quite the opposite is true of many other colonial chlorophyceans. For example, in *Scenedesmus quadricauda*, a member of the order Chloro-coccales, the walls of sister cells are fused, forming small, compact, four-cell colonies in which little or no mucilage is detected (Figure 2.5). In another chlorococcalean species, *Coelastrum microsporum*, sister cells are held in a very regular, spherical array by contacts formed between the cell walls of

Figure 2.6 A *Coelastrum* coenobium typically consists of a hollow sphere of 4, 8, 16, 32, 64, or even 128 cells that is held together by the contacts between the cell walls of adjacent cells and is filled with a mucilaginous matrix (black). (Adapted from Bold & Wynne 1985.)

nearest neighbors; however, the cells are less tightly packed in *Coelastrum* than in *Scenedesmus*, and the interior of the sphere and the spaces between cell walls are filled with a mucilaginous matrix (Figure 2.6).

The two species just illustrated share an important feature of many colonial chlorophyceans: The number of cells per colony is not random, but is some multiple of 2. In various species of *Coelastrum*, for example, there may be 4, 8, 16, 32, 64, or 128 cells per colony (Bold & Wynne 1985). This regularity in cell numbers is a consequence of the unique way in which these chlorophyceans divide. They do not just double in size and then undergo binary fission, as cells of so many other organisms do. Rather, each cell grows about 2^n-fold in volume, and then undergoes "multiple fission": a rapid, synchronous series of n divisions, while still inside the mother cell wall, to produce a cluster of 2^n daughter cells. In unicellular species such as *Chlamydomonas*, these 2^n cells separate from one another as they escape from the mother wall and swim away. But in colonial species, the 2^n cells produced by a round of cell divisions become cemented to one another by extracellular materials that they produce before hatching from their mother wall. As a result, the number of cells present in such a colony reflects the number of divisions that occurred during its formation and does not change until it initiates a new round of asexual reproduction. Such an organism, in which the number of cells is determined by the number of cleavage divisions that went into its initial formation, and in which cell number is not augmented by accretionary cell divisions, is called a "coenobium." *Volvox* and all of its close relatives, which we shall now discuss in some detail, are all coenobial.

2.3 The volvocine algae: a linear progression in organismic complexity?

Volvox, the rolling green globe that Antoni van Leeuwenhoek discovered in water he scooped from irrigation ditches, is a representative of one of about a dozen genera of green flagellates in the family Volvocaceae that are regularly found in such ditches and in warm roadside puddles, temporary pools, and stagnant ponds around the globe, as well as in many relatively deep eutrophic lakes.

Each volvocacean possesses 2^n Chlamydomonas-like, biflagellate cells held together in a coenobium of defined shape (usually spherical). Members of different volvocacean species and genera differ in cell number (owing to heritable differences in the modal value of *n*), the amount of extracellular matrix that binds the cells to one another, and the extent to which all of the cells are capable of participating in reproduction (Figure 2.7).

It has long been commonplace for textbooks to arrange *Chlamydomonas* plus five genera of volvocaceans (*Gonium, Pandorina, Eudorina, Pleodorina,* and *Volvox*) in a conceptual series in which there are progressive increases in size, cell number, volume of extracellular matrix, and the tendency to form terminally differentiated, sterile somatic cells (Figure 2.8). It then requires but a short extrapolation to suggest that such a series might well resemble the evolutionary progression by which the body plan of *Volvox*, with its germ–soma division of labor, evolved.

The sense that these organisms might really form a phylogenetic continuum has been reinforced by discovery of a number of intermediates that blur the boundaries between the standard genera. For example, an organism recently discovered in Poland blurs the distinction between *Chlamydomonas* and *Gonium*. Previously, *Gonium*, the smallest volvocacean, had been distinguished from *Chlamydomonas* principally by the fact that after mature cells had undergone a sequence of rapid divisions within the wall of the mother cell, the daughter cells of *Gonium* cohered to form a small colony, whereas those of *Chlamydomonas* invariably separated from one another and functioned independently (Figure 2.9). But in a recently discovered species that distinction has been lost, and when cells within a single colony divide, some produce new coenobia like the one from which they came, whereas others produce only free-swimming individual *Chlamydomonas*-like cells (Batko & Jakubiec 1989).[3] When unicellular and colonial individuals from that population were isolated and cultured separately, the two cultures produced uni-

---

[3] This difference involves a difference in the way the two types of cells divide; see Section 4.5.4.

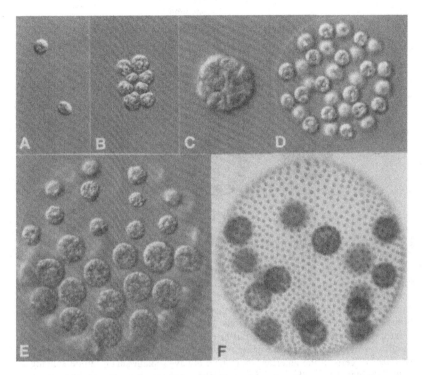

Figure 2.7 A photo gallery of volvocine algae. **A**: *Chlamydomonas reinhardtii*, the prototypical unicellular member of the group; note paired anterior flagella. **B**: An eight-cell colony of *Gonium pectorale*; note that *Gonium* forms a convex plate of cells, in distinction to the rest of the multicellular genera depicted here, which are spherical. **C**: A 16-cell colony of *Pandorina morum*; note the tight compaction of the cells and the lack of significant amounts of extracellular matrix. **D**: A 32-cell colony of *Eudorina elegans*; note the much greater amount of extracellular matrix than in *Pandorina*. **E**: A 64-cell spheroid of *Pleodorina californica*; note that the cells in the posterior (bottom) of the spheroid (which were originally indistinguishable from the anterior cells) have begun to enlarge in preparation for reproduction, while the anterior cells remain small, biflagellate, terminally differentiated somatic cells. **F**: A spheroid of *Volvox carteri* containing about 2,000 small, biflagellate, terminally differentiated somatic cells at the surface and about 16 large reproductive cells, or gonidia, in the interior. In distinction to *P. californica*, the somatic cells and gonidia of *V. carteri* are set apart very early in development, and the division of labor between them is complete. Magnification: A–E,×400X; F,×100X.

cellular and colonial offspring in similar proportions. Thus, although those authors decided to name the new isolate *Gonium dispersum*, they pointed out that an equally strong case might have been made for calling it a new species of *Chlamydomonas* (Batko & Jakubiec 1989).

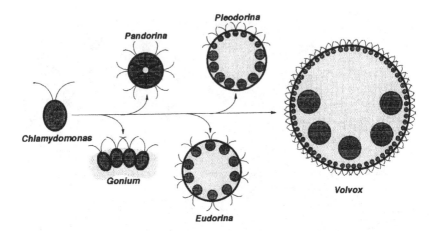

Figure 2.8 Diagrammatic representation of the volvocine lineage hypothesis.
Chlamydomonas and several genera of volvocaceans can be arranged in a
conceptual sequence in which there are progressive increases in the number of
cells, the amount of extracellular matrix, and the extent to which labor is divided
between reproductive cells and sterile somatic cells. In *Chlamydomonas, Gonium,*
and *Pandorina* (plus several other volvocacean genera not shown) all cells fulfill
both somatic and reproductive functions. In large *Eudorina* colonies, however, the
anteriormost four cells sometimes fail to enlarge; they continue to provide flagellar
motility while the other cells enlarge, redifferentiate, and divide. In *Pleodorina
californica*, cells in the anterior hemisphere always remain small, motile somatic
cells, whereas cells in the posterior hemisphere enlarge, become gonidia, and
divide. In *Volvox*, a similar pattern exists, but the ratio of somatic cells to gonidia
is much greater. Textbooks frequently imply that this conceptual series may
resemble the (presumably linear) phylogenetic pathway by which *Volvox* evolved
from a *Chlamydomonas*-like ancestor. However, modern evidence suggests that
the history of the group is more complex, that possibly a pattern resembling
this occurred several times, and also that some species within the genera
shown toward the left may have arisen by simplification of representatives of
genera shown toward the right.

Generic boundaries toward the other end of the range of volvocacean
complexity appear equally tenuous. The validity of a *Eudorina–Pleodorina*
distinction was called into question by Goldstein (1964) when he found that
two strains previously placed in these separate genera could mate and produce
viable hybrids.[4] Similarly, Vande Berg and Starr (1971) discussed the arbitrary nature of the *Pleodorina–Volvox* distinction when they discovered a
mutant of *V. powersii* that undoubtedly would have been classified as *Pleo-*

---

[4] Based on that observation, Goldstein argued that all species that others had partitioned between
Eudorina and *Pleodorina* should be combined in a single genus *Eudorina* (Goldstein

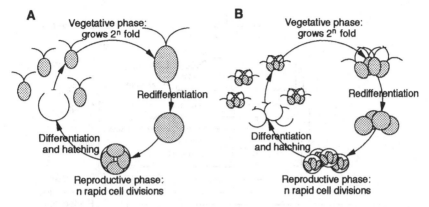

Figure 2.9 Typical patterns of asexual reproduction in the volvocales, as illustrated by *Chlamydomonas* (**A**) and *Gonium* (**B**). Each cell grows 2^n-fold during the vegetative portion of the life cycle and then rounds up, reorganizes its cytoplasm (often resorbing its flagella in the process), and divides rapidly n times to produce 2^n daughter cells within the mother-cell wall. After producing new flagella and their own cell walls, the progeny then escape from the mother-cell wall and swim away. In both *Chlamydomonas* and *Gonium* n may equal 2, 3, 4, or (rarely in *Chlamydomonas*) 5. The principal difference in the life cycles of unicellular and colonial volvocaleans is that in the latter the cells produced by cell division remain coherent after division, whereas in the former they do not.

dorina if it had been found in the wild. [That immediately summons up the much earlier opinion of Shaw (1916), who had concluded that the species now known as *V. powersii* was an intermediate between *Pleodorina* and *Volvox* and should be placed in a separate genus.] In a similar vein, a double mutant of *V. carteri* has been described more recently that, if found in nature, might be classified as a *Eudorina* (Tam & Kirk. 1991b).

In short, many observations support the idea that *Chlamydomonas* and the volvocaceans compose a nearly unbroken continuum of organisms varying in complexity from unicellular to multicellular with a complete division of labor, and suggest that the number of genetic differences that distinguish one level of complexity from the next may not be great. But does this constitute proof of what has been called ''the volvocine-lineage hypothesis'' (Figure 2.8) (Larson, Kirk & Kirk 1992), namely, the hypothesis that *Volvox* evolved from *Chlamydomonas* via a simple, monophyletic progression in organismic size and complexity?

1964). However, most algal taxonomists continue to place within the genus *Pleodorina* those species in which sterile somatic cells are invariably present, with *Eudorina* containing those species in which they are not invariably present.

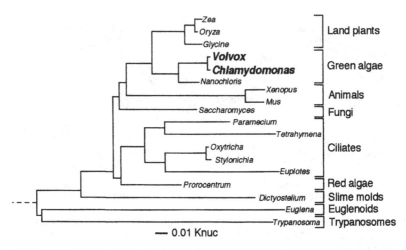

Figure 2.10 Dendrogram of selected taxa based on complete sequences of 18S rRNA molecules (Rausch, Larsen & Schmitt 1989). Lengths of horizontal lines are proportional to K_{nuc}, a measure of sequence divergence, but lengths of vertical lines are arbitrary. K_{nuc} would be a measure of the time since divergence if the rate of change (the "molecular clock") had been constant in all lineages, but the differing placements of modern taxa along the horizontal axis indicate that it has not been. In the plant and algal lineages, however, the rates of change appear to have been quite similar. In this regard it is noteworthy that there are fewer differences in rRNA sequences between *Chlamydomonas* and *Volvox* (specifically *C. reinhardtii* and *V. carteri*) than between *Zea* (maize) and *Oryza* (rice), two cereal grains that are estimated to have last shared a common ancestor about 50 million years ago. Thus it appears that *V. carteri* may have evolved from a *C. reinhardtii*-like ancestor during a time period not too different from that.

2.4 The volvocacean family tree appears to have a complicated branching pattern

The first published sequence of a nuclear-encoded rRNA of *Volvox* (Rausch et al. 1989) reinforced the inference drawn from morphology that *V. carteri* and *Chlamydomonas reinhardtii* are much more closely related to each other than they are to higher plants or members of other major taxa (Figure 2.10). Additional molecular-phylogenetic studies that will be discussed in more detail later strongly support the concept that the members of the Volvocaceae are all closely related to one another and to *C. reinhardtii* and certain other species of *Chlamydomonas*. However, such conclusions do not necessarily confer validity on the volvocine-lineage hypothesis. Indeed, many other considerations suggest that the evolutionary relationships within the volvocine lineage may be considerably more complex than this simplistic hypothesis suggests.

2.4.1   A single volvocacean species name can mask extensive genetic diversity

It is now increasingly clear that volvocaceans that are morphologically indistinguishable, and thus are classified as members of the same species, can
be extremely distant from one another genetically. This has been most thoroughly and convincingly demonstrated by Annette Coleman, who has studied
many strains of *Pandorina morum* isolated from ponds around the world
(Coleman 1959, 1975, 1977, 1979b, 1980, 1996a,b; Coleman & Zollner
1977; Coleman & Goff 1991; Coleman, Suarez & Goff 1994). She has established that this morphologically monotypic species contains at least 24
reproductively isolated mating groups, or "syngens," that differ from one
another in chromosome number (ranging from 4 to 12) (Coleman & Zollner
1977) and in various molecular markers such as the size and restriction patterns of their mitochondrial DNAs (Coleman & Goff 1991) and the sequences
of the internal transcribed spacers (ITS) of their nuclear rRNA-encoding
genes (Coleman et al. 1994; Coleman 1996a,b). It is particularly significant
that genetic distance within this so-called species is not related to the extent
of geographic separation in any simple way. A *P. morum* individual isolated
from a small pond in the American Midwest can be chromosomally indistinguishable from, and interfertile with, a strain isolated from Europe or Asia,
while being reproductively isolated and chromosomally very different from
a second *P. morum* individual taken from the same pond! It will be particularly interesting to see how these different syngens of *P. morum* appear to
be related to other genera and species of volvocaceans once a common set
of markers, such as ITS sequences, becomes available for comparative analysis.[5] As the next example to be considered will illustrate, it seems quite
possible that the different syngens of *P. morum* that Coleman has identified
will turn out to be related to quite different members of other genera within
the family Volvocaceae.

One of the most surprising findings in a preliminary comparative analysis
of rRNA sequences of 16 volvocaceans that we reported was the relatively
distant relationship that was inferred to exist between *Eudorina* and *Pleodorina* (Larson et al. 1992). That was disconcerting, because Goldstein had
reported that *Eudorina* and *Pleodorina* were so closely related that they

---

[5] At the time of this writing, Coleman reports that ITS sequences have been obtained for more
than 200 isolates, including all available genera, species, and strains of the family Volvocaceae, as well many other members of the order Volvocales. Once these data have been subjected to the necessary molecular-phylogenetic analysis, a picture of very great interest should
emerge. Based on her preliminary, visual analysis of some of the data, she states that "the
family as a whole is utterly fascinating, illustrating all levels of evolutionary divergence in
the different genera and species" (A. W. Coleman, personal communication).

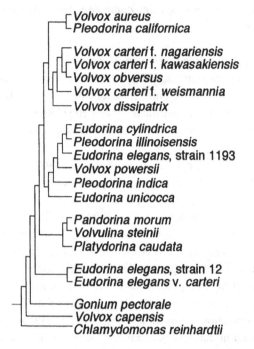

Figure 2.11 Phylogenetic relationships among selected volvocalean taxa inferred from analysis of variable regions of 18S and 28S rRNAs. Line lengths are arbitrary and are meant to indicate only the inferred branching order. Note, for example, that two different isolates of *Eudorina elegans* fall in very different clusters within the dendrogram. Similarly, the data indicate that the small number of *Volvox* isolates analyzed to date may fall within four separate lineages. (Based on Larson, Kirk & Kirk 1992, plus unpublished data.)

sometimes were able to mate and produce viable progeny; he had therefore concluded that the genus *Pleodorina* should be collapsed into *Eudorina* (Goldstein 1964). However, we quickly realized that the specimens that we happened to have available when our preliminary study was performed (*P. californica* and *E. elegans* strain 12) were not those that Goldstein had successfully hybridized (namely, *P. illinoisensis* and *E. elegans* strain 1193). When we subsequently analyzed the rRNA sequences of *P. illinoisensis* and *E. elegans* strain 1193, it turned out that they were nearly identical to one another, but extremely different from the two samples that we had originally analyzed (A. Larson, M. M. Kirk, and D. L. Kirk unpublished data) (Figure 2.11).[6] Indeed, in terms of rRNA sequences, *E. elegans* strain 1193 appears

---

[6] When we analyzed the hybrids derived from Goldstein's cross, which are still alive after 30 years (Starr & Zeikus 1993), we found that they possessed rRNAs identical with those of both parents, confirming Goldstein's conclusion based on karyotype (Goldstein 1964) that they are

to be more closely related to two other morphologically distinguishable species of *Eudorina*, and to members of five other volvocacean genera (*Pleodorina, Volvox, Pandorina, Volvulina,* and *Platydorina*), than it is to its presumed conspecific, *E. elegans* strain 12! Surprising as this finding may seem, it is actually consistent with the observations of Goldstein, who was able to mate *E. elegans* strain 1193 with *P. illinoisensis,* but not with most other strains of *E. elegans* (Goldstein 1964). We conclude that the name *Eudorina elegans,* like the name *Pandorina morum,* groups organisms that share a number of simple morphological features but may be quite unrelated to one another genetically. That opinion, and the data summarized in Figure 2.11 will be discussed in more detail shortly.

2.4.2 There are marked differences in developmental patterns within the genus Volvox

One obstacle to visualizing a simple monophyletic progression that would lead from *Chlamydomonas* to *Volvox* in a credible manner is the variability that exists within the genus *Volvox* itself. Although certain authors have occasionally referred to various species of *Volvox* as being "primitive," or others as "advanced" (e.g., Pocock 1933a; Cave & Pocock 1951b; Vande Berg & Starr 1971), no one has yet published a detailed and comprehensive analysis that has attempted to deduce how all the species within the genus might be related to one another or how they might all have been derived from a common ancestor. Indeed, many have interpreted the diversity within the genus *Volvox* as indicating that various of its members may have had different evolutionary roots.

The unifying feature of the genus *Volvox* is that in the asexual reproductive phase all members are organized as multicellular spheroids within which most cells are terminally differentiated as mortal somatic cells, with only a small minority functioning as "gonidia," or potentially immortal asexual reproductive cells. However, there are at least two very different patterns of spheroid organization within the genus, and these are associated with very different patterns of embryonic development, as can be illustrated by considering two well-studied species: *V. rousseletii* (Pocock 1933b; McCracken 1970; McCracken & Starr 1970) and *V. carteri* (Kochert 1968; Starr 1969,

---

pseudodiploids possessing both parental chromosome sets and thus must have been derived from zygotes that failed to execute meiosis (as all volvocacean zygotes normally do) before germinating.

1970a; Kochert & Olson 1970a; Green & Kirk 1981; Green, Viamontes & Kirk 1981).

The principal distinguishing feature that becomes apparent upon examination of various species of *Volvox* under a light microscope is the extent to which cells within the spheroid are interconnected. During embryonic development, cells of all volvocaceans are interconnected by cytoplasmic bridges that are the result of incomplete cytokinesis (Stein 1965a; Bisalputra & Stein 1966; Ikushima & Maruyama 1968; Pickett-Heaps 1970, 1975; Marchant 1976, 1977; Fulton 1978a,b; Green & Kirk 1981; Green et al. 1981). But in only a restricted number of *Volvox* species do such bridges persist in the adult (Smith 1944). *V. rousseletii* is one such species. All cells of an adult *V. rousseletii* spheroid are interconnected by such robust cytoplasmic strands that the cells appear stellate when viewed from their flagellar ends, and bell-shaped when viewed from one side (Figure 2.12). In *V. carteri*, in contrast (as well as in all other volvocacean genera and several other species of *Volvox*), cytoplasmic bridges break down at the end of embryogenesis, and cells of the adult lack any cytoplasmic communication with one another (Figure 2.13). This difference in cytoplasmic continuity is accompanied by a significant difference in the way in which gonidia differentiate and cleave.

In *V. rousseletii,* all cells are nearly the same size at the end of embryogenesis, and all cells develop flagella and contribute to spheroid motility. On closer examination, however, it is seen that a few of these cells are about twice the size of the others at the end of cleavage, presumably as a result of withdrawing from the cleavage program one cycle before their neighbors do (Pocock 1933b). These slightly larger cells later reveal their gonidial nature when they resorb their flagella, enlarge a bit more, and begin to divide, while still attached to neighboring somatic cells by bridges. Although gonidia of such species are only slightly larger than somatic cells when they begin to divide, each cleaving embryo grows between successive divisions, so that as cell number doubles at each division, cell size diminishes by much less than a factor of 2 (Figure 2.12).

Gonidial behavior in *V. carteri* is strikingly different. In this species, the presumptive gonidia and somatic cells of the next generation are visibly set apart during embryogenesis by asymmetric divisions and thus are substantially different in size in even the very youngest spheroids (Figure 2.13). In distinction to those of *V. rousseletii*, the gonidia of *V. carteri* never develop functional flagella or eyespots; rather, they put all of their energy into growth, eventually becoming 500–1000 times larger (in volume) than somatic cells before they begin to divide. Then, embryonic cleavage in *V. carteri* occurs in the absence of further growth, so that as cell number increases, cell size decreases proportionately (Figure 2.13).

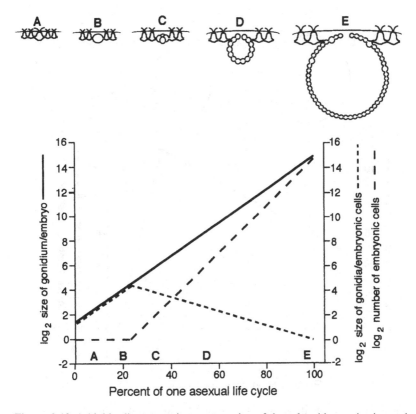

Figure 2.12 A highly diagrammatic representation of the spheroid organization and developmental pattern of *Volvox rousseletii*. Letters above the abscissa on the graph indicate the approximate stages of the life cycle that are represented by the diagrams. The values plotted on the graph are cellular volumes, relative to the volume of a somatic cell at the end of the embryonic cleavage period. Data were derived from Pocock (1933b), but were simplified by representing changes in cell size by lines connecting initial and final values. As in all *Volvox* species classified within the section Euvolvox (Table 2.1), all cells of a *V. rousseletii* spheroid remain connected throughout the life cycle by broad cytoplasmic bridges, giving the somatic cells a somewhat bell-shaped profile. Initially the prospective gonidia are only slightly larger than the somatic cells, and like the latter they develop functional flagella (**A**), but they grow more rapidly. After a short period of growth, the gonidia resorb their flagella and prepare to divide (**B**). Division cycles are lengthy, and embryonic cells grow between divisions, so that cell size is decreased by much less than a factor of 2 in successive divisions. The embryo remains connected to the somatic cells throughout division (**C–E**), but eventually the fully cleaved embryo will break these connections, undergo inversion (i.e., turn inside out to adopt the adult configuration), grow for a short time, and then hatch from the parental spheroid and swim away.

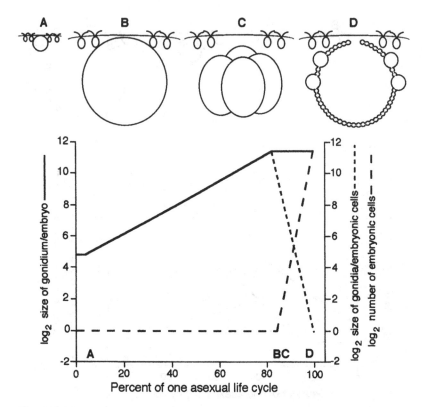

Figure 2.13 A highly diagrammatic representation of the spheroid organization and developmental pattern of *Volvox carteri* (compare with Figure 2.12). Graph generated as in Figure 2.12, with data taken from Kirk, Ransick et al. (1993). As in all other *Volvox* species that are classified within the section Merrillosphaera (Table 2.1), the cytoplasmic bridges that link all cells in the dividing embryo are broken at the end of embryogenesis, by which time the prospective gonidia of *V. carteri* are already about 30 times larger in volume than somatic cells, and they never develop functional flagella (**A**). However, the gonidia also grow much faster than the somatic cells for a prolonged period, becoming 500–1,000 times as large as somatic cells (**B**). Then they initiate a series of very rapid cleavage divisions, in the absence of further growth, so that cell size decreases by about a factor of 2 in each division cycle (**C, D**). As in *V. rousseletii*, the embryo then undergoes inversion, and after a period of further growth the juvenile spheroid hatches from the parental spheroid and swims away.

Are such distinctions phylogenetically significant? Some have believed they are and have postulated that species of *Volvox* with and without bridges as adults may have been derived from quite unrelated colonial forms (Crow 1918; Fritsch 1935).[7] Shaw considered such differences (and certain others)

[7] We and others have used rRNA sequence data to test the hypothesis of Crow (1918) that

so significant that he proposed that the genus *Volvox* be reserved for species (like *V. rousseletii*) that have robust cytoplasmic connections in the adult. The remainder of the species he proposed to subdivide among four new genera (Shaw 1916, 1919, 1922a–c). Shaw's proposal never became widely accepted in its original form, but it has become common practice to use certain of his terms to distinguish four "sections" of the genus *Volvox* (Smith 1944). In this taxonomic scheme, species lacking bridges in the adult (e.g., *V. carteri*) are placed in the section Merrillosphaera, species with stout bridges (e.g., *V. rousseletii*) are placed in the section Euvolvox, species with bridges of moderate size (e.g., *V. aureus*) are placed in the section Janetosphaera, and a species having very fine cytoplasmic bridges in the adult (*V. dissipatrix*) is placed in the section Copelandosphaera (Table 2.1). Each of these sections also has several other features (such as the structure of the extracellular matrix) that tend to distinguish it from the others (Smith 1944; Kirk, Birchem & King 1986).

Desnitski has taken an interesting and unique approach in trying to determine whether or not developmental differences among extant species of *Volvox* can be used to infer their phylogenetic relationships. Over the years, he has analyzed several species with regard to a range of morphological, physiological, and biochemical parameters related to gonidial development and the control of cleavage (Desnitski 1980, 1981a,b, 1982a–c, 1983a–d, 1984a,b, 1985a–c, 1986, 1987, 1990, 1995b). Combining those observations with information from the literature, he has concluded that there are four different developmental patterns in the genus. Program 1 (Table 2.2) involves prospective gonidia that are initially small, grow extensively, and then initiate several rapid, symmetrical divisions that occur in the absence of further growth. This clearly represents the primitive or ancestral program of *Volvox* development, because it closely resembles the asexual reproductive pattern that is observed in all other genera of volvocine algae (including *Chlamydomonas*) (cf. Figure 2.9). Program 2 (the *V. carteri* program described earlier) has the added feature of asymmetric division, which produces gonidia that start out considerably larger than their sibling somatic cells, and therefore mature more quickly than in program-1 species. Program 3 Desnitski inter-

whereas *Volvox* species that lack adult cytoplasmic connections are closely related to *Chlamydomonas* and the colonial volvocaceans, species with robust adult cytoplasmic connections are more closely related to entirely different unicellular and colonial forms, in what is now known as the family Haematococcaceae. Both sets of molecular studies have falsified Crow's hypothesis by showing that all species of *Volvox* thus far analyzed are much more closely related (by more than a factor of 10) to *C. reinhardtii* and the various genera of colonial volvocaceans than they are to the Haematococcaceae (Buchheim et al. 1990, 1994; Kirk, Kirk et al. 1990; Buchheim & Chapman 1991; Larson et al. 1992). Thus, in this coarse-grained sense, the genus *Volvox* clearly is monophyletic, with all species having shared a *C. reinhardtii*-like ancestor.

Table 2.1. *Morphological diversity and geographic distribution of the genus* Volvox

Section:[a] Species	Geographic distribution[b]	Cytoplasmic bridges in adults	Cell shape (polar view)	Cellular compartments
Euvolvox:				
V. amoeboensis	Africa			
V. barberi	Asia, NA			
V. capensis	Africa			
V. globator	Europe, Asia, NA	Broad	Stellate	Complete[c]
V. merrillii	Asia, Australia			
V. perglobator	North America			
V. prolificus	Asia			
V. rousseletii	Asia, Africa			
Janetosphaera:				
V. aureus	Af, As, Aus, Eur, NA	Intermediate	Round	Lack walls and floors
V. pocockiae	North America			
Copelandosphaera:				
V. dissipatrix	Asia, Australia	Fine	Round	Lack floors
Merrillosphaera:				
V. africanus	Af, As, Aus, NA			
V. carteri f. *kawas.*[d]	Asia			
V. carteri f. *nagar.*	Asia			
V. carteri f. *weism.*	Asia, Aus, NA	None	Round	Complete
V. gigas	Africa			
V. obversus	Asia, Australia			
V. powersii	North America			
V. spermatosphaera	North America			
V. tertius	As, Aus, Eur, NA			

[a]Adapted from Smith (1944); revised in light of subsequent observations.
[b]Data from Desnitski (1996a). Abbreviations: Af, Africa; As, Asia; Aus, Australia; Eur, Europe; NA, North America. (Observations from South America are lacking.)
[c]Each cell completely surrounded at a distance by a fibrous glycoprotein layer; walls of adjacent cellular compartments fused (Smith 1944; Kirk, Birchem & King 1986).
[d]Abbreviations of *V. carteri* formas: kawas., *kawasakiensis*; nagar., *nagariensis*; weism., *weismannia*.

prets as an intermediate between the ancestral program and program 4. Program 4 (in which gonidia begin to cleave while they are still very small, requiring light and continuous synthesis of macromolecules to continue cleaving, and in which embryos double in size between successive divisions) appears to characterize a majority of *Volvox* species, including *V. rousseletii*, which was described earlier. Indeed, this pattern of development is followed by 10 of the 11 species (in three sections of the genus) in which gonidia and cleaving embryos remain linked to adult somatic cells by cytoplasmic bridges; but it is not seen in any of the seven species in which such cytoplasmic connections are not retained in the adult.[8]

Although Desnitski obviously believes that programs 2–4 have evolved

---

[8] For the possible significance of this dichotomy, see Section 3.5.

Table 2.2. *Four types of developmental programs in the genus* Volvox[a]

Program number	Size of gonidial initials	Size of gonidia at cleavage	Rate of cleavage division	Growth between divisions?	Cleavage requires light?	Inhibitor sensitivity[b] of cleavage	Species exhibiting program
1	small	large	rapid	no	n.d.[c]	n.d.	*V. gigas* *V. pocockiae* *V. powersii* *V. spermatosphaera*
2	large[d]	larger	rapid	no	no	low	*V. carteri* *V. obversus*
3	small	large	slow	no	yes	intermed.	*V. tertius*
4	small	small	slow	yes	yes	high	*V. aureus* *V. globator*[e]

[a]Adapted from Desnitski (1995a).
[b]Rate and extent of cleavage arrest after addition of inhibitors of DNA, RNA, or protein synthesis.
[c]Not determined.
[d]As a consequence of asymmetric cleavage divisions.
[e]By all criteria studied, all other Euvolvox species and *V. dissipatrix* also exhibit program 4 .

from program 1, he has repeatedly indicated that the evolutionary relationships among these programs cannot be used to infer the evolutionary relationships among the particular species that presently exhibit them (Desnitski 1991, 1992, 1995a, 1996b). Only program 2 represents the kind of trait that evolutionary biologists would consider an informative character for phylogenetic reconstruction: a "synapomorphy," a derived feature shared by two or more species that appear on other grounds to be closely related (Duncan & Stuessy 1984). Program 1 represents a "symplesiomorphy" (an ancestral feature inherited without change from the common ancestor of the entire group); program 3 represents an "autopomorphy" (a derived feature possessed by only one species); and program 4, which is seen in three of the four sections of the genus, probably represents a "homoplasy" (an analogous, rather than homologous, feature that is the result of convergent evolution of similar traits in more than one line of descent). So, interesting as Desnitski's analysis is, it has provided us little in the way of clear insight into the phylogenetic relationships among the various extant species of *Volvox*.

2.4.3 *Molecular evidence indicates that most volvocacean taxa, including* Volvox, *are polyphyletic*

In the past there have been numerous suggestions that *Volvox* might be polyphyletic, but as far as I am aware there have been only two reports that have drawn the opposite conclusion: that the genus is monophyletic. Powers (1908) speculated that all the species of *Volvox* he was familiar with (7 of the currently recognized 18) could be arranged in "a more or less perfect series," implying a monophyletic relationship. However, neither his postulated evolutionary series nor the premises upon which he constructed it have won favor with subsequent authors. Much more recently, Nozaki and Itoh (1994) performed a cladistic analysis of selected morphological features in 25 species of volvocine algae, in an attempt to infer the phylogenetic relationships among them. In light of the limited number of *Volvox* species that were included in that analysis (one from each of the four sections recognized by Smith 1944) and the particular set of traits that were considered, it was not surprising that that study led to the conclusion that the four sections of the genus constituted a monophyletic group. What was surprising was that the study also led to the conclusion that members of the Euvolvox and Merrillosphaera sections – which usually have been thought to be the most disparate members of the genus – were the most closely related.[9]

The following year, however, when those same investigators collaborated with others to derive a molecular phylogeny of a similar set of volvocalean taxa, they came to dramatically different conclusions (Nozaki et al. 1995). The molecule used as the source of sequence information for the latter study was the chloroplast gene encoding the large subunit of ribulose bisphosphate carboxylase/oxygenase (*rbc*L). The most parsimonious tree derived from such data placed the four species of *Volvox* that they analyzed in three separate clades, and indicated that *V. rousseletii* (section Euvolvox) was significantly less closely related to *V. carteri* (section Merrillosphaera) than was either *V. dissipatrix* (section Copelandosphaera) or *V. aureus* (section Janetosphaera) (Figure 2.14). (Note that their molecular data also placed different representatives of *Pleodorina*, *Eudorina*, and *Gonium* in disjunct clades, affiliated with members of other genera.)

It is particularly significant that the tree derived by Nozaki and associates from analysis of a chloroplast gene (*rbc*L) indicated a relationship among the four sections of the genus *Volvox* that is congruent with the one we inferred from nuclear gene (rRNA) sequences (cf. Figures 2. 11 and 2.14). In both phylogenetic reconstructions the Janetosphaera section (*V. aureus*) is closely affiliated with *Pleodorina californica*, the Merrillosphaera and Copelandosphaera sections (*V. carteri* and *V. dissipatrix*, respectively) are closely affil-

---

[9] On the basis of that type of morphological analysis, Nozaki and his associates have proposed that species formerly grouped by others within the genus *Gonium* and the family Volvocaceae be removed from the Volvocaceae and placed in three genera and two new families (Nozaki 1993; Nozaki & Itoh 1994; Nozaki et al. 1996). Not all of those proposals were supported by their molecular-phylogenetic analysis, however (Nozaki et al. 1995).

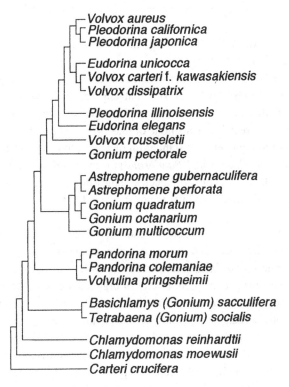

Figure 2.14 Phylogenetic relationships among selected green flagellates inferred from sequences of their chloroplast genes encoding the large subunit of ribulose bisphosphate carboxylase (*rbc*L) using the technique of maximum parsimony. Note that, as in Figure 2.11, this reconstruction affiliates *V. aureus* (section Janetosphaera) with *P. californica*, *V. carteri* (section Merrillosphaera) with *V. dissipatrix* (section Coplandosphaera), and the representative of section Euvolvox (*V. rousseletii* here and *V. capensis* in Figure 2.11) with *Gonium pectorale*. Note that this analysis also indicates that several other volvocacean genera, such as *Pleodorina*, *Eudorina*, and *Gonium*, are also polyphyletic. (Adapted from Nozaki et al. 1995, but the species that those authors identified as *Eudorina illinoisensis* is here given as *Pleodorina illinoisensis*, to conform with the convention that is used by most contemporary authors and elsewhere in this chapter.)

iated with one another, and the Euvolvox section (*V. rousseletii* in one study, *V. capensis* in the other) is off on a distant branch, next to a branch leading to *Gonium pectorale*. The congruence of these two trees strongly reinforces the opinion that the genus *Volvox* is polyphyletic. Moreover, our observation that *V. powersii* (arguably the most primitive member of the genus) (Vande Berg and Starr 1971; Desnitski 1995a) is placed by rRNA sequences in a clade distinct from the one containing other members of the section Merrillosphaera (with which it is usually grouped) suggests that the so-called genus

Volvox may include organisms representing at least four different lines of descent.

Volvox is far from being the only member of the family Volvocaceae exhibiting signs of polyphyly. Indeed, in each case in which we have sequenced the rRNA from more than one member of a genus or species, there is at least one member of that taxon that appears to be more similar to a member of another taxon than to one of the organisms with which it shares a Latin name (Figure 2.11). Thus none of these volvocacean genera or species appears to be monophyletic by standard molecular-phylogenetic criteria. The cases of *Eudorina elegans* and *Pandorina morum,* discussed earlier, provide particularly striking examples of species names that are applied to volvocaceans whose similarities in outward appearance belie their underlying genetic differences.

From such molecular-phylogenetic studies (which are as yet far from complete), we can draw the following important conclusions: Many, if not most, of the genus and species names now used to classify the Volvocaceae appear to cluster organisms at equivalent *grades* of organizational and developmental complexity, not *clades* of organisms that are related by common descent. Therefore, volvocaceans at a given level of organizational complexity cannot be inferred to have had a common phylogenetic history. The transition from organisms like *Eudorina*, which have only a single cell type, to organisms like *Volvox,* with two differentiated cell types, may have occurred repeatedly. There is also reason to suspect that certain extant volvocacean species may have arisen by "evolutionary reversal," that is, by a transition from a higher level of developmental complexity to a lower one.

2.4.4 There may have been several separate pathways leading from Chlamydomonas to Volvox

The picture of volvocacean phylogeny has been complicated further by the discovery that rRNA sequence data indicate that different species of *Chlamydomonas* (all of them members of the so-called *C. reinhardtii* lineage[10]) appear to be close relatives of different members of the Volvocaceae (Figure 2.15) (Buchheim et al. 1990; Buchheim & Chapman 1991). This raises the

---

[10] The "genus" *Chlamydomonas* is an extremely diverse assemblage of unicellular green flagellates consisting of about 450 named species (Ettl 1976), but modern molecular-phylogenetic evidence suggests that it is a paraphyletic assemblage representing at least four distinct lineages that are unified by their shared retention of certain primitive morphological features (Buchheim et al. 1990, 1996). Whereas the *C. reinhardtii* lineage appears to have given rise to all of the volvocaceans, other colonial green flagellates appear to have descended from quite different *Chlamydomonas*-like ancestors (Buchheim & Chapman 1991).

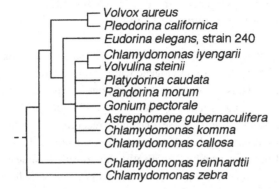

Figure 2.15 Phylogenetic relationships among selected volvocalean taxa inferred from analysis of variable regions of 18S and 28S rRNAs. As in Figure 2.11, line lengths are arbitrary. Note that although the available data do not permit statistically significant resolution of the branching order within a sizeable group of volvocacean taxa, three species of *Chlamydomonas* fall within this cluster, although *C. reinhardtii* does not. (Adapted from Buchheim & Chapman 1991.)

distinct possibility that different members of the Volvocaceae may have evolved from different *C. reinhardtii*–like unicellular ancestors, and hence are not really a monophyletic family after all. The available molecular data do not unambiguously resolve the relationships among these five species of *Chlamydomonas* (Buchheim & Chapman 1991), and there is as yet no way to rule out the possibility that one or more of these *Chlamydomonas* species may have evolved by simplification of a colonial volvocacean (Buchheim & Chapman 1991).[11]

In this context, it is important to note that molecular analysis suggests that *Volvox carteri* may have shared a common ancestor with *Chlamydomonas reinhardtii* as recently as 35 million years ago (Rausch et al. 1989), and with other species of *Chlamydomonas* even more recently than that (Buchheim & Chapman 1991; Buchheim et al. 1994). Thus, all of these experiments in volvocacean multicellularity and cellular differentiation appear to have begun far more recently (perhaps by a factor of more than 20) than similar experiments that led to the higher plants and animals.[12]

[11] A cladistic analysis that combined molecular and organismic data has grouped all of these *Chlamydomonas* species as a sister clade to all of the Volvocaceae (Buchheim et al. 1994), but that is not surprising, because the distinction between unicellular and multicellular organisms was weighted heavily in that analysis.

[12] Microfossils that have been called volvocaceans have been described from Devonian and even Precambrian strata (Kazmierczak 1975, 1976a,b, 1979, 1981). However, those fossilized organisms do not resemble any modern volvocaceans in overall morphology, and evidence that they were algal, as opposed to some other type of colonial eukaryotes, is lacking. Given the number of times that colonial flagellates appear to have evolved independently, and the relative paucity of other green-algal fossils in rocks of such age, the supposition that those

2.5   The genetic program for cellular differentiation in *Volvox* should be amenable to analysis

In short, the picture of the family Volvocaceae that is emerging from contemporary studies is that of a phylogenetically youthful and remarkably plastic group of organisms that have undergone numerous transitions in complexity, generating organisms with the capacity for germ–soma cellular differentiation (and possibly those that have recently lost such a capacity) repeatedly, in a relatively short time period. It seems unlikely that such transitions would have occurred so frequently and so rapidly unless the genetic changes required to generate a volvocacean with the capacity for germ–soma differentiation from an ancestor lacking that capacity were quite modest. Hence, we are increasingly optimistic that the molecular-genetic program regulating germ–soma cellular differentiation in a modern species of *Volvox*, such as *V. carteri*, may be simple enough to be analyzed in considerable detail by mere mortals.

Moreover, there is also reason to hope that once the molecular-genetic program for cell differentiation in a species of *Volvox* is understood in significant detail, it may be possible to trace the molecular-genetic pathway by which that program evolved. Because the entire volvocacean radiation appears to have taken place relatively recently, it can be hoped that modern organisms on any particular branch of the family tree may still contain within their genomes clues to the origins of the genes involved in regulating *Volvox* cytodifferentiation that will not have been sufficiently dimmed by the passage of time and the winds of genetic drift to have become illegible. Ultimately, of course, it should prove to be of even greater interest to determine the extent to which the genetic pathways leading to cytodifferentiation in *Volvox* species on separate branches of the family tree were similar or different, thereby addressing the question, Is there more than one way to skin this kind of evolutionary cat?

Before moving on to consider what is now known of the genetic, cytological, and molecular basis for germ–soma differentiation in modern *Volvox*, we will digress to consider what may have been the ecological forces and the cytological preconditions that fostered such rapid transitions among different levels of developmental complexity in the Volvocaceae.

fossils truly represent volvocaceans (let alone "Eovolvox," as some of them have been called) (Kazmierczak 1981) must be viewed with grave reservations.

3

Ecological Factors Fostering the Evolution of *Volvox*

... in the full blaze of Nebraska sunlight, *Volvox* is able to appear, multiply and riot in sexual reproduction in pools of rainwater of scarcely a fortnight's duration.

Powers (1908)

As the foregoing quotation so colorfully indicates, *Volvox* often appears in great abundance within days after the warm rains of early summer accumulate in depressions in the ground, and it is often joined by several of its colonial relatives. The ability of these volvocaceans to appear as soon as vernal pools are formed (or as soon as temperate lakes have thawed and stabilized in the spring) depends on a feature that they share with *Chlamydomonas* and many other green algae: When conditions begin to deteriorate toward the end of the growing season, they switch from asexual to sexual reproduction and produce dormant zygotes, or "zygospores," that are resistant to desiccation and freezing (Coleman 1983). These zygospores settle into the mud to wait out the adverse times; then, once favorable conditions return, they quickly germinate, and the germlings swim toward the surface and begin proliferating asexually.[1]

---

[1] Chapters 3 and 4 have been shaped in large measure by the work and thoughts of Graham Bell and Vassiliki Koufopanou and the literature in algal ecology and life-history biology to which they led me either through their writings or in personal communications. While their contributions are gratefully acknowledged, they obviously share no responsibility with me for any errors or distortions that may have occurred in my recasting of their insights.

Many volvocaceans (like many other algae) are astonishingly cosmopolitan, being found in similar environments around the world. For example, isolates of *Gonium pectorale* from Europe, Asia, and all parts of North America have been shown to be members of an interfertile population that is capable of sharing a single gene pool and is accordingly quite homogeneous at the DNA-sequence level (Stein 1966a,b; Coleman et al. 1994). Even "species" such as *Pandorina morum* that are extensively subdivided into reproductively isolated units, or syngens, are actually cosmopolitan, because members of a single syngen can be found on several different continents (see Chapter 2).

The secret underlying such cosmopolitanism apparently is that, lacking wings themselves, these algae have successfully exploited the wings of birds and the wings of the wind to carry them over long distances. Many studies have shown that volvocaceans and many other kinds of green algae can be transported in a viable state on or in the bodies of waterfowl (de Guerne 1888; Strøm 1925; Beger 1927; Smith 1933; Irénée-Marie 1938; Messikommer 1948; Schlichting 1960). Specifically, vegetative colonies of *Gonium, Pandorina,* and *Eudorina* have been shown to pass unscathed through the digestive tracts of migratory water birds, and it is estimated that by this means they can be transported between lakes as much as 150 miles apart in a single flight (Procter 1959). The presence of volvocacean zygospores in atmospheric dust can result in transportation over even greater distances.[2] This means that a single population of volvocaceans can be spread around the world, to be subjected to selective forces in a wide range of slightly different habitats.

3.1 Volvocaceans compete for resources in transitory, eutrophic waters

Three features characterize the environments in which the volvocaceans flourish: (1) They contain quiet, often relatively turbid, standing waters in which flagellar motility provides an advantage. (2) They are eutrophic and therefore provide, at least transiently, fairly abundant supplies of essential elements such as nitrogen, phosphorus, and sulfur. (3) They are subject to rapidly changing conditions as a result of a combination of biotic and abiotic factors.

As will be discussed in more detail shortly, certain species of *Volvox, Pandorina,* and *Eudorina* are frequently found as short-lived, early summer

---

[2] Coleman (1996a,b) has argued, on the basis of the current distribution patterns of volvocacean species and syngens, however, that dispersal by wind (which occurs preferentially in a latitudinal direction) probably does not occur nearly as frequently as dispersal by birds (which occurs primarily in a longitudinal direction).

"blooms" in certain types of permanent, eutrophic lakes. However, some of these same species and their close relatives are even more frequently found in much smaller, temporary bodies of water, particularly on or near farmlands, and they can even be found flourishing in tiny puddles formed where rainwater has collected in cattle hoofprints. Mary Agard Pocock has documented that what Powers said of *Volvox* in the quotation at the beginning of this chapter might equally well be said for many of its simpler relatives. She included Powers's Nebraska collecting grounds among the many sites she visited on four continents, and she reported that it was not at all unusual to find several genera of volvocaceans – and often two or three species of a single genus – in a single small pool (Rich & Pocock 1933; Pocock 1933a, 1938, 1953, 1955, 1962, and personal communication via letter to Gary Kochert, copied by him to me). Presumably that indicates that rather similar factors favor the growth and survival of various genera and species of volvocaceans, and hence competition among them for such resources must be quite keen. It can also be presumed that the ability of multiple species of volvocaceans to cohabit such environments year after year is a result of the rapidly changing nature of such environments, which prevents establishment of an equilibrium in which one species would become dominant.

The vast majority of places where the volvocaceans are found are small, temporary bodies of water (Iyengar 1920, 1933; Pocock 1933a,b, 1938, 1953, 1955; Rich & Pocock 1933; Smith 1933; Iyengar & Desikachary 1981), but because of their very ephemeral nature such sites are intrinsically difficult to study in a detailed, controlled way in successive years in an attempt to determine what factors appear to favor the proliferation of one type of alga over another. Hence, the best ecological information about the volvocaceans has been derived from studies of permanent ponds and lakes. However, because similar species are found as regular inhabitants of both permanent and transient bodies of water, and because it appears that the conditions that foster their growth are nearly as transitory in lakes as they are in rain puddles, it may be possible to extrapolate (with certain obvious precautions) from studies of lakes to the vernal pools where the volvocaceans are also found.

3.2   Even in permanent bodies of water, the volvocaceans flourish only briefly each year

The most consistent feature of eutrophic lakes and ponds in temperate zones is change. They are never at equilibrium (Hutchinson 1961), and "as environmental characteristics vary with season, so the selective advantage moves in train, from one category [of algae] to another" (Reynolds 1984b). Because

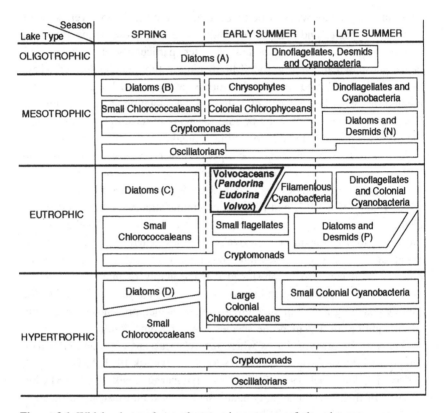

Figure 3.1 Widely observed annual succession patterns of algae in permanent temperate-zone lakes of different trophic (nutrient) levels. In eutrophic lakes in spring, as long as thermal mixing is occurring, diatoms and other small, nonmotile algae with high intrinsic reproductive rates dominate. But once thermal mixing has ceased and a thermocline has been established, the volvocaceans and other flagellates that are able to control their place in the water column dominate the community. However, once the fixed-nitrogen supply of the lake has been exhausted by such algae, they are succeeded by nitrogen-fixing cyanobacteria (which form gas-filled vesicles that provide the buoyancy required to float near the surface). Then as thermal mixing begins again in the late summer or early fall, new assortments of diatoms and desmids bloom. The letters in parentheses following "Diatoms" and "Diatoms and Desmids" identify different species complexes (Reynolds 1984a); they are included here to indicate that the diatoms, for example, that are found in abundance in lakes of different trophic levels and/or in a single lake at different seasons are different. The same is generally true of other taxonomic groups that are named at more than one point in this diagram. (Adapted from Reynolds 1984b, with permission of the author and Blackwell Science Limited.)

seasonal changes in temperature, light intensity, and other factors follow similar patterns from year to year, so also do seasonal patterns of algal growth. Although the details may vary from year to year and from lake to lake, the overall pattern in which one type of alga appears, flourishes briefly, and then disappears – to be succeeded by another type – is highly predictable for any given type of lake year after year (Pearsall 1932; Hutchinson 1967; Moss 1972, 1973a–c; Owens & Esaias 1976; Reynolds 1976, 1984a,b; Happey-Wood 1988).

When lakes thaw in the spring, the dense, cold meltwater released from the ice begins to sink, generating strong convection currents that stir the water to considerable depths, bringing nutrients and organisms up from the bottom toward the sunlight, and initiating a new season of algal growth. In such turbulent waters, flagellar motility provides little advantage, and small, non-motile algae with very high reproductive rates, including diatoms and chlorococcalean green algae, flourish (Figure 3.1). As solar heating of such a lake continues, however, the surface water expands, and a density gradient is created in the water column that abruptly arrests further convective stirring. That creates the thermal stratification that so many divers have experienced – with some shock to their nervous systems – as they have plunged through the warm upper layers of a lake to find themselves suddenly in water many degrees colder.

Once thermal stirring ceases and the lake becomes stratified, nonmotile algae find themselves sinking out of the euphotic zone (the zone where sunlight is intense enough to promote vigorous photosynthesis), and the advantage passes to species – notably volvocaceans such as *Pandorina, Eudorina,* and *Volvox* (Figure 3.1) – that can control their position in the water column and sustain their place in the sun. In the warm, quiet, well-lit surface waters, rich in mineral elements that have been brought up from the bottom by the earlier thermal mixing, the volvocaceans rapidly proliferate and form a ''bloom'' that often turns the water green. But in the process, they soon deplete the water in the euphotic zone of the resources necessary for their growth, most notably phosphorus and nitrogen.

In most lakes, phosphorus is the limiting element that determines what total biomass of algae (and hence small animals that feed on algae) the lake will sustain over a growing season (Moss 1973c; Owens & Esaias 1976; Schindler 1977). Nitrogen and carbon are required in greater amounts than phosphorus, but both of those elements can be obtained from the atmosphere by some or all algae, whereas phosphorus cannot.[3] Most of the phosphorus

---

[3] Obviously, the level of fixed nitrogen in the water affects the relative abundance of algae with

of an aquatic ecosystem is present in the sediments underlying the water, a great deal of which may be released each spring during the period of thermal stirring. However, once a lake becomes stratified, algae growing in the euphotic zone will invariably deplete that zone of phosphorus rather rapidly, routinely causing free-phosphorus concentrations to fall into the submicromolar range (Owens & Esaias 1976). It has repeatedly been shown experimentally that when inorganic phosphorus is added to the water (either under field conditions or in the laboratory), the suspended organisms sequester it rapidly, causing the free phosphorus concentration to fall with a half-time as short as 15 minutes (Owens & Esaias 1976). Although the growth rates of many species of green algae in natural situations can regularly be shown to be limited by phosphorus availability (Schindler 1971, 1977), it has also been shown that algal growth rates generally are not proportional to the external phosphorus concentrations, but rather to the concentration of phosphorus stored within the organism (Fuhs 1969; Fuhs et al. 1972; Rhee 1973). Indeed, measurements of free phosphorus in the water provide precious little insight into the dynamics of algal competition in such ecosystems, because algae can reproduce exponentially in water in which the free-phosphorus concentration is below measurable levels (Rhee 1973; Owens & Esaias, 1976); clearly they do this by exploiting internal phosphorus stores that were built up during periods when phosphorus was more abundant in the external medium (Rhee 1973, 1974). (The nature of such stores will be discussed shortly.) Thus, there can be little doubt that when phosphorus is available in the environment, competition among the algae to sequester it quickly must be intense.

We will return to the preceding theme in a moment. However, first it is worth pointing out that when phosphorus becomes limiting in the euphotic zone of a lake, *Volvox* is capable of demonstrating in a highly dramatic manner why it is selectively advantageous to have flagellar motility – and the capacity for phototaxis and chemotaxis – in an unstirred, nonhomogeneous environment. Gilbert Smith observed many years ago that lake populations of *Volvox* tend to accumulate near the surface during the day and then descend to considerable depths at night (Smith 1918). That observation has been confirmed (Utermöhl 1924, 1925) and extended to other volvocaceans, such as *Eudorina*, over the intervening years (Hutchinson 1967). Only more recently has it been discovered, however, just how incredibly far-ranging such diurnal vertical migrations can be. In a deep, clear lake in Africa, detailed

and without nitrogen-fixing ability, but it does not normally set an upper limit on the total algal biomass a lake will support (Schindler 1977).

observations demonstrated that *Volvox*[4] populations spent the daylight hours in the euphotic zone (where the soluble-phosphorus concentration was below the limit of detection, < 2 µg/L), but then swam downward at twilight to reach cool, dark regions (as deep as 30–40 meters below the surface!) where the soluble-phosphorus concentrations were 500–1000 µg/L (Sommer & Gliwicz 1986). Then, around dawn, the *Volvox* population swam back to the euphotic zone at rates in excess of 5 m/h! In that vertical migration, *Volvox* exceeded the rates and magnitudes achieved by marine dinoflagellates during their well-known diurnal migrations (Hasle 1950; Eppley, Holm-Hansen & Strickland 1968; Blasco 1978) and far exceeded the speed and range of any other migration reported for fresh-water algae (Berman & Rodhe 1971; Tizer 1973; Sibley, Herrgesell & Knight 1974; Frempong 1984). The function of this daily migration by *Volvox* almost certainly is to allow them to store photosynthesis products during the day and phosphorus during the night, as has been documented in the case of other algae that undergo a daily vertical migration (Salonen, Jones & Arvola 1984). Many other volvocaceans apparently follow a similar strategy, but usually in water of lesser depth and with migrations of more modest dimensions.

As successful as they appear for a time, within a matter of a few weeks after their first appearance the volvocaceans (and other organisms that overlap with them in time and space) will typically deplete the upper reaches of a lake of most of its fixed nitrogen (Moss 1973b,c). At that point the volvocaceans typically will engage in sexual reproduction, produce zygospores that settle to the bottom, and disappear from the water as abruptly they appeared, releasing the habitat for exploitation by cyanobacteria that are capable of floating at the surface (by virtue of gas-filled vesicles) and fixing atmospheric nitrogen (Reynolds 1984a,b) (Figure 3.1).

In temporary ponds and puddles on nutrient-rich soils, volvocaceans exhibit similar annual cycles of appearance and rapid asexual proliferation, followed by sexual reproduction and dormancy, but the factors regulating their life cycles in vernal ponds have been less well studied than those in lakes. Presumably, however, even in relatively shallow waters (which are frequently turbid) their flagellar motility is an important source of their success, as it permits them to move about to garner resources, such as sunlight and essential mineral elements, that are nonuniformly distributed (either intrinsically, or as a result of their own past consumption patterns).

This, however, brings us to a new question. Assuming that active motility provides green flagellates with a distinct biological advantage in unstirred

---

[4] Species not identified.

aqueous environments with patchy distributions of resources, why are there so many kinds of green flagellates in such environments? Where *Volvox* is found, one often also finds smaller colonial volvocaceans, *Chlamydomonas*, and other unicellular green flagellates. In many cases the unicells are initially the dominant members of the green-algal community. Given that the first green flagellates to appear in such sites historically were undoubtedly unicellular, and that unicells have persisted in such sites for many eons and often are the first green algae to become abundant in the spring, what environmental factors have favored the appearance and survival of progressively larger, multicellular green flagellates? There is another way to put this question: Given that *Chlamydomonas* is still found in such environments, why is there also *Volvox*? Alternatively, however, we might also ask this: Given that there is now *Volvox* in such environments, why is there also still *Chlamydomonas*?

3.3 Different aspects of the environment favor green algae of different body sizes

The preceding questions can be generalized, of course. The fossil record indicates that for many hundreds of millions of years there have been progressive increases in the sizes of the largest organisms on earth (Bonner 1965, 1988), but throughout that long period of time, tiny organisms have persisted and flourished. What accounts for these twin facts? It seems self-evident that there must be some combination of costs and benefits that accompany increased body size, such that under certain conditions the benefits of larger size outweigh its costs, whereas under other conditions the reverse is true. Let us first consider the costs of larger size.

3.3.1 Smaller organisms have an intrinsic advantage under "optimum" conditions

In the early 1960s, as François Jacob and Jacques Monod were challenging others to test the possibility that genetic regulatory mechanisms similar to the ones they had discovered in bacteria might be involved in eukaryotic cellular differentiation (and with tongue planted firmly in cheek), they coined the aphorism that "whatever is true of *E. coli* obviously must also be true of elephants." Among the many exceptions to that well-known and charming aphorism that might be listed, one surely ought to include the reproductive rates of those two species.

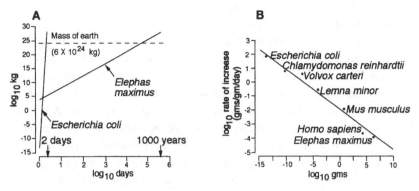

Figure 3.2 Size dependence of growth rates. **A**: Lines represent the theoretical changes in mass over time for a population of *E. coli* and a population of elephants if they could be fed adequately to maintain maximal reproductive rates. Although a pair of elephants weighs about 10^{16} times as much as a pair of *E. coli* cells, if both kinds of organisms were provided all of the nutrients that they and their progeny could utilize, it would take about 1,000 years for the elephants to produce the same mass of offspring that the bacteria would produce in two days – which would be many times the mass of our solar system. This is but one consequence of the relationship summarized by the Blueweiss line (**B**), which indicates that for organisms of different sizes, the maximum growth rate under optimum conditions (in grams of new biomass per gram of existing biomass per day) decreases with the 0.25 power of body weight. (**B** adapted from Bell 1985 and Blueweiss et al. 1978, except that the growth rates plotted for *Chlamydomonas* and *Volvox* are those observed in our laboratory.)

Under moderately favorable conditions it takes far less than an hour for *E. coli* to double its mass and reproduce itself, but rare indeed is the elephant that can reproduce in less than an hour! If we were to run an experiment in which we took two healthy young adult elephants (one of each gender) and two healthy *E. coli* cells and provided each with all of the nutrients that it needed to grow and reproduce at its maximum rate for just two days, we would quickly learn two things. The first would be that it is much cheaper to keep a pair of elephants well fed for a lifetime than it is to keep an *E. coli* culture well fed for just two days! The second would be that as long as two kinds of organisms are provided with all the resources that they need for optimum growth, the smaller one will always outreproduce the larger one.

As most schoolchildren learn these days, if an *E. coli* culture derived from one or two cells could be continuously supplied with all the nutrients required for optimum growth, it would produce a mass of progeny nearly 1,000 times greater than the mass of the earth in less than two days. However, it would take a pair elephants and their descendants about 10 centuries to generate the same mass of progeny, even though they would have had a 16-orders-of-magnitude head start (Figure 3.2A)!

Nevertheless, *E. coli* and elephants do have a certain commonality with respect to growth, because both follow a pervasive rule relating growth to body size. The "Blueweiss rule" (Blueweiss et al. 1978) states that the maximum rate of growth achieved by organisms under optimum conditions (or the intrinsic rate of increase, *r*, in units such as grams per gram per day) declines with approximately the 0.25 power of increasing body size (Figure 3.2B). This relationship indicates quantitatively the point that was made qualitatively earlier: As long as "optimum conditions" for growth of all species are maintained, small organisms always have a substantial reproductive advantage over large ones. But in nature, conditions seldom remain optimum for any organism for very long. Essential resources are finite in quantity. And predators lurk.

3.3.2 Predation favors organisms above a threshold size

The primary predators with which green flagellates must cope are invertebrate filter feeders, such as rotifers and small crustaceans (cladocerans and copepods) (Bell 1985). Such organisms have life cycles that resemble those of the volvocine algae, in that they are capable of rapid asexual proliferation during early summer, but when conditions begin to deteriorate they switch to sexual reproduction and produce dormant forms that overwinter in the substrate and become reactivated in the spring (Meglitch 1967; Brusca & Brusca 1990). Often such predators appear in vernal ponds and lakes shortly after the volvocine algae, and as they feed and proliferate rapidly, they can have a dramatic effect on the relative abundance of various species. For example, in one lake studied in France, three species of rotifers reached a combined population size exceeding a million per cubic meter, and at the peak of their grazing activities they were able to filter the total volume of water in the lake in less than a day and a half (Lair & Ali 1990). Meanwhile, the small crustaceans present were also capable of filtering all of the water in the lake about once every two days (Lair & Ali 1990).

The size range of the prey that such little herbivores can consume is governed by the dimensions of their filter-feeding apparati and certain behavioral adaptations; they seldom take prey exceeding 8,000 μm³ in volume (Burns 1968; McQueen 1970; Kryutchkova 1974; Pilarska 1977; Hino & Hirano 1980; Bogdan & Gilbert 1984; Bell 1985). Unicellular algae, including various species of *Chlamydomonas*, span about five orders of magnitude of size (Huber-Pestalozzi 1961; Bell 1985), but few of them exceed 8,000

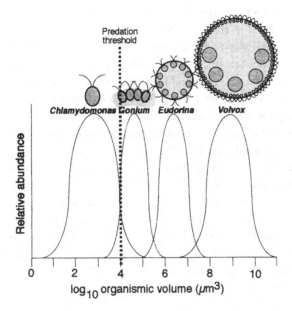

Figure 3.3 Size ranges for members of four volvocine genera. The dashed line represents the effective maximum size of prey consumed by filter feeders such as rotifers and caldocerans. Members of various *Chlamydomonas* species span four or five orders of magnitude in volume, but nearly all of them are potential prey for filter feeders, whereas all but the very smallest colonial volvocaceans escape such predation. The smallest *Volvox* spheroids, however, are more than 1,000 times that predation limit. The concept, the position of the predation threshold, and the size distributions of *Chlamydomonas* spp. are from Bell (1985). Data plotted for the other genera are for total organismic volumes (including extracellular matrix) and are taken from a variety of sources including Huber-Pestalozzi (1961), Iyengar & Desikachary (1981), and Smith (1944).

μm^3 in volume (Figure 3.3). Thus, nearly all such unicells are game for filter feeders. In contrast, all but the very smallest colonial algae are too large to be consumed by such filter feeders (Figure 3.3).[5]

In a lake such as the one in France mentioned earlier, a *Chlamydomonas*

---

[5] Although too large to be consumed by filter-feeding rotifers, *Volvox* is not free from rotifer predation. Many species of so-called raptorial rotifers have evolved a completely different feeding mechanism that permits them to attack much larger organisms than are accessible to their filter-feeding cousins. Among them are several (including, most notably, *Ascomorphella volvocicola*) that are able attack and consume *Volvox*. When such rotifers penetrate an adult *Volvox* spheroid, they lay eggs in the developing juveniles. Thus, when these juvenile spheroids hatch out, they are already infected with juvenile rotifers that feed on the spheroids from the interior and repeat the process. Such rotifer infections can be devastating to a *Volvox* population (Rich & Pocock 1933; Ganf, Shiel & Merick 1983; Heeg & Rayner 1988), but they are reported relatively infrequently. It has been suggested (Rich & Pocock 1933) that such infections are most likely to occur when a *Volvox* population has been abundant in a single pond for an unusually long period. Probably the factor limiting the success of such rotifers is their limited ability to initiate an infection of the potential prey in spring after both species have gone through a winter dormant period.

population could, in principle, be cleared from the water in much less than two days by filter feeders. Little wonder, then, that those authors reported that although small algae were present and proliferating rapidly in that lake, they did not accumulate in significant numbers, and the standing green-algal biomass of the lake was composed predominantly of large species (Lair & Ali 1990). Little wonder, also, that many *Chlamydomonas* species of modest size, including the well-known *C. reinhardtii*, are not truly aquatic, but live in moist soils that are inaccessible to invertebrate filter feeders (Iyengar 1920; Huber-Pestalozzi 1961; Harris 1989).

Although escape from predation appears to provide a plausible selective advantage for aquatic colonial algae that are too large to be consumed by filter feeders ($>8,000$ μm^3), it cannot readily be used to account for the evolution of colonial and multicellular volvocaceans that are orders of magnitude larger than this predation threshold (Figure 3.3). For a possible explanation of the success of these larger forms, including *Volvox*, we must turn to another aspect of algal biology.

3.3.3 The larger the colony, the more efficient the uptake and storage of essential nutrients

As mentioned earlier, essential minerals are initially abundant in the eutrophic environments inhabited by green flagellates, and hence the algae grow and proliferate rapidly. Briefly, the advantage may go to the smallest algae, because of their higher intrinsic growth rates, as explained earlier. But as the algae proliferate, they rapidly deplete the surrounding water of nutrients essential for their continued growth and reproduction (most rapidly and most notably phosphorus) (Schindler 1971; Owens & Esaias 1976). Once the level of free phosphates in the water drops into the submicromolar range, the reproductive rate of an alga will depend on the extent to which it has been able to capture and store this precious limiting resource.

The uptake, by algae, of nutrients such as phosphates[6] is a saturable enzymatic process (Rhee 1973; Owens & Esaias 1976); thus it can be described by the well-known Michaelis-Menten equation, $v_o = V_{max}[S]/K_m + [S]$, in which v_o is the initial rate of uptake of the nutrient, V_{max} is the maximum rate of uptake that can be achieved at saturating external concentrations of that nutrient, $[S]$ is the concentration of the nutrient at the time v_o is measured, and

---

[6] Although the concepts developed here probably apply, at least qualitatively, to the acquisition of other essential elements, such as nitrogen (N) and sulfur (S), phosphorus (P) is the element that is believed to be most often limiting in nature, and therefore it is the element for which uptake has been most widely studied.

K_m is the external concentration of the nutrient at which uptake occurs at half-maximal velocity. It has been argued that the rate of uptake of phosphorus when it is present at very high concentrations can be increased by changing either the K_m, or the V_{max} for the uptake process, or both (Bell 1985). However, the apparent values of K_m and V_{max} for transport that are measured for intact organisms are strongly influenced by the extent to which the nutrient is allowed to accumulate within the cytoplasm and exert feedback inhibition on the transport process by simple mass action. Therefore, the measured K_m and/or V_{max} for uptake of a nutrient can be increased not only by changing the nature and amounts of the transport enzyme present in the plasmalemma but also by somehow sequestering the nutrient that has reached the cytoplasm, so that it does not exert mass-action feedback. For a unicell, the principal way of doing this with phosphorus is to convert it to an intracellular store of polyphosphates (Harold 1966; Kulaev 1979). For example, when *Chlamydomonas* cells are fed isotope-labeled inorganic phosphates, they quickly take it up and accumulate it as polyphosphates, often with no significant increase in the internal concentration of orthophosphate, and at equilibrium more than 90% of the phosphorus potentially available for synthesis of nucleic acids may be present in the form of polyphosphates (Hebeler et al. 1992; Siderius et al. 1996). Moreover, it has repeatedly been shown that it is the magnitude of such internal polyphosphate stores, not the concentration of phosphorus in the environment, that determines the growth rate of a variety of algae (Fuhs 1969; Fuhs & Canelli 1970; Fuhs et al. 1972; Rhee 1972, 1973).

However, although polyphosphates do not exert direct mass-action feedback on phosphate uptake, as they accumulate they act as noncompetitive inhibitors of the transport system (Rhee 1973); thus the V_{max} for phosphate uptake falls as polyphosphates accumulate, imposing an upper limit to the amount of phosphorus that unicellular algae can accumulate when it is abundantly available in the surrounding water (Rhee 1972, 1973, 1974). In principle, a multicellular organism not only can develop such intracellular stores of polyphosphates but also can convert intracellular phosphorus compounds to chemical forms that can be transported to, and stored in, the extracellular matrix. That would have the twin effects of (1) increasing the amount of the nutrient that could be taken up without significant inhibition of the transport process and (2) building up a second kind of phosphorus store that could be used later on for growth and reproduction, after the concentration of free phosphorus in the surrounding water had fallen below measurable levels.

Graham Bell has reviewed and summarized the available data on phosphate uptake by organisms differing in size and cell number and has concluded that large multicellular algae do indeed have a distinct advantage with respect to phosphate uptake, particularly at very high external phosphate concentrations, such as exist at the time that competition among green flagellates for nutrients begins in early summer (Bell 1985). He states that because the efficiency of nutrient uptake at very high external concentrations can be increased by increasing either K_m or V_{max} or both, the product of K_m times V_{max} provides a useful measure of the capacity of an organism to take up and store that nutrient when it is abundantly available. When $\log(K_m \cdot V_{max})$ for phosphate uptake is plotted against log (total cytoplasmic volume)[7] for unicellular and multicellular organisms, marked differences are observed: For unicells, the value of $K_m \cdot V_{max}$ for phosphate uptake is only weakly size-dependent; however, multicellular algae exhibit $K_m \cdot V_{max}$ values that are orders of magnitude above the values predicted by the regression line obtained for the unicells (Figure 3.4A). Bell infers that this qualitative difference between the two types of organisms is due to the ability of multicellular volvocaceans to sequester the nutrient in the extracellular matrix (which, as mentioned earlier, increases in relative size with increasing cell number), and he speculates that such considerations may provide the explanation for the fact that when the maximal growth rates of various unicellular algae and colonial volvocaceans were compared under identical conditions, the volvocaceans exhibited growth rates significantly and substantially greater than those exhibited by the unicells (Figure 3.4B). For this reason, he muses that "paradoxically, one of the most important structures of the volvocacean colony may be the space in the middle" (Bell 1985).

As the concentration of phosphates in the surrounding water falls, the transport advantage of the large multicellular volvocaceans decreases proportionately, so that as the external concentration approaches zero, phosphate-uptake values for the multicellular species fall on the same regression line as for unicellular organisms (Bell 1985). However, at that point, large multicellular volvocaceans in an unstirred natural environment would realize a different advantage, as discussed in Section 3.3.2: They could swim more effectively to a different part of the habitat where phosphate levels had not yet been depleted. In principle, all green flagellates can swim back and forth between regions where different resources are available (storing photosynthate during the day and essential nutrients at night, for example), because all of them are capable of both chemotaxis and phototaxis. In practice, however, larger multicellular flagellates have a higher absolute swimming speed (Bonner 1965) and can therefore swim more rapidly between the regions where different essential resources are available. Moreover, once they reach

---

[7] Log transforms are used in such cases not only because they permit visualization of relationships between the dependent and independent variables over several orders of magnitude but also because they frequently linearize regressions.

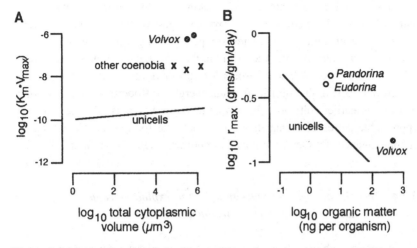

Figure 3.4 Maximum rates of phosphorus (P) uptake under saturating conditions (A)(given by $K_m \cdot V_{max}$) and maximum growth rates under identical conditions (B) as a function of size. **A:** The regression line shown was calculated from P-uptake measurements for 20 taxa of of unicellular green algae; its slope indicates that for such unicells the ability to capture and store P when it is extremely abundant increases only modestly with size. However, the corresponding values for two species of *Volvox* (*V. aurues* and *V. globator*; filled circles) and three other genera of multicellular green algae (*Coelastrum, Dictyosphaerium,* and *Pediastrum*; crosses) lie substantially above this line. These data indicate that when P is highly abundant, *Volvox* may be more than 1,000 times more efficient at capturing this resource than unicells are. **B:** The regression line shown was calculated from the maximum growth rates for 10 species of unicellular green algae under controlled laboratory conditions. Under the same set of conditions, the three volvocaceans examined all had significantly higher growth rates (by about a factor of 2) than would have been predicted from the regression line derived for the unicells. (Adapted from Bell 1985, in *The Origin and Evolution of Sex,* ed. H. O. Halvorson & A. Monroy, copyright© 1985 Alan R. Liss, Inc., adapted by permission of Wiley-Liss, Inc., a subsidiary of John Wiley & Sons, Inc.)

a region where a limiting nutrient is abundant, larger organisms again exhibit greater efficiency in taking up and storing that nutrient.

The data summarized in Figure 3.4A indicate that both lake-inhabiting species of *Volvox* studied (*V. aureus* and *V. globator*) have voracious appetites for P when it is abundantly available ($K_m \cdot V_{max}$ values three or four orders of magnitude greater than comparable values for unicells). Moreover, there is extensive evidence that the extracellular matrix of *Volvox* contains an abundance of sulfated and phosphorylated glycoproteins rich in N, P, and S (Sumper & Wenzl 1980; Wenzl & Sumper 1981; Gilles, Gilles & Jaenicke 1983). But is there any evidence that *Volvox* can use these elements that are stored in the extracellular matrix for reproduction and growth when external supplies dwindle? Adequate data are lacking, but the study from which the data in Figure 3.4A were obtained provide suggestive evidence that they can: Whereas the unicellular organisms stopped growing within about one day after P was withdrawn from the medium, the investigators found it necessary to withhold P from *Volvox* spheroids for up to five days before their growth ceased due to P limitation (Senft, Hunchberger & Roberts 1981), indicating that during earlier cultivation in normal medium they must have sequestered appreciable stores of P that they could mobilize for growth when external supplies waned. Further study of this important matter is warranted.

3.3.4 In field studies, volvocacean ability to exploit eutrophic conditions has been found to be correlated with size

If the superior ability of larger volvocaceans to hoard resources when they are abundantly available is of any selective advantage in the real world outside the laboratory, then it should be possible to demonstrate that as a pond in nature becomes more eutrophic, various volvocine algae will become increasingly abundant, in direct proportion to their size. The best data bearing on this point come from a two-year study of eight experimental ponds of similar sizes and shapes that were supplemented with different amounts of a mixed nitrogen-phosphorus fertilizer (O'Brien 1970; deNoyelles 1971). All of the deNoyelles-O'Brien study ponds had been allowed to equilibrate for several years after their construction, whereupon they were found to be equally oligotrophic and lacking in significant volvocacean populations (as is to be expected for oligotrophic ponds) (Figure 3.1). During the two-year study period, two ponds were left unfertilized as controls, and six were regularly fertilized (two ponds at each of three levels) such that the eight ponds ranged from oligotrophic to highly eutrophic. Extensive sampling was done throughout the study period to analyze the chemical composition and resident populations of algae (deNoyelles 1971) and invertebrates (O'Brien 1970) in each of the ponds over time.

Koufopanou and Bell (1993) analyzed the resulting data to compute the abundance of four categories of volvocine algae relative to the total algal biomass present in each of the ponds. Their analysis indicated that as nutrient levels were increased, not only did the total algal biomass in the ponds increase as anticipated but also each of the four categories of volvocine algae increased in *relative* abundance, becoming a progressively larger component of the algal community. Most importantly, however, that effect of nutrient availability on abundance varied strongly with the size and amount of extracellular matrix that the various organisms possessed: *Chlamydomonas* exhibited only a modest increase in relative abundance, the colonial volvocaceans showed intermediate effects, and *Volvox* exhibited the greatest effect (Figure 3.5). Other studies in which fish ponds were fertilized in various ways (Januszko 1971; Alam, Habib & Begum 1989), although less carefully controlled and detailed, led to similar conclusions (Koufopanou & Bell 1993).

Thus, increased efficiency of resource acquisition in a eutrophic environment provides a plausible selective advantage for volvocaceans of increased size. But how are we to account for the tendency to differentiate a mortal soma (a tendency that becomes increasingly stronger with increasing size in the Volvocaceae)?

3.4 In eutrophic environments, differentiation of soma and germ is reproductively advantageous

On superficial analysis, a germ–soma division of labor such as occurs in *Volvox* does not appear as though it should be selectively advantageous. It results in the great majority of the cells being sterile and eventually dying, with only a tiny minority of the cells producing offspring. How is it possible for an organism such as *Volvox*, which in some species produces only two or three offspring per 50,000-cell spheroid, to compete successfully with a simpler relative, such as *Eudorina*, that can produce 32 offspring per 32-cell adult? Bell has postulated that the answer lies in an extension of the resource-acquisition phenomena that have just been discussed (Bell 1985).

3.4.1 Hypothesis: The soma acts as a source for nutrients, and germ cells act as a sink

Findings summarized in the preceding section suggest that the extracellular matrix in the center of a volvocacean colony can be thought of as a "sink" into which cells at the periphery of the colony (the "source") can pump acquired nutrients so as to maintain a high rate of nutrient uptake and thereby outcompete unicells for these resources. Bell has postulated that an extension of the source–sink model may account for the success of organisms like *Volvox* with a germ–soma division of labor (Bell 1985). An organism in which one set of cells harvest essential resources while the cells of a second set use them to produce progeny should be successful only if the organism can thereby produce a greater mass of progeny than it would have if all cells had participated in both processes. Because the effect of a nutrient sink on rates of uptake is realized only when external concentrations of those nutri-

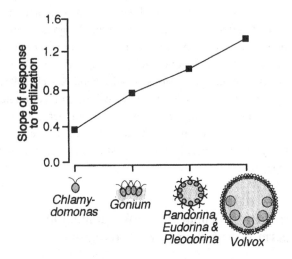

Figure 3.5 Changes in abundance (relative to all other algae) for four groups of volvocaleans in a set of experimental ponds treated with different levels of inorganic fertilizer. The entire algal biomass increased substantially with increasing levels of inorganic nutrients, but the fact that all of the slopes plotted here are positive indicates that the volvocaleans responded more strongly than most other algae and constituted a progressively larger fraction of the total biomass as nutrient levels were raised. Moreover, among these four groups of volvocaleans, this effect increased significantly with increasing organismic size and complexity. (The line connecting the data points is provided merely to guide the eye; it has no objective significance.) (Adapted from Koufopanou & Bell 1993, Soma and germ: An experimental approach using *Volvox*, *Proc. R. Soc. Lond. Ser. B. Biol. Sci.,* **254,** 107–13, with the permission of The Royal Society and the authors.)

ents are high, the Bell source–sink model predicts that a germ–soma division of labor should be advantageous only in a nutrient-rich environment.

3.4.2   Test: Gonidia produce larger progeny in the presence of an intact soma and rich medium

Koufopanou and Bell tested the predictions of the Bell source–sink model by measuring the reproductive performance of *V. carteri* gonidia cultured in three states and at four different concentrations of nutrients (Koufopanou & Bell 1993). The three states compared were gonidia in intact spheroids, gonidia associated with sheets of somatic cells in broken spheroids, and gonidia in complete isolation from somatic cells and extracellular matrix. Different concentrations of nutrients were provided via a series of dilutions of standard *Volvox* medium. In all states and all media, gonidia grew and divided to produce progeny spheroids (Figure 3.6). However, at all concentrations of

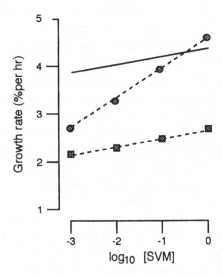

Figure 3.6 Growth rates in different dilutions of standard *Volvox* medium (SVM) for *V. carteri* gonidia tested in three different conditions. Circles: growth rates for gonidia in intact spheroids. Squares: averaged growth rates observed for gonidia under two conditions (isolated gonidia and gonidia of mechanically broken spheroids). (Because the gonidia of broken spheroids responded rather similarly to medium concentrations whether they were cultured in the presence or in the absence of somatic cells, the data for the two treatments have been averaged here.) Under all three conditions gonidia grew more rapidly with increasing nutrient concentrations. But the magnitude of this effect was much greater for gonidia in intact spheroids than for gonidia in the other two states. The dashed lines are regressions lines. The solid line is the calculated "compensation line"; it represents the total gonidial growth that one predicts should have been observed at each nutrient concentration if (1) the entire biomass of each spheroid had been devoted to gonidia (instead of a mixture of gonidia and somatic cells) and (2) these gonidia all grew at the same rates as isolated gonidia. It can be seen that at low nutrient levels the gonidia of intact spheroids fall below the compensation line, but at the highest level they fall above it. This indicates that the organism gains a reproductive advantage from the presence of sterile somatic cells only if the external nutrient concentration is high and the spheroids remain in their normal, intact state. The marked difference in growth rates for gonidia in intact versus broken spheroids provides compelling support for the source–sink hypothesis. (Adapted from Koufopanou & Bell 1993, Soma and germ: An experimental approach using *Volvox*, *Proc. R. Soc. Lond. Ser. B. Biol. Sci.*, **254**, 107–13, with the permission of The Royal Society and the authors.)

medium the gonidia in intact spheroids grew faster and produced a larger mass of progeny than did isolated gonidia or gonidia in broken spheroids. Moreover, the growth rates for gonidia in intact spheroids increased much more steeply in response to increasing nutrient availability than did the rates for the gonidia in either of the other two states, as predicted by the source–

sink model. It is particularly noteworthy that gonidia associated with somatic cells in broken spheroids did not fare significantly better at any concentration of medium than gonidia cultured in complete isolation from somatic cells. Thus it appears (again, consistent with the source–sink model) that the gonidia must be surrounded by an intact extracellular space (or "colony lumen" as Bell has called it) in order to gain any significant benefit from an association with somatic cells.

These experimental data indicate clearly that the gonidia benefit substantially from their association with somatic cells in intact spheroids. But is this benefit great enough to compensate for the resources that the organism devotes to somatic cells that make no direct contribution to the production of offspring? The answer to this important question comes from the same experiment. In all but the full-strength medium, gonidia of intact spheroids produced a smaller mass of progeny than would have been produced had all the cytoplasm of the spheroid been in cells that reproduced at the same rate as isolated gonidia (Figure 3.6). Thus, it appears that the germ–soma dichotomy of *Volvox* is distinctly disadvantageous in such "oligotrophic" environments. However, in a "eutrophic" environment, such as that simulated by standard *Volvox* medium, the germ–soma division of labor is reproductively advantageous, because in undiluted medium the mass of progeny produced by the gonidia of intact spheroids was greater than the mass that would have been produced had all the cytoplasm of the spheroid been in cells that reproduced at the same rate as isolated gonidia (Figure 3.6).

In short, the source–sink model appears to provide a cogent explanation for why *Volvox* is frequently found in great abundance in eutrophic waters, but is never found in oligotrophic waters. It also leads to the plausible hypothesis that it was under such eutrophic conditions that *Volvox* first evolved.

3.5 Two major forms of source–sink relationship between soma and germ have evolved in *Volvox*

As discussed previously (Section 2.4.2), two major patterns of organization and development in the genus *Volvox* are exemplified by *V. rousseletii* and *V. carteri*. In *V. rousseletii*, all adult cells are connected by cytoplasmic bridges, gonidia begin to divide while still quite small, and then the embryos (still connected to surrounding somatic cells) nearly double in mass between divisions. In *V. carteri*, in contrast, cells are not interconnected in the adult, gonidia grow to full size before initiating cleavage, and then they divide in the absence of further growth. These differences are interpreted by Bell (1985) as reflecting a difference in the trophic relationships between somatic cells and gonidia: The retention of cytoplasmic bridges by adult spheroids may have been fostered in the course of evolution because it increased the efficiency with which somatic cells could transfer their accumulated resources to the gonidia, where they could be used for reproduction.

The best evidence that somatic cells of Euvolvox species like *V. rousseletii* may actively transfer nutrients to the gonidia (and/or the embryos derived from them) is indirect and comes from careful observations made more than 60 years ago by Mary Pocock with respect to *V. rousseletii* development. Between divisions, *V. rousseletii* embryos nearly double in mass in about an hour, which is a faster growth rate than has ever been reported for any other eukaryotic cells.[8] Pocock observed, however, that once *V. rousseletii* gonidia begin to grow and divide, somatic cells cease growing altogether, and later actually decrease in size, particularly in the posterior of the spheroid – in the vicinity of the embryos (Pocock 1933b). As further evidence supporting the hypothesis that the rapid growth of *V. rousseletii* embryos might be the result of a flow of nutrients from somatic cells to embryos through the bridge network, Pocock observed that a somatic cell that apparently had become isolated from all others because of accidental severance of all of its cytoplasmic bridges was larger than all of its neighbors, and she suggested that its increased size may have resulted from its lost ability to export materials through the cytoplasmic-bridge system.

In *V. carteri* and other species that lack bridges between adult cells, nutrients can flow from somatic cells to gonidia only via the extracellular matrix. In this regard, it may be significant that in *V. carteri* the gonidia leave the surface monolayer very shortly after inversion and come to lie below the somatic-cell monolayer, where they are surrounded by extracellular matrix while they enlarge. In contrast, in *V. rousseletii* the developing gonidia and embryos remain closer to the surface (bulging inward somewhat) and remain connected to the somatic cells until cleavage is completed.

Although gonidia (and/or the embryos derived from them) grow more rapidly than somatic cells in all *Volvox* species, the gonidia and the somatic cells exhibit very different growth patterns in *Volvox* species with and without cytoplasmic connections between adult cells (Figure 3.7). Pocock reported that in *V. rousseletii* the gonidia are virtually indistinguishable from the somatic cells at the end of cleavage, but by the time they have enlarged enough

---

[8] Yeasts are generally considered to be about the most rapidly growing eukaryotic cells, but under the best of conditions yeast cells proliferate with a cell cycle time of about two hours (Alberts et al. 1989). Cells of *V. carteri* embryos – like those of many animal embryos – divide considerably more rapidly than do the cells of *V. rousseletii* or yeasts, but in both *V. carteri* and in rapidly cleaving animal embryos division occurs in the absence of significant growth (Starr 1970a; Gilbert 1994).

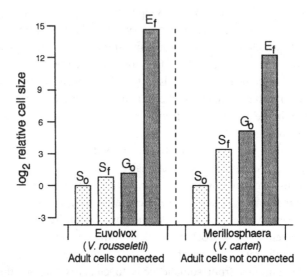

Figure 3.7 Two different forms of source–sink relationship in the genus *Volvox* (compare to Figures 2.12 and 2.13). S_o, young somatic cells at the end of cleavage; S_f, adult somatic cells in spheroids with fully cleaved embryos; G_o, young gonidia at the end of cleavage; E_f, total cellular volume of an embryo produced from a gonidium by the end of cleavage. All cell sizes are relative to the size of somatic cells at the end of cleavage (S_o). In *V. rousseletii*, somatic cells and gonidia start out life very similar in size, but the somatic cells grow little, and most of the new biomass of the spheroid is accumulated by the dividing embryos (Pocock 1933b), presumably as a result of the flow of photosynthetic products from somatic cells to gonidia and embryos via the cytoplasmic bridges. In contrast, in *V. carteri*, where such bridges are absent in the adult, the gonidia start out much larger than the somatic cells, but (relative to *V. rousseletii*) more of the total increase in biomass of the spheroid goes into somatic-cell growth, and less into growth of the gonidia/embryos.

to become clearly distinguishable in the juvenile spheroid, it can be seen that they have about twice as many connections to somatic cells as the somatic cells have to each other. From that she deduced that the presumptive gonidia probably forgo the last cleavage division and thereby start life twice as large as somatic cells (Pocock 1933b). However, by the time they finish cleaving, the *V. rousseletii* gonidia, or actually the embryos derived from them, will have grown about 5,600-fold, whereas the somatic cells will have grown no more than 2-fold (and will have actually reversed their growth during the period when the embryos are dividing, as indicated earlier). Thus, in *V. rousseletii*, the gonidia/embryos grow about 2,800 times as much as somatic cells (Figure 3.7). In contrast, in *V. carteri*, even though mature gonidia (and the embryos derived from them) are 500–1000 times as massive as the somatic cells in the same spheroid, the gonidia grow only about 16 times as much as somatic cells. This is because *V. carteri* gonidia start out 32-fold larger than somatic cells (as a result of asymmetric division), and then grow only about 140-fold before they cleave, whereas somatic cells increase in volume by about 9-fold during the same period (Figure 3.7). These differences are interpreted by Bell, quite cogently, to reflect the very different trophic relationships that exist between the two cell types in the two species (Bell 1985).

In short, *V. rousseletii* and *V. carteri* appear to represent two quite different approaches that have evolved (probably independently; see Section 2.4.4) to exploit the potential advantages of subdividing the labors of acquisition and resource utilization in a eutrophic environment. Moreover, as we saw in Section 2.4.2, several growth patterns intermediate between these two extremes are observed within the genus *Volvox*. *V. pocockiae* is of particular interest in this regard. Until Starr (1970b) described this species (which he named in honor of Mary Agard Pocock), it could be generalized that all species of *Volvox* that retained cytoplasmic bridges in the adult resembled *V. rousseletii* in the respect that their gonidia began cleaving while very small, and their embryos then grew between successive division cycles. Thus, the presence of gonidia that first enlarged extensively and then divided rapidly in the absence of further growth was one of the diagnostic features used to group species in the section Merrillosphaera. In *V. pocockiae*, however, cytoplasmic connections in the adult and other features that appear to identify it as a close relative of *V. aureus* are combined with a gonidial growth-and-division pattern similar to that of members of the section Merrillosphaera. The phylogenetic signficance of this combination of otherwise uncoupled features is not yet clear (but see Section 2.4.2 for a discussion of Desnitski's thoughts about it).

One attractive feature of Bells's source–sink hypothesis is that it attributes to most of the major evolutionary trends seen in the volvocaceans (i.e., increased size and cell number, increased relative volume of extracellular matrix, differentiation of germ from soma, and retention of cytoplasmic bridges in the adult in some species) the same type of selective advantage: increased efficiency of resource acquisition in eutrophic environments. This leaves open, however, the question of what may have been the cytological features that facilitated the development of such innovations, so that they could be tested by natural selection. Bells's former student, Vassiliki Koufopanou, has developed a plausible hypothesis in this regard, and it will be the subject of the next chapter.

4

Cytological Features Fostering the Evolution of *Volvox*

> In flagellated cells of green algae the flagellar apparatus is structurally connected to all major organelles. Most often the connection is provided by flagellar roots. . . . microtubular flagellar roots . . . determine the position of cell organelles with respect to the flagellar apparatus and . . . the plane of beat of the flagella. In some cases ([e.g. the] eyespot . . .) the positional relationship may be necessary for proper function, in others it may be necessary to ensure correct distribution of . . . organelles during cytokinesis.
>
> Melkonian (1984b)

There is a widespread belief that "developmental constraints" serve as boundary conditions that limit the types of morphological innovations that can arise within any group of organisms (Alberch 1982; Maynard Smith et al. 1985), though there is less than universal agreement about how developmental constraints are to be defined and recognized. The basic concept is that the range of morphological innovations that can be generated within any group of organisms is constrained by certain fundamental features of the cellular organization and developmental biology of that group of organisms.

The hypothesis to be developed in this chapter is that the extraordinary degree to which organelles within a green flagellate cell are interconnected by a highly regular cytoskeletal network, as outlined in the foregoing quotation, constitutes a fundamentally important developmental constraint that (when combined with the presence of a coherent cell wall and the selective pressures discussed in the preceding chapter) led, with a certain degree of inevitability, to the appearance of organisms, like *Volvox*, with a division of labor between somatic and reproductive cells.

Major portions of this hypothesis were originally articulated by Vassiliki Koufopanou, who postulated that three aspects of the biology of green algae combined to impose an important developmental constraint, namely, the flagellation constraint, that channeled volvocalean evolution in a way that eventually favored the appearance of multicellular organisms with a germ–soma division of labor (Koufopanou 1994):

1. An unusual, but nonetheless universal, feature of the division of a green-flagellate cell is that the basal bodies (BBs) of the flagella take positions near the spindle poles and behave like centrioles while they are still attached to the plasmalemma and many other organelles.

2. The type of cell wall that evolved in *Chlamydomonas* and its relatives prohibits the BBs from functioning in this way as long as they remain attached to flagella.

3. However, in the unstirred waters inhabited by many unicellular green flagellates and all volvocaceans, it could prove fatal to forgo flagellar motility during cell division.

Thus, the flagellation-constraint hypothesis postulates that although the tough, coherent wall developed by the chlamydomonads may have had other selective advantages (such as facilitating osmoregulation in a fresh-water environment), it created a fundamental dilemma: It meant that to gain the advantages of such a wall, organisms would have to find some new way to maintain flagellar motility while dividing or else move to environments where loss of flagellar motility during cell division was less disadvantageous.

Although the latter part of this chapter will be devoted to a discussion of Koufopanou's flagellation-constraint hypothesis, I will first push the analysis backward a step and present evidence that the unusual form of cell division that is observed in green flagellates (and that is the starting place for Koufopanou's analysis) is itself the result of a more fundamental and ineluctable developmental constraint, namely, the need to maintain fixed relationships among certain critical organelles in order to be able to move toward the light.

Before going further, however, we should inquire whether or not the concept of a flagellation constraint is credible: Just how important is it to volvocaceans living in quiet ponds to be able to maintain their preferred position in the water column via flagellar motility? Reynolds (1983) examined that question experimentally. Unable to deflagellate all of the volvocaceans in a natural environment, he did the next best thing: He created a situation in which flagellar motility was vitiated as a means of maintaining position in the water column. After a *Volvox* population had become well established in a large cylindrical tube within a stratified lake, Reynolds used an air lift to stir the water in the upper part of the tube intermittently. He observed that each time the water was agitated and the *Volvox* population was tumbled about, its reproductive rate plummeted (approaching zero when the water was stirred to a depth of 8 meters), but the reproductive rate of the population quickly rebounded each time that the stirring ceased, and the organisms were again able to achieve and maintain a position at a specific level within the water column (Reynolds 1983). *Quod erat demonstrandum.*

4.1 The need to find a place in the sun imposes a rigid constraint on the structure of green flagellate cells

The ability to move toward a moderate light source (positive phototaxis) and away from an intense one (negative phototaxis) is a ubiquitous and essential trait of green flagellates. The importance of positive phototaxis is obvious: Being photoautotrophs, they need to be able to reach a position where their rates of growth and reproduction will not be limited by the amount of light available for photosynthesis. The importance of negative phototaxis may be less obvious, but it is even greater: If the intensity of illumination is above the species optimum, toxic photooxidants will be produced in the chloroplast, where they will accumulate, bleach and destroy the chlorophyll, killing the cell. Indeed, direct summer sunshine will bleach and kill most species of *Volvox* within less than an hour (Mast 1907). Small photoautotrophs lacking the ability to move toward and away from the sun in an adaptive manner cannot survive in the environments where *Volvox* and its closest relatives live.

Lacking a nervous system to integrate and interpret environmental cues, green flagellates ensure that their movements in response to light are adaptive by maintaining a reliable spatial relationship between the organelle that detects light (the eyespot) and the two organelles that execute the responses to it (the flagella). The regularity of that spatial relationship is maintained within interphase cells by a complex set of cytoskeletal elements that link the organelles of the phototactic system in a highly precise and reproducible manner. Moreover, the continuity of that adaptive cellular organization from generation to generation is assured by using the very same cytoskeletal components to link critical parts of the dividing cell in an equally stereotyped manner.

The cytological basis for phototaxis has been most thoroughly examined in *Chlamydomonas reinhardtii*, as reviewed by Witman (1993), but the available evidence indicates that key features of that system are strongly conserved in all green flagellates, despite a wide range of variations on the basic recurring theme. In the accounts that follow, the emphasis will be on *C. reinhardtii*, on the assumption that it closely resembles the ancestral unicell from which the volvocaceans, including *Volvox*, evolved. However, information derived from studies of other green flagellates will be duly noted.

It now appears clear that the ability of *C. reinhardtii* to respond to light in a consistent and adaptive manner is based on three considerations: (1) The photosensitive organelle lies on one side of the cell and thus is particularly sensitive to light coming from that side. (2) The cell changes direction when a signal is generated by the photodetector, because the two flagella (although they appear similar) respond differently to any such signal. (3) A predictable spatial relationship of the eyespot to the two flagella of different response properties is established and maintained by a highly regular cytoskeleton that links organelles in a highly predictable manner during interphase and also controls their movements during cell division.

We will now analyze each of these components of the phototactic system of *Chlamydomonas* in considerable detail, because *it is a central premise of this book that an understanding of the basic cellular organization of the green flagellates is absolutely critical to an understanding of other aspects of their biology.*

4.1.1 *The photoreceptor uses calcium to signal light direction and intensity*

In its normal locomotory mode, *Chlamydomonas* swims forward by "doing the breast stroke," with the flagella executing ciliary-type effective strokes that swing them outward, away from one another, and then toward the posterior. But the waveforms executed by the flagella are neither precisely planar (Bessen, Fay & Witman 1980; Rüffer & Nultsch 1985) nor directed in precisely opposite directions. As a result, the cell rotates around its long axis about twice each second as it moves forward and thereby follows a left-handed helical path (Hoops & Witman 1983; Kamiya & Witman 1984; Witman 1993). Consequently, the eyespot, or "stigma," a conspicuous bright-red organelle that lies on one side of the cell, is also swept through a helical path, exposing it to the environment on all sides as the cell advances (Figure

4.1). The stigma of *C. reinhardtii* consists of two to four layers of hexago-
nally packed, carotenoid-rich granules that lie within an anterior lobe of the
chloroplast, just below a specialized region of the chloroplast envelope that
in turn is closely associated with a specialized region of the plasmalemma
containing an abundance of intramembrane particles (Melkonian & Robenek
1980).

Contrary to earlier assumptions, the photodetector of the cell resides not
in the stigma itself but in a specialized region of the plasmalemma above the
stigma (Melkonian & Robenek 1984; Lawson & Satir 1994; Zhang 1994)
that contains a unique kind of rhodopsin molecule. The spectral properties
of "chlamyrhodopsin," as it has come to be known, initially suggested that
it was very similar to the rhodopsin present in vertebrate eyes (Foster et al.
1984); however, its chromophore turned out to resemble archebacterial rhodopsin (an all-*trans* retinal that isomerizes to the 13-*cis* isomer on absorbing
light) (Witman 1993) rather than vertebrate rhodopsin (in which 11-*cis* retinal
is isomerized to all-*trans* retinal by light). Sequencing of the cloned gene
encoding the protein moiety of chlamyrhodopsin ("chlamyopsin") indicates
that although it is most similar in sequence to the opsins of various invertebrates and contains a conserved retinal-binding region, it differs from all
previously described opsins in that it has only four (not seven) transmem-
brane domains and contains an unusual abundance of polar and charged
amino acids (Deininger et al. 1995). The homologous gene has more recently
been cloned from *Volvox carteri* and shown to encode an opsin extremely
similar to chlamyopsin (W. Deininger, E. Ebnet & P. Hegemann, personal
communication). The structural features of these algal opsins, coupled with
the fact that a calcium-dependent photoreceptor current is generated within
50 μs after an eyespot is illuminated (which is too fast to involve a G-protein
signal-transduction chain of the sort employed in animal photoreceptors),
suggest that the algal opsins either are themselves the gated ion channels that
generate the photoreceptor currents, or are very closely coupled to those
channels (Deininger et al. 1995).

In the generally accepted current view, the function of the stigma is to
enhance directional sensitivity: When light strikes the photoreceptor side of
the cell, the stigma acts as a quarter-wave parabolic mirror that focuses light
on the photosensitive membrane, but when the light comes from the opposite
side, the stigma shades the photoreceptor (Foster & Smyth 1980). Thus, the
stigma increases the directional sensitivity of the photoreceptor manyfold.
Because the chloroplast can provide some shading to the photoreceptor, a *C.
reinhardtii* mutant that lacks a stigma is able to move phototactically, but its
responses are much less precise than those of a wild-type strain (Morel-
Laurens & Feinleib 1983). Because of these relationships, when the term

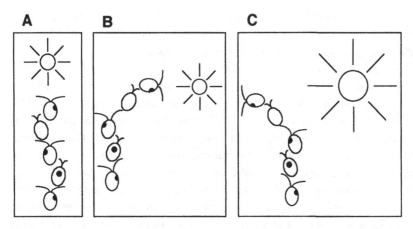

Figure 4.1 Phototactic swimming by *Chlamydomonas*. A *Chlamydomonas* cell follows a left-handed helical path as it moves forward. During each 0.5-second turn of this helix, the cell body makes one complete rotation around its long axis, with the eyespot (which is on one side of the cell) always facing the outside of the helix (**A**). This permits the eyespot to monitor the light intensity on all sides twice each second. When a cell is swimming directly toward a moderate light source (**A**), the eyespot receives the same illumination during all parts of the cell-rotation cycle, the two flagella beat with the same force, and thus the cell continues to swim toward the light. However, when the cell is illuminated from one side, the eyespot is stimulated more strongly in one half of the rotation cycle and less strongly in the other. If the light is of moderate intensity (**B**), when the eyespot is swung around to face the light, the flagellum closer to the eyespot (the *cis* flagellum) momentarily beats less forcefully. But then as the eyespot is rotated away from the light, it is the *trans* flagellum that momentarily beats less forcefully. Thus, in both halves of the rotation cycle the flagellum on the darker side is dominant, which causes the cell to turn toward the light, or exhibit "positive phototaxis." In contrast, if the light source is above some threshold intensity (**C**), the responses of the *cis* and *trans* flagella are reversed: as the eyespot is swung into an intense light beam, the *trans* flagellum beats less forcefully; and then as the eyespot is swung out of the light, the *cis* flagellum beats less forcefully. This causes the cell to turn away from a bright light, or exhibit "negative phototaxis." In a light gradient, *Chlamydomonas* cells accumulate in a specific region called the "neutral zone." Close observation indicates that in the neutral zone cells switch rapidly and repeatedly between positive and negative phototaxis, which causes them to tumble in place. The threshold for switching between positive and negative phototaxis – and hence the point within a light gradient where cells will accumulate – changes with the recent illumination history, time in the light–dark cycle, and many other physiological and environmental variables. But presumably these factors all act to ensure that the cells position themselves in that part of the environment where the light level is optimum for photosynthesis under the existing internal and external conditions.

"eyespot" is used in the following discussion, it should generally be taken to mean the visible red stigma plus the actual photoreceptor that lies just above it, because the two are so intimately associated.

It has long been known that phototaxis in flagellates is fully dependent on the presence of calcium in the extracellular environment (Stavis & Hirshberg 1973; Schmidt & Eckert 1976; Nultsch 1979; Nultsch, Pfau & Dolle 1986; Dolle, Pfau & Nultsch 1987), and it is now abundantly clear that in *Chlamydomonas*, as in other organisms from ciliates to mammals (Tamm 1994), flagellar activity is regulated by the concentration of calcium ions in the cytoplasm. The widely accepted current model is that a calcium pump operates continuously to keep intracellular calcium concentrations in the submicromolar range, but when light of moderate intensity strikes rhodopsin in the *Chlamydomonas* photoreceptor, calcium channels in the eyespot region open transiently, resulting in an influx of extracellular calcium. Over a certain range, the more intense the light, the greater the transient rise in intracellular calcium (Foster & Smyth 1980; Nultsch 1983; Harz & Hegemann 1991; Holland et al. 1996).[1] Thus, as the cell advances, intracellular calcium concentrations presumably rise as the eyespot rotates into a position where it is illuminated, but then fall again as it is rotated away from such a position.

4.1.2 The two flagella of Chlamydomonas respond oppositely to changes in calcium concentration

In response to moderate illumination from one side, *Chlamydomonas* cells execute a turn that causes them to swim toward the light (Boscov & Feinleib

---

[1] Following sudden exposure to intense light, *Chlamydomonas* and other green flagellates exhibit a "photophobic response" in which both flagella momentarily cease beating and then switch from the usual asymmetric, ciliary-type beat pattern to a symmetric, undulatory waveform similar to that typically observed in metazoan sperm; this causes the cell to swim backward briefly. Shortly thereafter, the cells begin swimming in the normal ciliary manner again, but usually in a different direction than before the photophobic response was evoked (Ringo 1967). The photophobic response is also regulated by calcium, but via a different set of calcium channels. When the eyespot is stimulated by light, local channels open in a graded manner, causing a calcium influx and membrane depolarization that is proportional to the light intensity. However, if photostimulation suddenly passes some critical value, membrane depolarization also passes a threshold, and voltage-sensitive calcium channels located all along the flagellar membrane (Beck & Uhl 1994) open in an all-or-none response, flooding the axoneme with calcium and triggering the transition from the ciliary pattern to the flagellar pattern of beating. The transition from the ciliary to the flagellar waveform has been shown to occur in isolated or demembranated, reactivated flagella of *Chlamydomonas* at a Ca^{2+} concentration of 10^{-4} M (Hyams & Borisy 1978; Bessen et al. 1980; Kamiya & Witman 1984). Although this response of the cell is no less interesting or adaptive than the phototactic responses to steady illumination, it is not emphasized in the text because it does not depend on or reveal the intrinsic differences between the *cis* and *trans* flagella, nor does it cause the cell to orient itself with respect to the location of the light source.

1979) (Figure 4.1). It is intuitively obvious that such a turn could not be executed if both flagella responded identically to photostimulation of the cell, and indeed they do not (Rüffer & Nultsch 1991). In any biflagellate alga, one of the flagella (called *cis*) is always located closer to the eyespot than the other (*trans*). It has long been known that in many other biflagellates the *cis* and *trans* flagella are very different in size, structure, and function, as reviewed by Moestrup (1982), but in *Chlamydomonas* and all of its close relatives the *cis* and *trans* flagella appear identical, even under the closest scrutiny in the electron microscope. Nevertheless, it is now clear that they are very different physiologically, and their differences appear to account for the phototactic behavior of the cell. As the illumination level rises (within the range that will cause positive phototaxis), the *trans* flagellum transiently beats more strongly, and the *cis* flagellum transiently beats less strongly; then, as the illumination level falls, the *trans* flagellum transiently beats less strongly and the *cis* flagellum transiently beats more strongly (Rüffer & Nultsch 1991; Nultsch & Rüffer 1994). As we will see later, those two sets of responses both have the effect of causing the cell to turn toward the light.

In a classic study, Kamiya and Witman (1984) demonstrated that the differences in behavior of the two flagella are intrinsic to the flagellar axonemes themselves and are neither dependent on differences in ion permeability between the two flagellar membranes, nor on any other cellular mechanism that would cause the ionic environments within the two axonemes to be different. Kamiya and Witman (1984) showed that when demembranated *Chlamydomonas* cells were reactivated by adenosine triphosphate (ATP), the behaviors of the *cis* and *trans* flagella were strongly dependent on the concentration of Ca^{2+} in the reactivating solution. At a Ca^{2+} concentration of 10^{-8} M, the two flagella beat with the same force, and the cells swam forward, in the typical helical path, with the eyespot always facing the outside of the helix. However, when the Ca^{2+} concentration was lowered to 10^{-9} M, the *trans* flagellum became less active than the *cis* flagellum, and therefore the cell turned away from the eyespot side. In contrast, when the Ca^{2+} concentration was raised to 10^{-7} M, it was the *cis* flagellum that became less active, and thus the cell turned toward the eyespot side.

All three of these calcium-dependent flagellar responses are parallel to responses that were observed by Rüffer and Nultsch (1991), who studied the effects of rapid changes in light intensity on beat patterns, and all three presumably will have the same overall effect: to head the cell toward the light (Figure 4.1). When a cell swims forward in a helical path, with a moderate light lying off to one side, the eyespot will become maximally illuminated only once in each 360° helical turn. It is believed that at that point the intracellular Ca^{2+} concentration will rise above 10^{-8} M, the *trans* flagellum will become dominant, and the cell will turn in the direction that the eyespot is facing at that instant, that is, toward the light. But as the cell rotates so that the eyespot is no longer facing the light, the intracellular Ca^{2+} concentration will fall below 10^{-8} M, the *cis* flagellum will become dominant, and the cell will turn toward the *trans* side, which now is the illuminated side![2] Then, once the cell has fine-tuned its path so that it is heading directly toward the light, the eyespot will be equally stimulated in all parts of its helical trajectory, the intracellular Ca^{2+} concentration will remain constant (presumably around 10^{-8} M), and the cell will continue to swim forward. Strong support for this model to explain the phototactic behavior of *C. reinhardtii* came with the demonstration that a mutant strain that lacks the capacity for calcium-dependent switching of flagellar dominance also lacks the capacity for phototaxis (Horst & Witman 1993).

4.1.3   The difference between the two flagella is due to a difference in basal-body age

The studies just described indicate that the phototactic ability of *C. reinhardtii* depends on *intrinsic* differences between its two flagella: In the presence of ATP, the *cis* flagellum beats less strongly, whereas the *trans* flagellum beats more strongly, when the intracellular Ca^{2+} concentration rises from 10^{-9} to 10^{-7} M. This difference is now known to be a consequence of a predictable difference in developmental status between the basal bodies (BBs) that lie at the bases of, and act as the organizing centers for, the *cis* and *trans* flagella.

In *Chlamydomonas* and other green flagellates, as in mammalian cells, new BBs (or centrioles) are formed in proximity to older ones (Johnson & Porter 1968; Gould 1975; Wheatley 1982; Johnson & Rosenbaum 1992). These BBs are distributed at the time of cell division in a semiconservative fashion: Each daughter cell receives one older ("parental") BB and one newly formed ("daughter") BB (Adams, Wright & Jarvik 1985; Aitchison & Brown 1986; Melkonian, Reize & Preisig 1987; Holmes & Dutcher 1989). Since 1980 there had been clear evidence that in a very different kind of biflagellate unicell the parental and daughter BBs were functionally very different (Beech, Wetherbee & Pickett-Heaps 1980), but the first indication that such might be the case for *Chlamydomonas* came with the description of a number of uniflagellar, or *uni*, mutants of *C. reinhardtii* in which it was

---

[2] As noted in Figure 4.1C, above some threshold intensity of light, the responses of the *cis* and *trans* flagella to light direction are reversed; the physiological basis for this reversal is not at all understood, but it causes the cell to execute negative phototaxis and swim away from a light source that is above some optimum intensity.

observed that following cell division a flagellum was regularly formed by the *trans* BB, but not by the *cis* BB (Huang, Ramanis et al. 1982). In a flash of insight, those authors proposed the following working hypothesis to account for that regularity of the *uni* mutant: (1) The *cis* BB is always the younger one. (2) The *cis* BB in any given generation becomes a *trans* BB in all subsequent generations. (3) Full maturation of a BB takes two cellular generations. (4) The defect in a *uni* mutant affects BB maturation in such a way that it takes two cellular generations for the BB to be able to template the formation of a flagellar axoneme (Huang, Ramanis et al. 1982). Subsequently, an elegant and detailed light-microscope study of cell division in *C. reinhardtii* verified the most critical of those assumptions, namely, that in wild-type cells the *cis* BB is always the younger one (Holmes & Dutcher 1989). Implicitly, that (when combined with the known physiological differences between the *cis* and *trans* flagella) confirmed the prediction that it takes two cellular generations for a BB to mature and that the physiological differences between the *cis* and *trans* flagella of *Chlamydomonas* are consequences of the maturational differences between their BBs. Simultaneously, studies of cell division in a related biflagellate (*Brachiomonas*), using entirely different methods, came to remarkably similar conclusions about the regular segregational behavior of parental and daughter BBs during cell division (Segaar & Gerritsen 1989). Moreover, the conclusion that maturation of a BB requires more than one cell cycle has been generalized to other flagellates by study of cell division in other green algae (Melkonian, Reize & Preisig 1987), golden algae (Wetherbee et al. 1988), and cyanophorans (Heimann, Reize & Melkonian 1989) that have two flagella of grossly dissimilar lengths and motility patterns.

   Thus, it now becomes clear that the functional nonequivalence of the two flagella of *Chlamydomonas* (which makes phototaxis possible) is a reflection of a developmental nonequivalence that lasts only one cell generation.[3] The BB that is in the *cis* position in any particular cell generation and that nucleates an axoneme that beats *less* strongly as the calcium concentration rises will in all subsequent generations be located in the *trans* position and will nucleate an axoneme that beats *more* strongly in response to the same increase in intracellular calcium. The molecular basis for this behavioral difference between flagella nucleated by young and older BBs remains entirely

---

[3] It is important to emphasize that flagellar maturation is a cell-cycle-dependent phenomenon, not a time-dependent or function-dependent phenomenon. A BB of the *cis* type is transformed into a *trans* type when, and only when, it goes through an additional cell-division cycle, whether that cell-division cycle takes less than one hour or more than 24 hours to complete, and whether or not that BB was involved in organizing a flagellum or a complete basal apparatus in the period between cell divisions (Holmes & Dutcher 1992).

mysterious; nevertheless, it appears to underlie the ability of these cells to move adaptively with respect to light, and hence it imposes a substantial developmental constraint that probably has operated throughout the evolutionary history of green flagellate unicells (Holmes & Dutcher 1992).

4.1.4 Proper orientation of the eyespot and the BBs must be maintained in each generation

It follows from what has just been described that to maintain the ability to move toward the light, green flagellates like *Chlamydomonas* must have a way to assure that the eyespot and the younger BB will end up on the same side of the cell after each round of cell division. A cell that divided in a way such that the parental and daughter BBs were located at random with respect to the eyespot clearly would produce daughter cells in which the response to light would be nonadaptive; such cells would be very likely to disappear from the population without leaving progeny.

But there is a second requirement for directional swimming in *Chlamydomonas*: The two BBs must be oriented in a very specific manner with respect to one another. Each flagellar axoneme of *Chlamydomonas* has an intrinsic polarity that determines the direction of the effective stroke: Of the nine outer doublets characteristic of all eukaryotic flagella, those that are now conventionally numbered 1, 5, and 6 have structural specializations (not only in *Chlamydomonas* but also in a wide range of other organisms), and as the flagellum begins its effective stroke, doublets 5 and 6 are always on the leading edge, and doublet 1 is on the trailing edge (Hoops & Witman 1983). Mutants lacking these structural specializations of doublets 1, 5, and 6 are unable to execute the normal ciliary ("breast-stroke") type of beat, and are therefore restricted to swimming backward (Segal et al. 1984). Similar axonemal asymmetries exist in *Volvox carteri* (Hoops 1993) and probably in all other volvocaleans. Moreover, there is compelling evidence to suggest that this structural and functional polarity of the axoneme is a direct reflection of a corresponding polarity within the BB that templates it (Hoops & Floyd 1983; Hoops 1984, 1993; Greuel & Floyd 1985; Taylor, Floyd & Hoops 1985).

In wild-type *Chlamydomonas* and most other biflagellate unicells, the two flagella are oriented with what is called 180° rotational symmetry. That is to say, they are arranged so that their number-1 doublets face each other, and their number-5 and number-6 doublets face outward (Floyd, Hoops & Swanson 1980; Hoops & Witman 1983; Mattox & Stewart 1984). As a consequence, their effective strokes are directed away from one another (Figure

4.2). This relationship is required for directional swimming: A *C. reinhardtii* mutant in which the normal 180° rotational symmetry of the axonemes is not maintained is incapable of normal swimming behavior (Hoops, Wright et al. 1984).

In a particularly elegant study, Holmes and Dutcher (1989) showed that the proximity of the eyespot to the daughter BB, as well as the 180° rotational symmetry of adjacent BBs, resulted from the precise structure and stereotyped behavior of the cytoskeleton in interphase and dividing cells.

4.2 Three kinds of cytoskeletal elements in the basal apparatus link BBs to other organelles

The BBs of green flagellates do more than act as organizing centers for the flagellar axonemes; they lie at the focal point of an extremely complex and highly regular "basal apparatus" (BA)[4] in which bundles of microtubules and several other kinds of fibers connect the BBs in a highly predictable manner to one another, to their daughter "probasal bodies," and to virtually all other major organelles in the cell, including the rest of the cytoskeleton, the plasmalemma, the eyespot, the nucleus, the chloroplast, and so forth.[5] Indeed, the BA appears to act as the organizing center for the entire green-algal cell, both in interphase and during cell division, and thus it is the functional equivalent of the centrosome of an animal cell (Johnson & Rosenbaum 1992). However, at the electron-microscopic level the algal BA is seen to be much richer in structural elements and (within any given species) much more regular in its architecture than any animal centrosome.[6]

---

[4] Many authors prefer the term "flagellar apparatus" for this complex of structural elements (Andersen et al. 1991). But because, as we shall see, the complex often executes critically important functions in cell division in the absence of attached flagella, I prefer the term "basal apparatus."

[5] Additional details of the structure and function of this apparatus in the volvocine algae and their close relatives (many of which are beyond the scope of this review), are avilable (Ringo 1967; Johnson & Porter 1968; Olson & Kochert 1970; Pickett-Heaps 1970, 1972, 1975, 1976; Cavalier-Smith 1974; Coss 1974; Gould 1975; Lembi 1975; Stewart & Mattox 1975, 1978, 1980; Mattox & Stewart 1977, 1984; Floyd 1978; Hyams & Borisy 1978; Salisbury & Floyd 1978; Katz & McLean 1979; Green et al. 1981; Hoops 1981, 1984; Mattson 1984; Melkonian 1984a,b; O'Kelley & Floyd 1984; Greuel & Floyd 1985; Hoops & Witman 1985; Wright, Salisbury & Jarvik 1985; Doonan & Grief 1987; Melkonian, Reize et al. 1987; Gaffal 1988; Salisbury, Baron & Sanders 1988; Holmes & Dutcher 1989; Segaar & Gerritsen 1989; Wright et al. 1989; Segaar 1990; Kirk, Kaufman et al. 1991).

[6] Despite its precisely defined organization within any given species, the algal BA exhibits enough subtle structural variation at higher taxonomic levels to make it one of the most widely used and most powerful tools in modern attempts to discern phylogenetic relationships among algal groups. For this aspect of BA biology, see Stewart & Mattox (1975, 1978, 1980), Moestrup (1978, 1982), Floyd et al. (1980), Melkonian (1980, 1982, 1984a, 1990), Hoops & Floyd (1982a,b, 1983), Hoops, Floyd & Swanson (1982), O'Kelley & Floyd (1983, 1984), Hoops

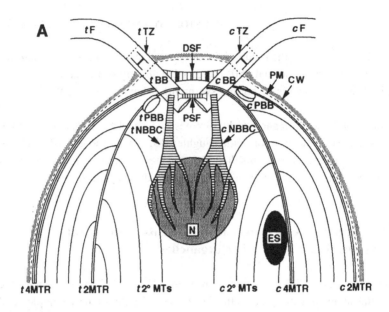

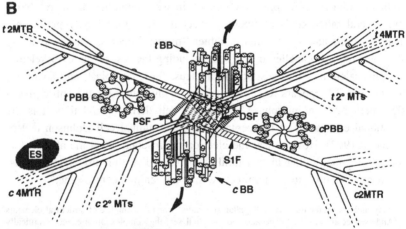

Figure 4.2 Two schematic drawings of the basal apparatus (BA) of an interphase *Chlamydomonas* cell. **A**: Lateral overview of the relationships of the basal bodies (BBs) to certain other components of the cell. **B**: A more detailed view (from above and slightly to one side, and rotated 90° relative to A) of the BBs themselves and some of their attachments. These diagrams have been drawn to emphasize connections between parts, but not to accurately represent the relative dimensions of those parts. The BBs are intrinsically asymmetric and are arranged in 180° rotational symmetry with respect to each other; as a result, their flagella beat with effective strokes (indicated by bold, curved arrows in B) in opposite directions. Because the MTRs and other components of the BA are attached to specific MT triplets of each BB, the whole BA can be thought of as being composed of two half-BAs arranged with 180° rotational symmetry: the *cis* half, which is associated with the eyespot (ES), and the *trans* half, which is not. Here the various components have been given prefixes (*c* or *t*) to indicate which half-BA each of them is associated with. The long axes of the BBs of an interphase cell are tipped at about

4.2.1 Microtubular rootlets link the BBs to the cortical cytoskeleton and the eyespot

BBs are asymmetric structures: Each has a distal end that can nucleate formation of an axoneme and a proximal end (containing a characteristic "cartwheel structure") that cannot. Moreover, when viewed from the distal end, the nine microtubule triplets of the BB are invariably seen to be imbricated in the clockwise direction; the enantiomorphic form that theoretically might exist has never been seen (Floyd et al. 1980; Melkonian 1984a). This fundamental asymmetry is highlighted by the high degree of regularity with which particular accessory structures are attached to particular triplets on the BBs in any given species.

One of the most characteristic taxon-specific features of any green-algal BA is a well-defined set of microtubules (MTs), the MT rootlets (MTRs), that are attached to the BBs at specific positions and that extend outward under the plasmalemma (Melkonian 1984a). In all chlorophyceans (a group that includes *Chlamydomonas* and all of the volvocaceans) each BB has one

(1984), Mattox & Stewart (1984), Greuel & Floyd (1985), Hoops & Witman (1985), and Hoek, Stam & Olsen (1988).

Caption to Figure 4.2 *(cont.)*
90° to one another and are connected to each other and to the plasma membrane, the nucleus, and most other organelles by a complex array of microtubules (MTs) and fibrous elements, many of which have been omitted here for clarity. Among the most conspicuous and regular of these features is a cruciate array of MT rootlets (MTRs) in which rootlets containing two and four MTs (2MTRs and 4MTRs) alternate and act as organizing centers for an extensive array of secondary MTs (2° MTs) that connect to other organelles. (Note that in the BB-proximal region, the MTs of the 4MTRs are in a 3-over-1 configuration, but in more distal regions they switch to a side-by-side configuration.) In interphase, the two half-BAs are connected across the midline by proximal (PSF) and distal (DSF) striated fibers that link the BBs and by system I fibers (S1F) that link MTRs of like type. However, during cell division the links between the two halves of the BA will be broken, permitting the BBs to move in opposite directions, as will be described in Figures 4.4 and 4.5. Each BB is associated with (and is connected by fine filaments to) a probasal body (PBB) that was formed during the previous mitotic cycle and that now lies on the other side of the 2MTR. Key to the spatial coordination of cytoplasmic and nuclear events throughout the cell cycle are a pair of striated nucleus-BB connectors (NBBCs) that are composed predominantly of the contractile protein centrin and are intimately associated with the BBs at one end and the nucleus (N) at the other (see also Figure 4.3). The apparent symmetry of the cell and its BA is broken by the stigma, or eyespot (ES), which lies within the chloroplast (omitted for clarity) in a region that is immediately clockwise of the 4MTR that is connected to the younger of the two BBs (the cBB). Other abbreviations: CW, cell wall; F, flagellum; PM, plasma membrane; TZ, transitional zones of flagella.

rootlet containing four MTs (a 4MTR) attached in the vicinity of its number-2 and number-3 triplets and a rootlet containing two MTs (a 2MTR) on the other side, near its number-8 and number-9 triplets (Figure 4.2). Because the two halves of the BA are arranged with 180° rotational symmetry, as the MTRs angle outward under the plasmalemma, they form a highly character-istic "4-2-4-2 cruciate array."[7]

The regularity with which the two kinds of MTRs (and various other cytoskeletal elements, some of which will be described later) are attached to specific BB triplets and also the regularity with which structurally and func-tionally distinguishable outer-doublet MTs of the flagellar axoneme emanate from particular BB triplets (Hoops & Witman 1983; Hoops 1993) provide compelling evidence that each of these nine triplet MTs must have a unique molecular signature. To date, we can only speculate on the nature of the molecules that distinguish one triplet from the next, but the fact that isolated BBs contain more than 300 proteins (Dutcher 1986) leaves much room for the imagination. In any case, there can be little doubt that the BBs possess an extraordinary amount of "positional information" and that many of the distinctive organizational features of green flagellates that underlie their suc-cess in aqueous environments can be traced back to this fundamental asym-metry of their BBs. And many such features can be traced back to the BBs via the MTRs.

In *C. reinhardtii*, as in other green flagellates (Bouck & Brown 1973; Melkonian 1978; Brown, Stearns & Macrae 1982; Segaar, Gerritsen & De Bakker 1989; Ehler, Holmes & Dutcher 1995), each of the MTRs (or material adhering to it) functions as a microtubule-organizing center (MTOC), and as a result, additional MTs (which constitute the "cortical" or "secondary" MT cytoskeleton) emanate from the MTRs and extend toward the posterior end of the cell, between the plasmalemma and the chloroplast. The MTRs can be readily distinguished from the secondary MTs by immunocytology: The rootlet MTs, like the MTs of the BBs and the axoneme (but not the secondary MTs) are thermostable MTs that contain acetylated α-tubulin and stain with a monoclonal antibody directed against this epitope (LeDizet & Piperno 1986).

Of truly central significance is the fact that the eyespot is always associated with a defined site in this MTR array (Gruber & Rosario 1974; Moestrup 1978; Melkonian 1984b; Gaffal, el-Gammal & Friedrichs 1993). In *C. rein-hardtii* it has been shown that the eyespot lies immediately clockwise of the 4MTR that emanates from the younger BB (Holmes & Dutcher 1989), and

---

[7] In more distantly related algae, an *X*-2-*X*-2 array is seen in which the value of *X* varies greatly, but in a taxon-specific way.

considerable evidence indicates that it lies in a homologous location in all green algae (Moestrup 1978, 1982; Melkonian 1982), which makes eyespot location one of the most conserved features of green-algal cellular organization.

It apparently is this particular 4MTR, then, unimposing as it appears, that assures that an adaptive spatial relationship will exist between the eyespot and the younger BB. We will return later to consider the way in which this special relationship is reestablished in each cellular generation.

4.2.2   Centrin-rich fibers connect the BBs to one another and to the nucleus

It has long been known that the BA in many green flagellates includes a pair of striated, fibrous rootlets that pass from the BBs to the surface of the nucleus, and in some species beyond the nucleus to the plasmalemma at the posterior end of the cell (Kater 1929; Manton 1952, 1956; Parke & Manton 1965; Pitelka 1969, 1974; Melkonian 1984b).[8] Some such rootlets have been known to exhibit rapid and dramatic contraction in the presence of calcium (Salisbury & Floyd 1978). A novel eukaryotic motility system was uncovered when the first such rootlets were isolated and their major protein component, a 20-kDa calcium-binding protein now called centrin,[9] was purified and characterized (Salisbury, Baron et al. 1984).

Centrin is a member of the EF-hand family of calcium-binding proteins that includes calmodulin; in the presence of submicromolar concentrations of calcium, centrin fibers contract with extraordinary speed and force, apparently by undergoing extensive supercoiling (Salisbury 1989a; Melkonian, Beech et al. 1992). The discovery of centrin and the use of specific antibodies to search for centrin in other sites not only led to novel insights into many aspects of algal biology (Salisbury, Sanders & Harpst 1987; Schulze et al. 1987; Melkonian, Schulze et al. 1988; Melkonian 1989; Salisbury 1989a,b; Melkonian, Beech et al. 1992) but also led to the recognition that centrin is a nearly universal feature of eukaryotic cells, particularly in the centrosomal area (Salisbury, Baron et al. 1986; Schiebel & Bornens 1995).

Among the many locations in which centrin was detected with anti-centrin

---

[8] These have often been called "striated rootlets" or "rhizoplasts" when they are particularly massive (Dangeard 1901; Hamburger 1905; Manton 1952, 1956; Parke & Manton 1965). It is now proposed that they and all other BA fibers that react with anti-centrin antibodies be called "system II fibers" (Melkonian, Schulze et al. 1988) to distinguish them from "system I fibers" that contain the protein assemblin (Lechtreck & Melkonian 1991).

[9] Synonym: caltractin (Huang, Watterson et al. 1988).

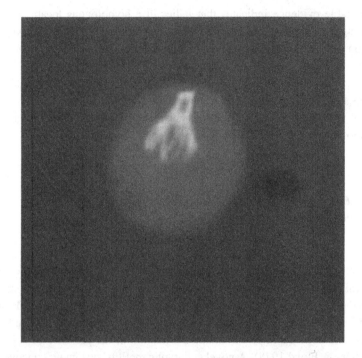

Figure 4.3 NBBCs of an interphase *C. reinhardtii* cell, as revealed by staining with an anti-centrin monoclonal antibody and a fluorescein-labeled secondary antibody. Low level autoflourescence reveals the outline of the cell body (grey disc); the flagellar end is at the top. The horizontal element at the top of the "diamond ring" configuration has been shown to reflect the location of the (centrin-containing) BBs and the distal striated fiber; the vertical elements are the NBBCs that descend from the BBs as two parallel fibers until they reach the surface of the nucleus, where they branch into a set of fibrils that are tightly associated with the nuclear envelope (and probably through it to the nuclear matrix). (Photograph courtesy of Mark A. Sanders, Imaging Center, College of Arts and Sciences, University of Minnesota, and Jeffrey L. Salisbury, Laboratory for Cell Biology, Mayo Clinic Foundation.)

antibodies was a pair of fine fibers connecting the nucleus to the BBs in *C. reinhardtii* (Wright et al. 1985). These fibers are attached to the sides of the BBs and extend (in the relaxed state) about 5 μm through the cytoplasm until they contact the nucleus, whereupon they branch extensively to form a series of fibrils that extend out over much of the nuclear surface (Figure 4.3) (Salisbury 1988). These nucleus-BB connectors, or NBBCs, had been detected by light microscopy as much as 50 years earlier (Kater 1929; Pitelka 1969, 1974). However, in contrast to the more robust versions evident in certain other flagellates, the NBBCs of *Chlamydomonas* are not preserved during conventional electron-microscopic (EM) preparative techniques (Salisbury

1988), so they were seldom visualized by the first generation of cell biologists who studied *Chlamydomonas* ultrastructure (however, see Cavalier-Smith 1967). Nevertheless, they are sufficiently stable that when cells of *C. reinhardtii* are lysed, the nucleus, BBs, and NBBCs can readily be isolated as a structural unit; moreover, in the presence of calcium the NBBCs contract dramatically, reducing the separation between the nucleus and the BBs by more than 95% (Wright et al. 1985). Cyclic contraction and relaxation of the NBBCs (and corresponding movement of the nucleus toward and away from the BA) can also be effected in permeabilized *C. reinhardtii* cells by exposing them alternately to high and low concentrations of calcium in the presence of ATP (Salisbury, Sanders & Harpst 1987). With respect to the role of these connectors in cell division (which will be discussed later), it is important to note that the NBBCs remain firmly associated with the nuclei even after the nuclear envelope has been dissolved with detergent (Wright et al. 1985); thus it seems likely that these centrin-rich fibers are somehow connected to the nuclear matrix, perhaps via integral membrane proteins of the nuclear envelope.

Centrin is found in at least two other locations in the BA of *C. reinhardtii*, in addition to the NBBCs, namely, in the distal striated fiber that links the two BBs to one another (Figure 4.2) and in the transition zone between the BB and the flagellar axoneme. Different amino acid substitutions resulting from point mutations in the gene encoding centrin were shown to affect these three structural regions differentially (Taillon et al. 1992; Taillon & Jarvik 1995), suggesting that centrin probably engages in different types of molecular interactions in these three different cytological locations. This conclusion is supported by the observation that centrin appears to have quite different functions in these three regions.

Centrin in the transition zone has been implicated in the process of flagellar autotomy (shedding), which occurs in response to noxious stimuli and requires calcium in the medium. Under these conditions, the centrin-containing fibers that connect the outer doublets to the central cylinder of the transition zone contract as the flagellum is severed just distal to the transition region (Lewin & Lee 1985; Sanders & Salisbury 1989). It may be, however, that this contraction is a result, rather than a cause of autotomy, because mutants lacking centrin in the transition region are capable of autotomy (Jarvik & Suhan 1991).

The distal striated fiber (the centrin-rich fiber that connects the BBs to one another) is apparently formed when two half-fibers that have been initiated at equivalent points on adjacent BBs (on triplets 9, 1, and 2) link up to form a complete bridging structure (Gaffal 1988). A completed distal fiber clearly is essential for maintaining the 180° rotational symmetry of the BBs, because in certain mutants that form half-fibers that fail to connect with one another the BBs and flagella do not establish the proper 180° rotational relationship (Wright, Chojnacki & Jarvik 1983; Hoops, Wright et al. 1984; Adams et al. 1985). It is particularly important to note that such mutants have neither the ability to swim in a directional manner nor the ability to segregate BBs or other organelles properly at cell division.[10]

4.2.3 Assemblin-rich fibers connect the two halves of the BA to one another

Connecting elements of a third type found in algal BAs are "system I fibers," which typically are associated with the MTRs and serve to link MTRs of like type in the two halves of the BA to one another across the midline during interphase (Melkonian 1980). System I fibers are now known to be composed of a 34-kDa fibrous protein that has been called "assemblin" because of the ease with which it can be disassembled with ionic detergents or chaotropic agents and then reassembled in vitro into paracrystalline arrays with the same periodicity of cross-striations as are observed in situ (Lechtreck, McFadden & Melkonian 1989; Lechtreck & Melkonian 1991). Recent analysis indicates that the striated system I fiber that connects, for example, the two 2MTRs from opposite halves of a *Chlamydomonas* BA (Figure 4.2) is actually (like the centrin-rich distal striated fiber described earlier) composed of two half-fibers that have 180° rotational symmetry and that associate with one another at their proximal ends to form what appears (in a conventional EM image)

---

[10] The distal striated fiber has an additional function in some green algae. In the naked biflag-ellate *Spermatozopsis similis* (as in many other naked or scale-covered green-algal cells), as the cell swims forward with the conventional ciliary type of beat, the BBs are oriented in an end-to-end, antiparallel manner. But upon exposure to an intense light source, the two BBs are rapidly swung upward toward one another, so that they lie in a side-by-side, parallel manner as the cell executes the backward-swimming flagellar waveform characteristic of the photophobic response (Watson 1975; Melkonian 1978; Melkonian & Preisig 1984). Isolated flagellar apparatuses of *S. similis* were used to show that this change in BB orientation is due to a rapid, extensive, and forceful calcium-activated contraction of the proximal striated fiber (McFadden et al. 1987). This contraction normally occurs in synchrony with the change in flagellar beat pattern, but it has been nicely shown that these two effects of elevated calcium concentration are mechanistically independent (McFadden et al. 1987). Such reorientation of flagella would not be possible in species like *Chlamydomonas* that have a coherent cell wall through which the flagella emerge and which restricts movement of their BB ends. It is therefore not surprising that the BBs of *Chlamydomonas* are always oriented at a 90° angle to one another (halfway between the angles observed in forward- and backward-swimming naked cells) and that when *Chlamydomonas* flagella are beating in a backward-swimming mode, each exhibits a sharp reverse bend just distal to the point where it exits the cell wall, so that the proximal ends of the two flagella lie parallel to one another, in a similar manner as (but by an entirely different mechanism than) they do in backward-swimming naked cells (Ringo 1967; Hyams & Borisy 1978; Hoops & Witman 1985).

to be a single continuous fiber (Lechtreck & Melkonian 1991). It is postulated that the primary function of the system I fibers may be to hold the two halves of the BA together in the face of the stresses generated by flagella beating. The observation that system I fibers (like distal striated fibers) are much more robust in naked biflagellates like *Spermatozopsis* than they are in walled biflagellates like *Chlamydomonas* (in which the cell walls are thought to absorb most of the stresses generated by the flagella) is consistent with this view (Lechtreck & Melkonian 1991).

There can be little doubt that these three structural elements – MTs, centrin-rich fibers, and assemblin-rich fibers – constitute only the largest and most easily studied of the many cytoskeletal elements that contribute to the structure and function of the algal BA. Many other structural components – such as proximal striated fibers linking adjacent BBs at their proximal ends, and an assortment of filaments linking the BBs (in some cases directly, but in many cases indirectly) to the plasmalemma and to probasal bodies, contractile vacuoles, chloroplasts, and so forth – can be readily seen in the electron microscope, but in most cases the compositions of such connectors remain unknown. Moreover, there undoubtedly are numerous other components of the BA whose existence has yet to be detected in any way. Nevertheless, it is already apparent that the BA is a cytoskeletal assemblage of extraordinary complexity and regularity that plays a central role in the biology of the organism during interphase. Its role during cell division is neither less complex nor less important.

4.3 Cell division in *Chlamydomonas* and related green algae has some unusual features

Cell division in *Chlamydomonas* has been studied repeatedly, and in a variety of ways, for nearly a century,[11] and yet new aspects continue to be uncovered. Although it has long been obvious that in several respects cell division in *Chlamydomonas* differs substantially from division in the more familiar higher plants and animals, it is equally obvious that it serves as a starting

[11] For example, see Dangeard (1899), Belar (1926), Kater (1929), Schaechter & DeLamater (1955), Buffaloe (1958), Bernstein (1964, 1966, 1968), Cavalier-Smith (1967, 1974), Ringo (1967), Johnson & Porter (1968), Howell & Naliboff (1973), Coss (1974), Triemer & Brown (1974), Gould (1975), Pickett-Heaps (1975), Wetherell & Kraus (1975), Ettl (1976), Lien & Knutsen (1979), Coleman (1982), Craigie & Cavalier-Smith (1982), Huang, Ramanis et al. (1982), Donnan & John (1983), Harper & John (1986), Doonan & Grief (1987), John (1987), Gaffal (1988), Salisbury, Baron et al. (1988), Holmes & Dutcher (1989, 1992), Gaffal & el-Gammal (1990), Gaffal, el-Gammal & Friedrichs (1993), Ehler et al. (1995), and Gaffal, Arnold et al. (1995).

place for understanding division in all other chlorophyceans, including *Volvox* and its kin.

The most distinctive feature of chlamydomonad division was mentioned in Chapter 2: All green flagellates that have a *Chlamydomonas*-type cell wall undergo "multiple fission," rather than the more familiar binary fission (Figure 2.9). Whereas plant and animal cells typically grow twofold before dividing once, *Chlamydomonas* and its relatives regularly grow 2^n-fold, and then divide rapidly n times (replicating their nuclear DNA between divisions) to produce 2^n progeny cells. In *C. reinhardtii*, the value of n can be as high as 5 under optimum growth conditions (Lien & Knutsen 1979), but under the conditions more commonly employed in *Chlamydomonas* research labs it is usually 2 or 3. Under conditions in which $n = 3$, for example, three rounds of rapid division produce eight daughter cells tightly packed within the mother-cell wall; these daughter cells then construct new walls and flagella, produce an "autolysin" that digests the mother wall (Schlösser 1976, 1984; Jaenicke, Kuhne et al. 1987), and then hatch and swim away (Figure 2.9).

It is noteworthy that under illumination conditions resembling those in nature (i.e., a 24-hour light–dark cycle), most of the daylight hours are devoted to growth, and division events in *C. reinhardtii* are confined to the dark period. The initiation and number of these division events are thought to be controlled by the joint action of a "timer" and a "sizer," in which the timer determines when the commitment to divide will occur, and the sizer determines how many divisions will ensue once the cell has become committed to divide (Craigie & Cavalier-Smith 1982; Donnan & John 1983; Harper & John 1986; John 1987).

Division in the chlamydomonads also has several distinctive cytological features. For one thing, the mitotic spindle is of the "closed" type; it develops within the nuclear envelope, which remains largely intact throughout mitosis (Johnson & Porter 1968). Moreover, the spindle forms not near the center of the cell, as in many other eukaryotes, but near the cell surface, with the spindle curved in such a manner that its poles are held in close proximity to the BBs, which remain attached to the plasmalemma (Coss 1974). Finally, like all other chlorophycean algae, *Chlamydomonas* has an unusual type of cytokinetic apparatus, called a "phycoplast," in which the mitotic spindle collapses completely at telophase as two new sets of MTs form (perpendicular to the former spindle axis, and midway between the daughter nuclei) to define the plane in which cytokinesis will occur (Pickett-Heaps 1972, 1975, 1976). These features will be described in more detail later.

4.4   The BA plays a key role in segregation of organelles during division of *C. reinhardtii*

During interphase, the *C. reinhardtii* BA functions as a unitary structure in which two rather similar halves are held together by at least two sets of connectors described earlier: the centrin-rich distal striated fiber that links adjacent BBs, and the assemblin-rich system I fibers that link MTRs of like type across the midline. In preprophase, however, both of these sets of connectors break down, which allows the two halves of the BA to act independently (but symmetrically) during division. At about the same time, both flagella are disassembled and resorbed, and BB replication is completed, so that by the beginning of prophase the BA consists of two similar halves, each of which contains two full-length BBs, one parental and one daughter, neither of which is associated with a flagellum (Figures 4.4 and 4.5).

Despite all of these dynamic changes, the two halves of the BA continue to interconnect many other parts of the cell. During division they interact closely with the mitotic spindle, play an important role in cytokinesis, and participate in the orderly assortment of organelles to the daughter cells and in the establishment of predictable symmetry relationships within those cells (Johnson & Porter 1968; Pickett-Heaps 1972, 1975, 1976; Coss 1974; Floyd 1978; Hoops & Witman 1985; Wright et al. 1985; Doonan & Grief 1987; Gaffal 1988; Salisbury, Baron & Sanders 1988; Holmes & Dutcher 1989, 1992; Segaar & Gerritsen 1989; Gaffal & el-Gammal 1990; Segaar 1990; Kirk, Kaufman et al. 1991; Gaffal, el-Gammal & Friedrichs 1993; Ehler et al. 1995).

4.4.1   BBs are essential for coordination of karyokinesis with cytokinesis

As the descriptions that follow will indicate, during cell division various components of the BA execute highly stereotyped, dynamic activities that are temporally and spatially linked to various aspects of karyokinesis and cytokinesis. These activities are so thoroughly integrated in wild-type cells that until recently it had been impossible to discern with any clarity which components of the BA take the lead role in mediating various critical aspects of the division process. However, important insights in this regard have now been provided by a superb recent study conducted by Ehler et al. (1995). Those authors were able to dissect the roles of several components of the division apparatus by performing a detailed cytological analysis of division

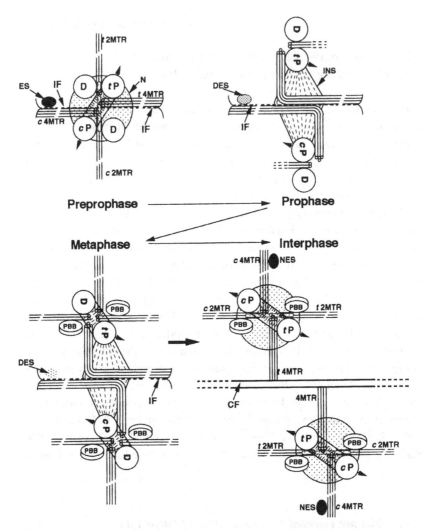

Figure 4.4 Behavior of the BBs and MTRs during cell division (compare to Figures 4.2 and 4.5). **Preprophase:** One of the first signs of imminent mitotic activity is the formation of an incipient division furrow (IF) in the plasmalemma immediately adjacent to the 4MTRs and the eyespot (ES). At that time probasal bodies (pro-BBs) formed in the preceding mitotic cycle elongate into full-length "daughter" BBs (D), and each remains attached to the same 2MTR as its "parental" BB (*c*P or *t*P). The *cis* parental BB (*c*P) is the younger of the two parental BBs, having been transformed from a pro-BB to a functional BB in the preceeding mitotic cycle. **Prophase:** Connections between the two halves of the BA have broken down, and as the intranuclear mitotic spindle (INS) forms and elongates the two P-D BB pairs separate, remaining near the spindle poles. During this movement each BB pair rotates clockwise through 90°. The MTRs remain attached to the BBs, but the 2MTRs shorten while the 4MTRs elongate, making a sharp bend in the region of the incipient furrow as they do so. The connections between the MTRs and the BBs assure that the two members of each BB pair will be held in a fixed relationship to one another as they rotate and move and that the

in *bld2*, a *C. reinhardtii* mutant in which more than 99% of all cells lack all traces of BBs (and thus flagella, whence the name "bald").[12]

As is evident from the fact that *bld2* is still in culture 20 years after its isolation (Goodenough & St. Clair 1975), basal bodies are not required for either karyokinesis or cytokinesis. However, as the recent study shows, BBs must be essential for *integrating* these two processes, because in the absence of BBs *bld2* cells are unable to coordinate cytokinesis with karyokinesis in either space or time: A substantial majority of *bld2* cells will exhibit serious mispositioning of either the mitotic spindles or the cleavage furrow or both, which at best produces sister cells that are significantly different in size, and at worst results in the formation of anucleate and binucleate daughter cells, neither of which will be able to divide again. Temporal coordination is equally disturbed in *bld2*: Whereas in wild-type cells a cleavage furrow that has ingressed far enough to become visible in the light microscope never

---

[12] In less than 1% of *bld2* cells, a circle of singlet MTs, rather than the normal triplet MT arrays, can be observed in the region where BBs are normally located. On this basis, it is postulated that the defect in the *bld2* mutant is in the ability to form doublet and triplet MTs, not in the ability to initiate BB formation (Goodenough & St. Clair 1975).

Caption to Figure 4.4 *(cont.)*
daughter BBs will always be swung away from the cleavage furrow. DES, disintegrating eyespot. **Metaphase:** The BBs in each P-D pair have moved apart slightly as new striated fibers have formed between them, and as new MTRs have begun to form next to the daughter BBs. By now, new pro-BBs (PBB) have also formed next to each full-length BB, and each pro-BB is attached to the same 2MTR as its parental BB. **Interphase:** As the cleavage furrow (CF) was completed, and a complete BA was reconstructed in each daughter cell, two aspects of BB maturation have occurred: (1) Having passed through a second complete cell-division cycle, what was previously a full-length but immature BB capable of templating only a *cis*-type flagellum (cP in the previous drawings), has now matured into a *trans*-type BB (tP) capable of templating a *trans*-type flagellum. (2) The BBs that grew to full length during prophase (D in the previous drawings) have become functional but immature BBs (cP) capable of templating formation of *cis*-type flagella (with effective strokes that will be oriented as indicated by arrow heads). New eyespots (NES) are now formed in each of the daughters next to the 4MTRs that are distal to the cleavage furrow. Because of the highly choreographed behavior of the BBs and the rootlets, these will always be the 4MTRs emanating from the younger (cP) BBs. If (as is usually the case) two or more divisions occur in rapid succession, the same stereotyped movements of BBs and rootlets occur at each division. At the end, each daughter cell will have one BB proximal to the last cleavage furrow that has been through at least two complete cell-division cycles and is therefore a mature (tP-type) BB, as well as a BB distal to the cleavage furrow that was formed from a pro-BB in the last division cycle and that therefore is an immature (cP-type) BB. Most importantly, in every daughter cell the eyespot will form next to the 4MTR emanating from its younger BB; this invariant relationship underlies the ability of the cell to orient with respect to light.

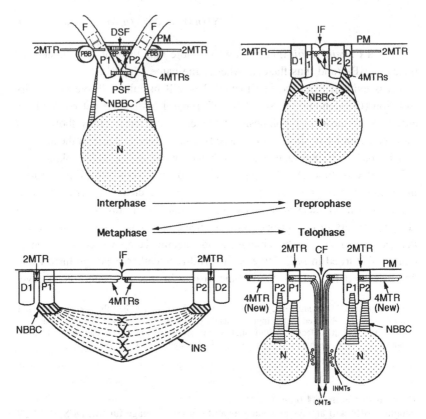

Figure 4.5 Coordinated behavior of the BBs and nuclei during cell division (abbreviations as in Figures 4.2 and 4.4). **Interphase:** A more stylized and simplified version of what is shown in Figure 4.2A. **Preprophase:** Five changes occur in preprophase: (1) The flagella, including their transition zones, are resorbed. (2) The proximal and distal fibers connecting BBs break down, and in the process the BBs assume a more nearly parallel orientation. (3) Pro-BBs elongate to become full-length D BBs. (4) NBBCs contract, drawing the nucleus up into the vicinity of the BBs. (5) An incipient division furrow (IF) forms. **Metaphase:** As the spindle elongates, the BB pairs separate and rotate 90°, with parental BBs attached to the spindle poles by the contracted NBBCs. Meanwhile, the 4MTRs elongate (bending in the vicinity of the furrow), and the 2MTRs shorten. Note that the metaphase plate now lies directly beneath the incipient furrow. (New pro-BBs and MTRs shown in Figure 4.4 have been omitted for clarity.) **Telophase:** The BBs and the newly formed daughter nuclei to which they are attached move back to the vicinity of the incipient furrow. By now, the *c*P BB has matured to a *t*P type, and both daughter BBs have developed into new *c*P-type BBs. As the spindle disassembles, two new sets of MTs are formed: two parallel arrays of cleavage MTs (CMTs) that begin next to the 4MTRs flanking the furrow and that penetrate deeply into the cell in the space between the nuclei, and two sets of internuclear MTs (INMTs) that are orthogonal to both the CMTs and the former spindle axis. Together, these two sets of MTs constitute the cytokinetic structure called a "phycoplast" that is found only in certain green algae, and that defines the plane in which the cleavage furrow (CF) now ingresses to divide the cell.

appears before late anaphase (and much more frequently appears first in early telophase), in *bld2* visible ingression of the furrow begins as early as prophase. On the other hand, because mitotic spindles and furrows do form in *bld2* cells, albeit in an uncoordinated manner, the mutant permitted Ehler et al. (1995) to deduce the roles that other components of the cytoskeleton play in controlling these important aspects of the division process. These new insights will be mentioned in appropriate locations in the discussions to follow.

4.4.2   The NBBCs couple the movements of the BBs to that of the mitotic spindle

Of the three centrin-rich structures attached to the *Chlamydomonas* BBs during interphase, only the NBBCs persist to play a role during division. As the distal striated fiber and the transition zone (along with the rest of the flagellum) are being disassembled during preprophase, the NBBCs contract, drawing the nucleus from its interphase position deep within the cell into close apposition to the BA, where it remains throughout division (Salisbury, Baron & Sanders 1988). As the spindle forms and elongates during prophase, the BBs move apart, in the plane of the membrane, remaining closely associated with the spindle poles (Coss 1974; Gaffal, el-Gammal & Friedrichs 1993) via their centrin linkages (Salisbury, Baron & Sanders 1988) (Figures 4.4 and 4.5).

This centrin-mediated association of the BBs with the spindle poles during cell division appears to be essential for coordinating the movements of the BBs with the rest of the division apparatus. In the *vfl-2* mutant of *C. reinhardtii*, which has a point mutation in the gene encoding centrin (Taillon et al. 1992), BBs are connected to the nuclei during interphase, but the NBBCs are deficient in centrin and are weak and readily broken (Wright 1985). As a result, BBs often fail to segregate normally to the daughter cells. This results in the variable-flagellar-number phenotype and a loss of the capacity to swim directionally and execute phototaxis (Wright et al. 1989).

In the *bld2* mutant, although centrin is associated with the spindle poles during division, in the absence of BBs it fails to connect to the BA in more than 99% of the cells (Ehler et al. 1995) (see fn. 12 above). There can be little doubt that this absence of centrin-mediated connections between the spindle and the BA accounts for the loss of spatial coordination between karyokinesis and cytokinesis in *bld2*. Thus we can conclude that the function of the association between BBs and spindle poles that is normally seen in the green flagellates is not to permit the BBs to play an essential role as

MTOCs for the spindle, but is to provide an essential structural and regulatory link between the karyokinetic apparatus and the cytokinetic apparatus.

4.4.3 MTRs maintain the spatial relationship between old and new BBs and define the cleavage plane

Although it is an NBBC that normally holds each parent–daughter BB pair near a spindle pole, it is the MTRs that hold the members of each of these BB pairs in the precise spatial relationship that assures that the younger BB will always be on the side of the daughter cell where a new eyespot will be formed after the completion of cell division (Holmes & Dutcher 1989).

BB replication in *C. reinhardtii* begins during one mitotic cycle but is not completed until just before the beginning of the next (Gould 1975; Gaffal 1988). During metaphase and/or anaphase, a short new probasal body is formed adjacent to each of the existing full-length BBs and is connected to it by fibers that persist during interphase. Then, early in preprophase, as the flagella are being resorbed in preparation for division, the MT triplets of the probasal bodies are elongated until each reaches full length, so that by prophase each half of the BA contains two full-length BBs. Importantly, each parent–daughter pair is connected to the 2MTR that lies between them, and it is this connection that keeps them in a constant orientation with respect to one another as they move with the spindle pole (Figure 4.4).

Several different types of cytological studies have converged in recent years to give us an integrated view of the rather extraordinary behavior of the MTRs during division in the chlamydomonads and their role in orienting the BBs and the eyespot properly in the next cell generation (Doonan & Grief 1987; Gaffal 1988; Holmes & Dutcher 1989, 1992; Segaar & Gerritsen 1989; Segaar et al. 1989; Gaffal & el-Gammal 1990; Segaar 1990; Gaffal, el-Gammal & Friedrichs 1993; Ehler et al. 1995). The following summary is a synthesis of the conclusions drawn from those studies and is illustrated diagrammatically in Figures 4.4 and 4.5.

One of the first externally visible signs that division is imminent is the formation of an incipient cleavage furrow at the anterior end of the cell when the flagella begin to be resorbed and the NBBCs contract. This groove (which at this stage is detectable only at the EM level) always forms adjacent to the 4MTRs and bisects the BA. During prophase, the nucleus elongates, and the spindle forms at approximately a right angle to the incipient furrow, and as this happens the two halves of the BA separate, keeping pace with the spindle poles while remaining attached to their MTRs. As the BBs move, the 2MTRs shorten, but the 4MTRs elongate, bending sharply in the vicinity of the furrow. Thus, by metaphase, the BB-proximal end of each 4MTR lies parallel to the spindle axis, whereas its distal end remains perpendicular to the spindle and associated with the incipient furrow.[13]

The continuing association of the BBs with the MTRs during their prophase-to-anaphase movements always results in the younger BB lying farther from the cleavage furrow than the older BB does. This has a fundamentally important consequence, as we shall see.

At metaphase, the "parental" 4MTR attached to the older BB persists, but additional MTRs begin to be formed, so that by late anaphase each BB pair lies at the center of a developing new 4-2-4-2 cruciate array of MTRs. At about the same time, each BB pair will become connected by a new distal striated fiber, a new NBBC will be formed to connect the younger BB to the adjacent spindle pole, and a new pair of probasal bodies will be formed (in association with the existing BBs). Thus, before the end of mitosis each prospective daughter cell has a complete BA very similar to that which was present in the mother cell during interphase (Figures 4.4 and 4.5).

4.4.4   The role of the BA in cytokinesis assures orderly segregation of organelles

At telophase, the mitotic spindle collapses quickly and completely, and the daughter nuclei, with their associated BAs, move back toward one another until they come to straddle the incipient cleavage furrow. Then two new sets of MTs form: The first is an array of "cleavage" MTs that emanate from the vicinity of each BA and extend deep into the cytoplasm between the daughter nuclei. The second set of MTs, the "internuclear" MTs, form perpendicular to both the former spindle axis and the cleavage MTs (Figure 4.5). Together, the cleavage and internuclear MTs constitute the "phycoplast," the distinctive cytokinetic apparatus of chlorophycean algae.

In the *bld2* mutant, in which the locations of the MTRs are essentially random (owing to the absence of BBs), cleavage MTs form (and cytokinesis subsequently occurs) wherever the 4MTRs happen to be located (Ehler et al. 1995). This observation confirms for *C. reinhardtii* what had previously been established for a related biflagellate unicell (Segaar & Gerritsen 1989; Segaar

---

[13] In the first detailed ultrastructural study of *C. reinhardtii* division, only the distal ends of the 4MTRs were detected. These MTs were termed "the metaphase band," in the mistaken belief that they had been assembled de novo above the spindle equator at metaphase (Johnson & Porter 1968). It is now clear, however, that such is not the case (Doonan & Grief 1987; Gaffal 1988; Holmes & Dutcher 1989, 1992; Gaffal & el-Gammal 1990).

et al. 1989; Segaar 1990): It is the distal portions of the 4MTRs that line the incipient cleavage furrow that act as the MTOCs for the cleavage MTs.

Once the phycoplast has formed, the cleavage furrow (which has been incipient since prophase) ingresses in the plane defined by the two parallel arrays of cleavage MTs. Because one BA lies on each side of this furrow and is connected directly or indirectly to virtually all of the organelles in that half of the cell, an orderly segregation of organelles to each daughter cell is assured. For example, even the Golgi stacks, of which there are two in the interphase cell, are segregated to the daughter cells dependably, presumably as a result of their connections to opposite halves of the BA (Holmes & Dutcher 1992).

The subdivision of the chloroplast is more complex and considerably more interesting than the preceding paragraph suggests. Two decades ago, Ettl (1975) concluded from his light-microscopic examination of division in two other species of *Chlamydomonas* that the chloroplast develops deep longitudinal furrows very early in the mitotic cycle, and that its division ("plastokinesis") is completed before karyokinesis (let alone cytokinesis) has been completed! This conclusion has recently been verified and expanded with respect to *C. reinhardtii* by an elegant computer-aided reconstruction of electron micrographs of serial-sectioned cells (Gaffal, Arnold et al. 1995). The furrows described by Ettl develop in the chloroplast as early as prophase, and because they form along the 4MTRs (which extend for a considerable distance in the thin space between the plasmalemma and the chloroplast) they invariably lie in the same plane as the incipient cytokinetic furrow. The chloroplast furrows deepen during mitosis, reaching the base of the chloroplast by metaphase, and by the time that cytokinesis occurs the chloroplast has already been bisected, with one half going to each presumptive daughter cell. It appears that the 4MTRs identify the sites at which chloroplast furrowing should be initiated, but it is far from clear where or how the motile force for plastokinesis is generated, that is, whether it results from the activities of motile elements that are internal or external to the chloroplast. Equally mysterious is the mechanism by which the newly separated daughter chloroplasts undergo rapid and dramatic shape changes at telophase, transforming from highly asymmetric half-bowls to bilaterally symmetrical hollow ellipsoids (Gaffal, Arnold et al. 1995). Is there a previously unrecognized motile apparatus within the chloroplast? This question clearly deserves further attention.

4.4.5 A new eyespot forms next to the new 4MTR emanating from the younger BB

If (as is usually the case) the daughter cells produced in the first division cycle are to divide again, the sequence of events just described will be quickly repeated. In the process, the younger BB of each daughter cell will mature into a parental one – within an hour of its formation, and without benefit of ever having constructed a flagellum or having passed through a typical interphase (Figure 4.4) (Gaffal 1988). As a consequence of its newly established rootlet connections, as this now-mature BB participates in its second division cycle, it will end up *proximal* to the division plane, rather than *distal* to it as in its first encounter with a spindle pole. And, of course, its associated daughter BB (which will have elongated in preprophase) (Gaffal 1988) will end up on the distal side (Holmes & Dutcher 1989; Segaar & Gerritsen 1989).

During division, the stigma of the mother cell (which is always located adjacent to the first cleavage furrow) (Figure 4.4) disintegrates. Then, following the last division, each daughter cell produces a new stigma. Each of these is always formed on the side of the cell distal to the last cleavage furrow, beside the 4MTR that lies in this region (Holmes & Dutcher 1989). Because the movements of the BBs and MTRs have been so stereotyped during each division cycle, this 4MTR will predictably be the newer one: the one that emanates from the younger BB. As has been repeatedly mentioned, this association between the younger BB and the new eyespot is crucial: It assures that when the eyespot is subsequently stimulated by light, the responses of the *cis* and *trans* flagella will be adaptive.

A recent study (Deininger et al. 1995) has added a new dimension to our understanding of eyespot formation. By performing indirect immunofluorescence with an antibody to "chlamyopsin" (the protein moiety of the chlamyrhodopsin discussed in Section 4.1.1), those authors were able to follow the behavior of the photodetector part of the eyespot during the cell-division cycle in *C. reinhardtii*. They observed that the opsin spot in the plasmalemma of the mother cell (in contrast to the stigma, which disappears early in the division cycle) (Ettl 1976; Holmes & Dutcher 1989) persists until the end of the first division cycle and is still visible at the time that two new opsin spots are being formed in the daughters. As would be predicted from previous studies showing that the stigma lies on one side of the cleavage furrow (Holmes & Dutcher 1989; Gaffal, el-Gammal & Friedrichs 1993), the maternal opsin spot is not subdivided by the cleavage furrow, but lies on one side of it. Most interestingly, the new opsin spots of the daughter cells also lie near the cleavage furrow initially; only later do they migrate to the distal sides of the cells, to the region where new stigmas will eventually be formed

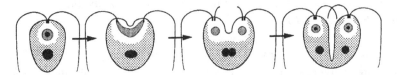

Figure 4.6 Cell division in a naked or scaled green flagellate. Green algae that lack a coherent wall divide by a form of simple binary fission in which the BBs serve as centrioles at the spindle poles (much as diagrammed in Figures 4.4 and 4.5) while remaining attached to actively beating flagella. (Adapted from Iyengar & Desikachary 1981.)

de novo. Moreover, the opsin spot formed in the daughter cell containing the maternal opsin spot appears to be significantly larger than the opsin spot formed in the sister cell. And at the end of the second division cycle, two of the daughter cells have larger, and two have smaller, opsin spots (Deininger et al. 1995). These observations raise the distinct possibility that in each division cycle the opsin molecules of the mother cell are at least partially conserved and transferred into the photodetector that is being assembled in one of the daughter cells. However, the fact that opsin spots also form in the other daughter cell indicates that the photodetector, like the stigma, can be produced de novo.

4.5   The BA and cell wall established the conditions that led to germ–soma differentiation

The essential message of the foregoing sections is that it is the highly stereotyped behavior of the BA during division that assures that in each generation a green flagellate will be able to move toward the light, a response upon which it depends for survival.

In naked or scale-covered green flagellates, the BBs remain firmly attached to their flagellar axonemes while the BA participates in all of these cell-division activities. In such species, as the mitotic spindle forms between sister BBs, the BBs simply move apart in the plane of the plasmalemma – with flagella attached and actively beating. Then, at telophase, a cytokinetic furrow forms between the BBs and the cell divides, having remained fully motile throughout mitosis and cytokinesis (Figure 4.6) (Floyd 1978). It is important to note that binary fission is observed not only in those green flagellates in which the absence of a *Chlamydomonas*-like coherent cell wall is believed to be a primitive feature, but also in those naked flagellates, derived from *Chlamydomonas*-like ancestors, in which the wall has been lost secondarily

(Melkonian 1990; Koufopanou 1990, 1994; Buchheim et al. 1996). The fact that binary fission is seen in naked flagellates that are descendants of walled flagellates that divided by multiple fission gives added credence to the concept that *multiple fission is an evolutionary adaptation that is coupled to the presence of a coherent cell wall.*

4.5.1 A coherent cell wall created a dilemma for green flagellates: "the flagellation constraint"

When the coherent glycoprotein wall (see Section 2.1 and Figure 2.2B) of *Chlamydomonas* and its relatives evolved, possibly by fusion of the loose glycoprotein scales covering ancestral forms (Mattox & Stewart 1984), a dilemma arose. In all green flagellates with a coherent wall, the flagella of the interphase cell pass through the wall via specialized flagellar channels (Figure 4.7). This precludes any lateral movement by the proximal ends of the flagella. Thus, the flagella cannot simply move apart (as they do in naked flagellates) as a mitotic spindle forms between the BBs. Cells living inside such a wall cannot retain the ancestral relationship of the BBs to both the flagella and the spindle during cell division; *one association or the other must go.* This is the source of Vassiliki Koufopanou's "flagellation constraint."[14]

4.5.2 Small algae with coherent walls exhibit three solutions to the flagellation constraint

The chlorococcaleans (see Section 2.2) resolved the dilemma imposed by the flagellation constraint by making two kinds of strategic retreats: (1) They abandoned flagella in the asexual phase of the life cycle. (2) They abandoned the coherent cell wall in the (flagellated) sexual stage (Mattox & Stewart 1984; Bold & Wynne 1985). As might be anticipated from what has gone before, the absence of flagella in the chlorococcalean asexual phase was ac-

---

[14] A very different kind of flagellar constraint has been proposed as an important stimulus to cellular differentiation in the ancestors of the metazoans. Margulis noted that as far as is now known, no metazoan cell divides while it is flagellated (Margulis 1981). Therefore, she postulated, the evolution of the multiple cell layers and the cellular differentiations that characterize metazoans may have begun when cells of an ancestral colonial flagellate lost their flagella and moved to the interior to divide. This idea has gained considerable currency (Buss 1987). Nevertheless, it is completely unclear why there should have been such a constraint against simultaneous flagellation and division in the ancestor to the metazoans, because many modern protists (including the ciliates and many colorless flagellates, as well as the wall-less green flagellates discussed here) divide while flagellated or ciliated.

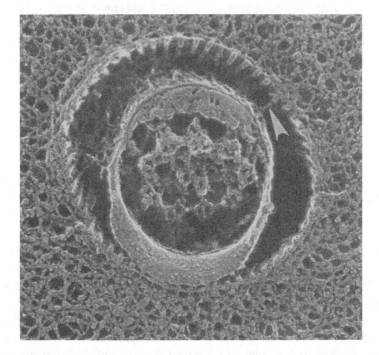

Figure 4.7 Quick-freeze, deep-etch preparation revealing some details of a flagellar channel of *V. carteri* that surrounds each flagellum where it emerges from the cell body. The channel is a highly specialized component of the extracellular matrix, one layer of which is seen surrounding the channel. In life the flagellar base is tethered to the wall by a series of radiating filaments that attach to each of the ridges of the channel, but in this preparation all but one of these filaments (at arrowhead) have broken and collapsed down on the flagellar membrane. Similar flagellar channels are present in all green flagellates with a coherent cell wall. These channels prevent the flagellar bases from moving laterally and therefore preclude the simple type of binary fission diagrammed in Figure 4.6. (Micrograph courtesy of U. W. Goodenough and J. E. Heuser.)

companied by an extensive reorganization of both interphase and dividing cells; cytologically these algae reveal their affinities to the chlamydomonads only in the flagellated sexual phase (Mattox & Stewart 1984; Melkonian 1990). Moreover, the resulting loss of motility has (as might also be antici- pated) largely restricted the chlorococcaleans to stirred environments where they contact essential resources passively (see Figure 3.1, for example).[15]

---

[15] In the lineage leading to the higher plants, the evolution of the cellulosic cell wall was accompanied by a similar loss of flagellar motility, and eventually by the loss of centrioles altogether.

Flagellate unicells like *C. reinhardtii* that have left the open water to live in moist, fertile soils (Sack et al. 1994) have dealt with the flagellation constraint in a second way. As we have just seen, as a *C. reinhardtii* cell prepares to divide, it resorbs its flagella. At the same time, it adds a sticky coating to its wall that permits it to attach to the substrate; then it divides in a temporarily immotile state (Schlösser 1984; Harris 1989). Lacking attached flagella, the BBs of *C. reinhardtii* (and other species that resorb the flagella before dividing) are free to move in the plane of the membrane during mitosis and cytokinesis, as discussed in detail earlier. Presumably because it normally lives in thin films of water coating soil particles, the temporary loss of motility during division is not seriously disadvantageous to a mud-dwelling alga like *C. reinhardtii*.

In pond- or lake-dwelling flagellates, however, loss of motility during division presumably would be much more disadvantageous. All green algae are negatively buoyant and rapidly settle toward the bottom of an unstirred water column in the absence of flagellar motility (Reynolds 1984a; Koufopanou 1994).

An essential component of Koufopanou's flagellation-constraint hypothesis is that green flagellates that live in a truly aqueous environment, where flagellar motility is necessary for remaining in the euphotic zone, must find some way to retain flagellar motility while dividing.[16] The way that this is accomplished in many pond-dwelling species of *Chlamydomonas* and many related walled unicells, as well as in all volvocacean species up to the size of *Eudorina*, constitutes a third solution to the flagellation-constraint dilemma. In these species, the flagella become detached from the BBs during prophase, but the flagella are not resorbed; they remain inserted in the cytoplasm and continue beating throughout division (Gaffal 1977; Hoops 1981; Hoops & Witman 1985). Meanwhile, the BBs that have detached from the flagella are now free to separate from one another in the plane of the membrane as the spindle forms between them. However, as we will see in a bit, there appears to be a limit to how long flagellar motility can be maintained in the absence of an attached BA.

The chlamydomonad wall not only imposed a flagellation constraint but also had a second, equally important consequence that was mentioned earlier:

---

[16] See the introductory section of this chapter for data validating this assumption.

4.5.3 The presence of a Chlamydomonas-*type wall is also accompanied by multiple fission*

Naked and scaled green flagellates divide by simple binary fission (Floyd 1978). In contrast, all of the green algae with coherent glycoprotein cell walls undergo multiple fission of the sort described in Chapter 2 and earlier in this chapter (see Figure 2.9). As noted, the fact that binary fission is seen in naked flagellates that are descendants of walled flagellates that divided by multiple fission gives added credence to the concept that multiple fission is an evolutionary adaptation that is coupled to the presence of a coherent cell wall.

Why these two phenomena are so tightly coupled is not certain. However, a credible working hypothesis is that the flagellation constraint was first resolved (as it is today in *C. reinhardtii*) with loss of motility during division, and thus variants that combined several divisions in one immotile period per day would have had a survival advantage over ancestral forms that had to become immotile several times per day in order to complete the same number of divisions. In this regard, it may well be significant that most species that lose their flagella while dividing do all of their dividing at night and are motile when the sun is shining (see Section 4.3). Whether or not this working hypothesis about its origin has any validity, multiple fission is currently a fact of life for all walled chlamydomonads and all volvocaceans.

4.5.4 Multiple fission facilitates colony formation

The practice of dividing rapidly several times within the maternal cell wall has the effect of holding all sister cells that are formed in any given generation in close association until they have developed their own new walls and flagella and are ready to hatch out and swim free. It seems highly probable that *multiple fission was an important precondition facilitating the formation of volvocacean colonies.*

The first steps toward multicellularity most likely were taken by chlamydomonads in which cytokinesis was not always completed. Incomplete cytokinesis (which results in the formation of cytoplasmic bridges that hold sister cells in a fixed spatial relationship until cell wall deposition has begun) is observed today in all volvocaceans that have been carefully examined (Janet 1923a; Pocock 1933b; Bisalputra & Stein 1966; Darden 1966; Ikushima & Maruyama 1968; Kochert 1968; Deason et al. 1969; Starr 1969, 1970a; Pickett-Heaps 1970, 1975; Marchant 1976, 1977; Gottlieb & Goldstein 1977; Viamontes & Kirk 1977; Fulton 1978a; Birchem & Kochert

1979b; Viamontes, Fochtmann & Kirk 1979; Green & Kirk 1981; Green et al. 1981).

A recently discovered species of *Gonium* (*G. dispersum*) (Batko & Jakubiec 1989) that was briefly discussed earlier (Section 2.3) gives us a clear insight into how this step toward multicellularity may have been taken. *G. dispersum* embryos within a single colony exhibit two different forms of cell division. Some cells divide so completely in each cycle that (like dividing *Chlamydomonas* cells) they are free to move about within the mother wall to assume an optimum packing arrangement. Others within the same parental colony divide so that all daughter cells are held in a concave plate, as is characteristic of embryos of other *Gonium* species. The former hatch out and swim as unicells, whereas the latter hatch out as more or less typical *Gonium* colonies. Those authors noted that it was only because the latter form outnumbered the former that they opted to call this organism *Gonium* rather than *Chlamydomonas*.

It is noteworthy that the walls of adjacent cells within a *G. dispersum* colony are so loosely joined that individual cells frequently break free after hatching. A second step toward multicellularity very likely involved coalescence of walls between adjacent cells, as is now seen in all other species of *Gonium* (Figure 2.10). From there it would have been a small step to the formation of wall elements that joined the outer ends of adjacent cell walls into a single "colony boundary," as occurs today as a second, discrete step during the development of each juvenile colony of *Pandorina* (Fulton 1978b). At that point, variants of increasing size and cell number could have begun to be selected for if (but only if) those larger variants could find some new way to remain motile while their cells divided.

4.5.5 Larger algae circumvent the flagellation constraint in a new way: with sterile somatic cells

Analysis of a unicellular relative of *Chlamydomonas* that retains BB-less flagella during division has shown that separation of the BBs from the flagellar axonemes need not compromise flagellar motility, at least initially (Hoops & Witman 1985). Small colonial volvocaceans, like many green flagellate unicells, retain the flagellum of the parental cell in a BB-less state during embryonic cleavage. However, Marchant noted that in a strain of *Eudorina elegans* with 16 or 32 cells, "parental flagella continue to beat, although not as strongly or as well coordinated as when the cells are not dividing" (Marchant 1977). Moreover, three other species of colonial flagellates in which dividing cells initially retain BB-less flagella have been analyzed, and in each case it was reported that flagella were always lost eventually, sometimes as early as the third division, and in no case later than the fifth division (Iyengar & Ramanathan 1951; Rayburn & Starr 1974; Nozaki 1986b). Apparently, flagella that continue beating after having lost their firm connections to the rest of the cell via the BA eventually tend to tear themselves loose from the cell.

In the laboratory, such loss of flagellar motility merely results in the dividing organisms accumulating on the bottom of the culture vessel (Rayburn & Starr 1974). But in a pond it could result in a termination of photosynthesis, as the organisms fell out of the euphotic zone. The larger the organism, and hence the more rounds of division it needs to complete, the more serious this consideration becomes, because the rate at which nonmotile volvocaceans fall through an unstirred water column increases with increasing size (Koufopanou 1990). Thus, retention of flagella that have lost their BBs appears to have provided a viable solution to the flagellation constraint only for unicells and colonies up to about 32 cells in size.

By Koufopanou's hypothesis, if larger volvocaceans were to succeed, they would have to resolve the flagellation constraint in some different way in order to keep their place in the euphotic zone while completing more than five rounds of cell division.

And so they did. Koufopanou (1994) has collected, summarized, and analyzed in considerable detail data indicating that among the volvocaceans, as cell number increases, there is an increasing tendency for a subset of the cells in the anterior end of the organism to remain small, continue beating their flagella, and thereby provide the colony with a continuous source of motility while the rest of the cells enlarge, divide, and produce progeny. Such a cellular dichotomy is observed sporadically in 32-cell volvocaceans, frequently in 64-cell ones, and invariably in those having more than 64 cells in the adult (Goldstein 1967; Kochert 1973; Starr 1980; Iyengar & Desikachary 1981; Koufopanou 1994). By the time an organism is produced that contains more than 500 cells, this has become, of course, the full-blown germ–soma division of labor that characterizes *Volvox*. These relationships will now be considered in slightly more detail.

4.6 The extent of germ–soma differentiation increases with increasing size

In *Eudorina* cultures in which individuals contain 16 cells, all cells regularly enlarge in concert and divide at the same time to produce progeny (Goldstein 1964; Marchant 1977). However, under richer growth conditions, in which individuals of the same strains contain 32 cells, the anteriormost 4 cells usually enlarge less than posterior cells, and if they ever divide (which they do not always do), they do so later and produce smaller progeny than the posterior cells (Goldstein 1964). In *Pleodorina californica*, a certain number of anterior cells always remain terminally differentiated as flagellated somatic cells. Moreover, within a single *P. californica* clone, the relative abundance of somatic cells increases with total cell number: from about 25% in individuals with 32-cells to as high as 50% in individuals with 128 cells (Gerisch 1959; Goldstein 1964; Kikuchi 1978).[17] Thus, within the *Eudorina-Pleodorina* group there clearly is a tendency – sometimes expressed even within a single clone – to increase the ratio of somatic cells to gonidia as total cell number rises.

This trend continues in the genus *Volvox*. Species of *Volvox* (such as *V. powersii* or *V. gigas*) that have as few as 500–2,000 cells may have as many as 70–80 gonidia (Powers 1907; Pocock 1933b; Smith 1944; Vande Berg & Starr 1971), whereas species (such as *V. barberi* and *V. amoeboensis*) that have as many as 50,000 cells regularly have fewer than 10 gonidia (Shaw 1922b; Rich & Pocock 1933; Smith 1944; Iyengar & Desikachary 1981). Thus, *as total cell number increases, the ratio of somatic cells to gonidia tends to rise,* from less than 15 for *V. gigas* to more than 5,000 for *V. barberi*.

A correlated trend is revealed when gonidial and somatic-cell volumes are plotted as functions of the total number of cells per spheroid (Bell 1985; Koufopanou 1994): As total cell number increases, gonidia get progressively larger, and somatic cells get progressively smaller (Figure 4.8A). This trend, which is particularly evident in *Volvox*, actually begins in *Pleodorina,* in which the somatic cells are smaller, but the gonidia are larger, than the cells of *Eudorina* (Figure 4.8A). The significance of this trend in gonidial size is obvious: In a lineage in which cell division is restricted to the period of embryonic cleavage, it takes a larger germ cell to produce an organism with more cells. However, as the size of the gonidia increases with organismic

---

[17] Interestingly, Gerisch has shown that if the small, normally somatic anterior cells of *P. californica* are isolated from the rest of the spheroid, they will grow, divide, and produce progeny, indicating that they have reproductive potentials that are normally repressed within the intact spheroid (Gerisch 1959). The situation is qualitatively different in *Volvox*: No one has ever reported any conditions under which wild-type *Volvox* somatic cells can be stimulated to divide. (However, as we will see in Chapter 6, a full set of reproductive potentials can be restored to *V. carteri* somatic cells by mutation of a single locus.) A second indication that the reproductive behavior of *P. californica* cells is controlled to a large extent by interspheroidal signaling came from the work of Kikuchi (1978), who showed that if developing gonidia of *P. californica* were released from the spheroid mechanically, they immediately stopped enlarging and began to cleave. This also represents a marked difference from *Volvox*: If *V. carteri* cells are released from a spheroid mechanically, they continue to grow, eventually reaching essentially normal size before beginning to divide (Starr 1970a; Koufopanou & Bell 1993).

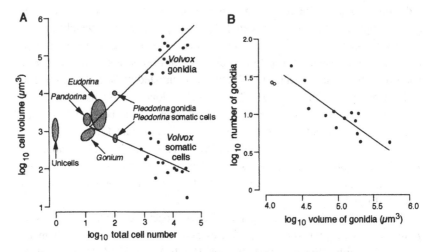

Figure 4.8 Allometric relationships between organismal size and cell size in selected volvocaleans. **A:** Cell size as a function of organismal size and complexity. The range of cell sizes in colonial volvocaceans (*Gonium, Pandorina,* and *Eudorina*) is not greatly different from that in their unicellular relatives. In *Pleodorina,* with its partial division of labor, however, the somatic cells remain near the low end of this size range, whereas their gonidia enlarge to beyond the high end of this range. This trend continues in *Volvox:* As the organisms get larger (in terms of total cell number), the somatic cells tend to get smaller, while the gonidia get larger. **B:** Relationship between number and size of gonidia in various species of *Volvox.* As gonidia get larger (in organisms of increasing size), they also tend to be less numerous. Data for two species of *Pleodorina* are included (open circles) for reference. (A adapted from Bell 1985, in *The Origin and Evolution of Sex,* ed. H. O. Halvorson & A. Monroy, copyright © 1985 Alan R. Liss, Inc., adapted by permission of Wiley-Liss, Inc., a subsidiary of John Wiley & Sons, Inc.: B adapted from Koufopanou 1994, The evolution of soma in the Volvocales, *American Naturalist,* **143**, 907–31, copyright © 1994 The University of Chicago Press, by permission of the publisher, The University of Chicago Press.)

size, the average number of gonidia produced per individual decreases (Figure 4.8B). This apparently indicates some limit to the rate at which the total amount of germ tissue can be increased.[18]

In short, it is clear that as total cell number increases in the volvocaceans, gonidia decrease in number but increase in size, whereas somatic cells increase in number and decrease in size. This is a rephrasing of a concept that has been obvious to biologists for more than a century: Division of labor

---

[18] Because the slope of the line relating gonidial size to gonidial number is −0.67 (less than −1), it follows that the total volume of germ cells produced per *Volvox* spheroid increases with increasing cell number. However, the rate at which the total reproductive-cell volume increases with increasing cell number is not significantly greater in the genus *Volvox* than it is in the colonial volvocaceans in which all cells reproduce (Koufopanou 1994).

increases as size increases. It is also a prime illustration of a principle that was discussed extensively by Adam Smith (1776) in a quite different context and that John Tyler Bonner (1952) has called "the principle of magnitude and division of labor":

... it is impossible to become large and survive in selective competition without becoming complex. A division of labor is essential to support and manage the size. ... It is a principle at least as old as Archimedes, and it was put forth with total clarity by Galileo. The volume or weight of an organism is a cubic function of the linear dimensions. ... However, many of the key activities of an organism vary as the square of the linear dimensions. ... For this reason, with increase of size there has been *an elaboration of surfaces and levers* to keep pace with the increase in bulk. This elaboration is the complexity, the division of labor [Bonner 1974, italics added].

As size increases in the larger volvocaceans, an "elaboration of surfaces and levers" occurs by the production of more and smaller somatic cells. These numerous small cells provide more absorptive surface (for acquiring nutrients) and more levers (flagella) for manipulating the environment than would be provided if the same mass were divided among a smaller number of larger cells. Thus, the size-dependent trends in cellular composition among the Volvocaceae are readily understood in terms of both the source–sink model and the flagellation-constraint hypotheses that have been discussed.

4.7 Environment provided the stimulus and cytology provided the mechanism for germ–soma differentiation

From the elegant analyses of Bell and Koufopanou, two rather different hypotheses regarding the origins of the germ–soma dichotomy of *Volvox* have emerged: the source–sink hypothesis discussed in Chapter 3 (Bell 1985) and the flagellation-constraint hypothesis considered here (Koufopanou 1994). Are these two hypotheses to be considered as alternatives to one another? I think not. They appear to me to be highly complementary.

An impressive array of biochemical, ecological, and experimental studies support the hypothesis that flagellates of larger size have an advantage over smaller ones in capturing the resources of a temporarily rich environment, like that found in eutrophic lakes and puddles, and that ultimately an additional reproductive advantage accrued to organisms like *Volvox* that divided the labors of resource acquisition and utilization between soma and germ (Bell 1985; Koufopanou & Bell 1993). However, such a potential advantage would have been of little significance in evolutionary history if the biology of the primitive green flagellates had not lent itself to the production of organisms of increased cell number and larger size that could maintain their place in the euphotic zone.

Similarly, a convincing case has been made that the flagellation constraint (which arose when the necessity for maintaining a division mechanism that would assure that progeny would be capable of phototaxis was combined with a coherent cell wall) constituted a precondition that facilitated the appearance initially of small colonial volvocaceans with a single cell type, and then of larger forms with a germ–soma division of labor (Koufopanou 1994). But would such a potential ever have been realized if the reproductive costs of becoming larger and setting aside a cohort of sterile somatic cells had outweighed the benefits? Very likely not.

The following would seem to be a reasonable synthesis of these two hypotheses: The transient abundance of nutrients in quiet ponds and lakes each spring constituted an ecological factor that would provide a selective advantage to any green-flagellate variants that happened to cohere to form a multicellular body. And the characteristic organization of the *Chlamydomonas*-like cell, with its unusual pattern of cell division (multiple fission), provided the cytological features that made the production of such a multicellular organism a relatively simple accomplishment. Moreover, when the powerful selective advantage of further increases in size was combined with the stringent requirement for continuous motility in the unstirred lakes and ponds that the volvocaceans inhabited, the eventual appearance of *Volvox*, with its germ–soma dichotomy, seems in retrospect to have been almost inevitable.

Indeed, when the problem is viewed in this combined ecological-cytological context, it seems not in the least improbable that (as postulated in Chapter 2) different species of *Volvox* may have evolved independently, from slightly different volvocacean ancestors, and by different genetic pathways. In any case, the remainder of this book will be devoted to a review of the biology of the species of *Volvox* with which the greatest progress has been made in analyzing the genetic pathway by which an organism with a germ–soma division of labor may have arisen from ancestral forms in which a single cell type executed all vegetative and reproductive functions.

5

Volvox carteri: A Rosetta Stone for Deciphering the Origins of Cytodifferentiation

Progress in understanding a biological process and its control is dependent upon, and at the same time limited by, the nature of the organism which provides the system under investigation. . . . The genus *Volvox* offers a variety of species which may serve in varying ways as experimental material for studies in differentiation of a simple multicellular organism with only two kinds of cells, somatic and reproductive. . . . *Volvox carteri* f. *nagariensis* has been more thoroughly investigated than other species and has been shown to possess an unusual combination of characteristics which make it especially adapted to [such] studies.

Starr (1970a)

The foregoing statement is as valid now as it was when it was written 27 years ago. Modern studies of *Volvox* biology began in Richard Starr's laboratory in the 1960s, when William Darden, then a graduate student, demonstrated that axenic cultures of *Volvox aureus* could be maintained indefinitely in a chemically defined medium that had been developed a few years earlier for culturing other types of algae (Provasoli & Pintner 1959). In such cultures he was able to observe and study all stages in the asexual and sexual life history of *V. aureus* under controlled conditions (Darden 1966). Later, he and his associates would use this culture system to examine a variety of aspects of *V. aureus* biology (Darden 1968, 1970, 1971, 1973a,b; 1980; Deason et al. 1969; Darden & Sayers 1969, 1971; Ely & Darden 1972; Tucker & Darden 1972).

For more than 250 years before that time most biologists who had admired freshly isolated samples of *Volvox* had failed in their attempts to maintain the organism in the laboratory for more than a few days or weeks. There had been rare exceptions. Two methods for relatively long term cultivation of *Volvox* had been reported many years before Darden's time (Uspenski & Uspenskaja 1925; Pringsheim 1930), but for some reason neither of those methods had led to publication of any detailed investigations of *Volvox* biology. Indeed, the two authors who had published the most extensive studies of *Volvox* biology in the decades preceding Darden's report had drawn their samples from populations that had become established in outdoor tanks or ponds near their own laboratories, and both reported failure in their repeated attempts to establish permanent laboratory cultures in pond water, synthetic media, or mixtures of the two (Pocock 1933b; Metzner 1945a,b).

Starr and his students quickly capitalized on Darden's success by bringing other species of *Volvox* under long-term cultivation and laboratory scrutiny. In 1965 a second student, Gary Kochert, reisolated from Kansas and Nebraska farm ponds the species that had so impressed Powers with its potential for analyzing germ–soma differentiation that he had named it after August Weismann[1] (Powers 1908). Kochert promptly described the full asexual and sexual life history of *Volvox carteri* forma *weismannia* (Kochert 1968). While Kochert's work was in progress, Starr isolated a morphologically similar organism, *V. carteri* forma *nagariensis* (Figure 5.1), from a pond near a rice paddy in Japan, and proceeded to describe its life history (Starr 1969). Soon, Starr's group would analyze development in many other *Volvox* species and strains (McCracken & Starr 1970; Starr 1970b, 1971, 1972a; Vande Berg & Starr 1971; Karn et al. 1974; Starr, O'Neil & Miller 1980; Miller & Starr 1981), but none of those would prove to be more suitable for detailed developmental studies than *Volvox carteri* forma *nagariensis*.

5.1 Some features recommending *V. carteri* f. *nagariensis* as a developmental model

Within a year after he had first described its life history, Starr had established clearly that *V. carteri* f. *nagariensis* had the following features that would

---

[1] Not realizing that a very similar form of *Volvox* had previously been described by Carter in India (Carter 1859), Powers had named his isolate *V. weismannia* (Powers 1908). Although its status as a separate species has been a matter of some dispute over the intervening decades (Shaw 1922c; Pascher 1927; Printz 1927; Iyengar 1933; Smith 1944; Metzner 1945a,b), it is now generally considered a forma of *V. carteri* rather than a separate species.

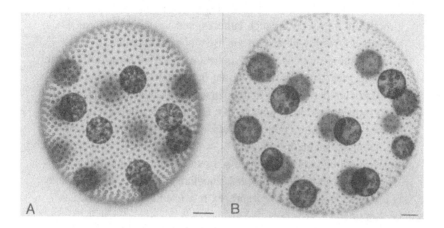

Figure 5.1 Two spheroids of *Volvox carteri* forma *nagariensis*. **A:** An adult spheroid with 16 mature, uncleaved gonidia lying just below the surface monolayer of about 2,000 small somatic cells. **B:** A parental spheroid (about 18 hours older than the one shown in A) containing 16 fully cleaved and inverted juvenile spheroids. Note that the new generation of gonidia can be clearly distinguished in each of the juveniles that lies near the focal plan. Bars = 50 μm.

be useful in adapting it as a model system for analyzing the genetic and cytological basis of cellular differentiation (Starr 1970a).

- In laboratory culture it completes a cycle of asexual reproduction and thereby undergoes about a 16-fold increase in mass in as little as two days. Thus, large isogenic populations can be established quickly from a single individual.
- Development can be synchronized with a light–dark cycle, yielding cultures in which all individuals are at essentially the same point in the life cycle at any given time.
- Visible germ–soma differentiation begins with a stereotyped cleavage program in which certain embryonic cells divide unequally to produce a large germ-cell initial and a small somatic initial.
- The species consists of genetically distinct male and female strains that are morphologically indistinguishable in the asexual phase, but that will, under certain conditions, undergo modified programs of asymmetric division and produce sexual individuals containing small, motile sperm and large immotile eggs, respectively.
- Sexual males produce a potent pheromone that will induce asexual cultures of both mating types to initiate sexual reproduction. Thus, sexual reproduction can be triggered at will by exposing asexual cultures to a filtrate from sexual male cultures.

- Spontaneous mutants with morphological abnormalities appear with a moderate frequency in such cultures, and large clones of such mutants are readily generated via asexual reproduction.
- Many of these mutants exhibit interesting aberrations of asymmetric division, germ–soma differentiation, the switch from asexual to sexual reproduction, or other fundamental developmental processes. Such mutants not only indicate that all of these processes are under rather direct genetic control, but also provide tools for analyzing the nature of such controls.
- Because of the ease with which sex can be induced and the resulting zygotes can be germinated, Mendelian analysis of these mutants is possible. Both sex-linked and sex-restricted autosomal loci can be identified.

Subsequently the following traits would be added to the list by others (to be cited later):

- Conventional mutagenesis can be used to increase recovery of variants with morphological or biochemical defects, and the loci these mutations define can be combined with various kinds of DNA polymorphisms to construct a linkage map for the species.
- The organism is amenable to a wide range of cytological, biochemical, and molecular analyses.
- Transposons exist that can be used for tagging and recovery of genes regulating development.
- Stable nuclear transformation with exogenous DNA is possible, unselectable markers can be efficiently co-transformed with selectable markers, and gene replacement is feasible.
- A cloned, inducible *Volvox* gene can be used to engineer constructs carrying either an inducible promoter or a reporter gene.
- Genes from other green algae can be incorporated and expressed in *V. carteri*, thereby changing its physiological properties.

5.2 Two formas of *V. carteri* differ greatly in their genetic accessibility

As their names imply, *V. carteri* f. *weismannia* and *V. carteri* f. *nagariensis* are very similar in gross morphology and life history (Kochert 1968; Starr 1969, 1970a). Indeed, Iyengar, who first grouped them as different formas of a single species, distinguished them exclusively on the basis of the number of gonidia they tend to contain when isolated from nature (Iyengar 1933).

However, in both species gonidial number varies with culture conditions (Kochert 1968; Starr 1969, 1970a), and overlapping ranges of gonidial numbers are obtained when wild-type strains of both formas are grown under identical conditions (D. L. Kirk unpublished observations).

Despite their gross morphological similarities and the fact that their ranges overlap in India (Iyengar 1933; Apte 1936; Starr 1970a; Iyengar & Desikachary 1981),[2] it is clear that forma *nagariensis* and forma *weismannia* are completely isolated from one another reproductively: The sperm of one forma fail to impregnate sexual females of the other under any conditions yet tested (Starr 1970a; D. L. Kirk unpublished observations). Moreover, the two formas are readily distinguished at the ultrastructural level (Kirk, Birchem & King 1986).

Knowledge that these two formas of *V. carteri* are reproductively isolated did not prepare the *Volvox* research community for what must be considered one of the most puzzling (and still unexplained) observations to emerge in the first decade of modern *Volvox* research: the discovery that whereas spontaneous and induced mutants can be recovered from *V. carteri* f. *nagariensis* with ease, the same is not true for *V. carteri* f. *weismannia*. When the two formas are grown side by side and subjected to a wide range of mutagenic treatments, mutants are recovered in considerable abundance from forma *nagariensis*, but none are recovered from forma *weismannia* (G. Kochert and R. Huskey personal communications; D. L. Kirk & M. M. Kirk, unpublished observations). Indeed, not a single mutant of forma *weismannia* has yet been described in print. This difference cannot be ascribed to pseudodiploidy in forma *weismannia*, because the two formas have virtually indistinguishable karyotypes (G. Kochert personal communication) and nuclear DNA levels (Kirk & Harper 1986); it remains to be determined whether or not the difference results from a qualitative difference in the DNA repair systems of the two forma. In any case, whereas forma *nagariensis* has proved accessible to genetic analysis and manipulation,[3] forma *weismannia* has not.

This difference is at least one important reason why, although forma *weis-*

---

[2] However, whereas *V. carteri* forma *weismannia* is also found in Australia and North America, forma *nagariensis* has been found only in Japan and India (see Table 2.1).

[3] For examples of studies in which mutants of *V. carteri* f. *nagariensis* were described, analyzed, and/or utilized, see Starr (1970a), Sessoms & Huskey (1973), Meredith & Starr (1975), Pall (1975), Kurn, Colb & Shapiro (1978), Huskey (1979), Huskey & Griffin (1979), Huskey et al. (1979b), Callahan & Huskey (1980), Zeikus & Starr (1980), Kurn & Sela (1981), Kirk, Viamontes et al. (1982), Kurn (1982), Mattson (1984), Kirk & Harper (1986), Harper, Huson & Kirk (1987), Kirk, Baran et al. (1987), Harper & Mages (1988), Kirk (1988, 1990, 1994), Mages et al. (1988), Starr & Jaenicke (1989), Adams et al. (1990), Kirk, Kaufman et al. (1991), Koufopanou & Bell (1991), Tam & Kirk. (1991b), Schmitt, Fabry & Kirk (1992), Miller, Schmitt & Kirk (1993), Sumper et al. (1993), Schiedlmeier et al. (1994), Godl et al. (1995), Hallmann & Sumper (1994b, 1996), and Gruber, Kirzinger & Schmitt (1996).

mannia was brought into continuous cultivation first and was more frequently employed in early cytological and biochemical studies of *V. carteri* development,[4] many more studies of *V. carteri* have utilized forma *nagariensis*, particularly in the past decade and a half.[5] In the vast majority of cases, the cultures of *V. carteri* f. *nagariensis* that have been studied have been direct descendants of the two strains Starr isolated from Japan, namely, strains HK 10 (female) and strain HK 9 (male). A few studies, however, have employed *V. carteri* f. *nagariensis* male and female strains isolated more recently from a pond near Poona, India, and/or progeny derived from a cross between the Japanese and Indian strains (Adams et al. 1990; Tam & Kirk 1991b; Miller et al. 1993; Schiedlmeier et al. 1994). Even more recently, a second set of *V. carteri* f. *nagariensis* male and female strains have been isolated in Japan and brought into culture; these have turned out to be extremely similar genetically to the HK 10 and HK 9 strains, although there are some differences that are discernible at the molecular level (Miller et al. 1993).

A third reproductively distinct forma of *V. carteri*, namely, forma *kawasakiensis*, was isolated from ponds in Japan and described in the late 1980s

---

[4] See, e.g., Kochert & Olson (1970a,b), Kochert & Yates (1970, 1974), Olson & Kochert (1970), Hutt & Kochert (1971), Kochert (1971, 1973, 1975, 1978, 1981), Kochert & Sansing (1971), Yates et al. (1975), Margolis-Kazan & Blamire (1976), Yates & Kochert (1976), Birchem & Kochert (1979a,b), Coggin, Hutt & Kochert (1979), Kochert & Crump (1979), Hagen & Kochert (1980), Pommerville & Kochert (1981, 1982), Weinheimer (1983).

[5] For example, in addition to the genetic studies cited in footnote 3, see Pall (1973, 1974), Bradley et al. (1974), Karn et al. (1974), Roberts (1974), Starr (1975), Kirk & Kirk (1976, 1983, 1985, 1986), Margolis-Kazan & Blamire (1976, 1977, 1979a,b), Kelland (1977), Viamontes & Kirk (1977), Kurn et al. (1978), Bause & Jaenicke (1979), Jaenicke & Waffenschmidt (1979, 1981), Kurn & Sela (1979, 1981), Sumper (1979), Viamontes et al. (1979), Wenzl & Sumper (1979, 1981, 1982, 1986a,b, 1987), Caplen & Blamire (1980), Dauwalder, Whaley & Starr (1980), Mitchell (1980), Sumper & Wenzl (1980), Desnitski (1981a, 1982a, 1983b, 1984b, 1985c, 1986, 1987, 1990, 1992), Gilles, Bittner & Jaenicke (1981), Green & Kirk (1981, 1982), Green et al. (1981), Kurn (1981, 1982), Müller, Bause & Jaenicke (1981, 1984), Gilles & Jaenicke (1982), Jaenicke (1982), Jaenicke & Gilles (1982, 1985), Kirk, Viamontes et al. (1982), Kurn & Duskin (1982), Pommerville & Kochert (1982), Willadsen & Sumper (1982), Bause, Müller & Jaenicke (1983), Gilles, Gilles & Jaenicke (1983, 1984), Mattson (1984), Wenzl, Thym & Sumper (1984), Gilles, Moka, et al. (1985a), Baran & Huskey (1986), Coggin & Kochert (1986), Kirk, Birchem et al. (1986), Kirk & Harper (1986), Schlipfenbacher et al. (1986), Adair et al. (1987), Gilles, Balshüsemann & Jaenicke (1987), Jaenicke, Kuhne et al. (1987), Tschochner, Lottspeich & Sumper (1987), Goodenough & Heuser (1988), Kaska et al. (1988), Kirk (1988, 1990, 1994, 1995), Mages, Tschochner & Sumper (1988), Müller & Schmitt (1988), Adair & Appel (1989), Ertl et al. (1989, 1992), Rausch et al. (1989), Balshüsemann & Jaenicke (1990a,b), Cresnar et al. (1990), Kirk, Kirk et al. (1990), Müller, Lindauer et al. (1990), Waffenschmidt, Knittler & Jaenicke (1990), Haas & Sumper (1991), Jaenicke, van Leyen & Siegmund (1991), Tam & Kirk (1991a), Tam, Stamer & Kirk (1991), Fabry, Nass et al. (1992), Gruber et al. (1992, 1996), Mengele & Sumper (1992), Schmitt et al. (1992), Schmitt & Kirk (1992), Fabry, Jacobsen et al. (1993), Jaenicke, Feldwisch et al. (1993), Kirk, Ransick et al. (1993), Lindauer et al.. (1993a,b), Dietmaier & Fabry (1994), Hallmann & Sumper (1994a), Huber & Sumper (1994), Schiedlmeier & Schmitt (1994), Fabry, Müller et al. (1995a), Fabry, Steigerwald et al. (1995b), Huber, Beyser & Fabry (1996), Selmer et al. (1996), Choi, Przybylska & Straus (1996), Liss et al. (in press).

(Nozaki 1988). Forma *kawasakiensis* appears to be more closely related to forma *nagariensis* than is forma *weismannia* (Nozaki 1988) (see Figure 2.14), but to date nothing has been published that would indicate which of the other formas *kawasakiensis* may more closely resemble in terms of its accessibility to genetic analysis.[6]

Because such a preponderance of recent studies have employed forma *nagariensis*, in the rest of this book wherever the name *V. carteri* is used without further modification, it will refer to forma *nagariensis*, and more specifically to strains HK 10 and HK 9 and their descendants. In this regard, discretion demands that a point that should be obvious by now must be made explicit: There are numerous variations in organization and development within the genus *Volvox*, and hence it should not be assumed that anything that is said about *V. carteri* in the following account necessarily applies to any other species. As interesting and important as the differences observed in other species of *Volvox* may be to many, they will be largely ignored in the remainder of this book, not only because of space constraints but also out of concern that inclusion of such details could easily obscure the issues upon which I prefer to focus attention.

5.3 An overview of *V. carteri* anatomy and reproduction

A young adult asexual spheroid[7] of *V. carteri* contains some 2,000–4,000 small somatic cells near the surface and about 16 large gonidia that lie slightly

---

[6] Three other formas of *V. carteri* have been described in the past (reviewed in Nozaki 1988), but are not presently represented in culture collections.

[7] An aside regarding terminology is in order. Traditionally, zoologists have laid claims to *Volvox* and other green flagellates as "animals" (albeit green ones) in the "sub-kingdom" Protozoa (Grell 1973), whereas botanists, of course, have claimed them as archetypal algae, and hence "plants" (Bold & Wynne 1985). Such competing claims of proprietorship resulted in a conflicting and hence confusing set of terms being used by different authors to describe *Volvox*. Early in the century, Powers attempted to strike a new path: "I shall deliberately use, in this paper, such expressions as 'somatic cells,' 'reproductive cells,' etc., and I shall also use, with equal freedom, the terms 'colony,' 'coenobium' and the like. Such expressions are flatly contradictory, in a sense; but so are the facts. They confuse no one familiar with the different points of view from which *Volvox* may be considered. The very beauty of *Volvox* and its group lies in the happy way in which they override 'fundamental distinctions.' Out of the seeming chaos, therefore, of terms ... old and new, terms botanical and terms zoological, terms metazoan and terms protozoan, ... I choose those most convenient and useful for the context ..." (Powers 1907). Starr broadened the path opened by Powers. For example, although he was himself a "card-carrying" botanist, Starr found less than satisfying the two traditional botanical terms for the unit organism in a *Volvox* culture that Powers had used, namely "colony" and "coenobium." He has told me that he considered "colony" entirely appropriate for an organism, such as *Pandorina*, that has a single cell type, but inappropriate for a multicellular organism like *Volvox* that has two highly differentiated and fully interde-

deeper within the transparent extracellular matrix that holds all the cells in fixed relationships to one another (Figure 5.1A). The spheroid exhibits a distinct anterior–posterior polarity that is revealed not only by the direction in which it swims (anterior pole first, of course) but also in several morphological features. The organism initially enlarges somewhat more rapidly along the anterior–posterior axis than along the equatorial axis, so that prior to gonidial cleavage it is a prolate spheroid. Several somatic-cell features are graded so subtly along the anterior–posterior axis that differences are imperceptible when one examines neighboring cells, but they become readily apparent when cells at the two poles are compared: Somatic cells near the anterior pole are slightly larger, have much larger eyespots, and are more widely spaced than cells near the posterior pole. Although all somatic cells have their eyespots located on the side facing toward the posterior of the spheroid, closer examination reveals that in cells near the anterior pole the eyespots are located quite a distance from the flagella (as much as a third of the way down the side of the cell), whereas cells in more posterior regions have their eyespots near the cell apex, closer to the flagella. The gonidia are arranged in four tipped rhombohedrons in the posterior portion of the spheroid.

5.3.1 Somatic cells are polarized, but gonidia are radial in organization

V. carteri somatic cells (Figure 5.2A), like *Chlamydomonas* cells (Figure 2.2), are organized in a highly polarized manner: Major organelles are arrayed in a predictable sequence between the paired flagella that dominate the anterior end of each cell and the large, cup-shaped chloroplast (with its conspicuous basal pyrenoid, or starch-forming organelle) that fills the posterior end of the cell. At the bases of the flagella lies a basal apparatus (BA) that differs in only one important way (to be described shortly) from the BA of *Chlamydomonas* described in the preceding chapter. Lateral to the basal bodies (BBs) lie a pair of contractile vacuoles, and posterior to them lie an array of mitochondria, Golgi complexes, and vesicles that surround the nucleus,

pendent cell types, and that while the term "coenobium" was clearly applicable to *Volvox*, he felt it was too specialized to convey meaning to many outside the botanical community (R. Starr personal communication). So he standardized the use of a term that had previously been used as an alternative to "colony" and "coenobium" (albeit only sporadically) by Mary Pocock (1933b; 1938), namely, "spheroid." Most subsequent workers have used this term, as well as several other terminological conventions adopted or established by Starr (1969, 1970a). Additional conventions have been established since then that will be discussed as they occur.

which, like the nucleus of *Chlamydomonas,* lies near the center of the cell, cupped by the chloroplast.

As in the vegetative cells of *Chlamydomonas* (Holmes & Dutcher 1989), the apparent bilateral symmetry of a *V. carteri* somatic cell is broken by a unitary eyespot, or "stigma," that lies on one side of the cell (the side that faces the posterior of the spheroid) in an anterior part of the chloroplast (Figure 5.2A). As mentioned earlier, this critical organelle is smaller and closer to the cell apex in posterior somatic cells than in anterior ones.

In contrast to the highly polarized organization of somatic cells, gonidia are organized in a radial manner, around a central nucleus that contains a prominent nucleolus (Figure 5.2B). The periphery of each gonidium is dom-inated by a chloroplast that possesses numerous pyrenoids and that sends projections centripetally into radial strands of cytoplasm that subtend the nucleus. Some of these radial strands of cytoplasm also contain an assortment of other organelles (mitochondria, Golgi elements, secretory vesicles, etc.). But perhaps the most striking feature of a developing gonidium is its highly vacuolate appearance: The numerous radial strands of cytoplasm are all sep-arated by equally numerous large vacuoles (Figure 5.2B).

5.3.2 The BA is arranged very differently in somatic cells and gonidia

As discussed at length in the preceding chapter, one thing that distinguishes green flagellates from most other organisms is the role that is played in cell structure and function by the BA, and it is impossible to understand the structure and function of green flagellates in modern terms without reference to the structure and function of the BA. The gonidia and somatic cells of *V. carteri* constitute no exception.

V. carteri gonidia never possess functional flagella. However, presumptive gonidia, like presumptive somatic cells, develop short flagellar rudiments (about 1–2 μm long) in the late stages of embryogenesis. Very soon thereafter, however, as cytodifferentiation and deposition of extracellular ma-trix (ECM) begin, the flagella of somatic cells begin to elongate, whereas the rudimentary flagella of the gonidia are resorbed, leaving behind a telltale pair of empty flagellar tunnels in the ECM. Just beneath those empty flagellar tunnels each gonidium retains a BA (Figure 5.3A,B) that is organized much like that of *Chlamydomonas* (i.e., with 180° rotational symmetry), and that similarly connects the BBs to each other, to the plasmalemma, to the nucleus, and (directly or indirectly) to most other organelles in the cell. As the eyespot breaks the apparent bilateral symmetry of a *Chlamydomonas* cell, the BA

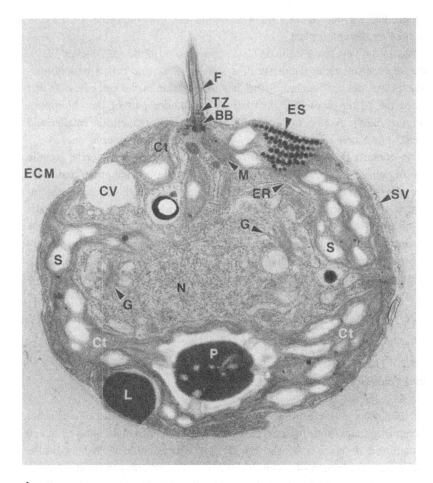

A

Figure 5.2 Electron micrographs of *V. carteri* cells. The somatic cell (**A**) has a polarized organization similar to that of *Chlamydomonas* (Figure 2.2), with a pair of prominent anterior flagella (F) (one of which is out of the plane of section) and a single posterior pyrenoid (P) that lies within the chloroplast (Ct). The chloroplast dominates the posterior end of the cell, but also extends along the sides of the cell nearly up to the basal bodies (BB), thereby surrounding a central cytoplasmic domain containing the nucleus (N), endoplasmic reticulum (ER), Golgi complexes (G), and other organelles. In addition to the pyrenoid, the chloroplast contains many starch granules (S), a few lipid-storage droplets (L), and a prominent stigma, or eyespot (ES). Mitochondria (M) are preferentially associated with the BBs, but are less numerous than in species, such as *C. reinhardtii* (Figure 2.2), that are capable of heterotrophic growth. Other abbreviations: CV, contractile vacuole; ECM, extracellular matrix; SV, secretory vesicles discharging ECM components to the extracellular space; TZ, transition zone of the flagellum. In contrast to the somatic cell, the gonidium (**B**) is organized radially about a central nucleus that has a prominent nucleolus (Nu) and is subtended by radial strands of cytoplasm that pass between the numerous large vacuoles (V). The chloroplast dominates the entire

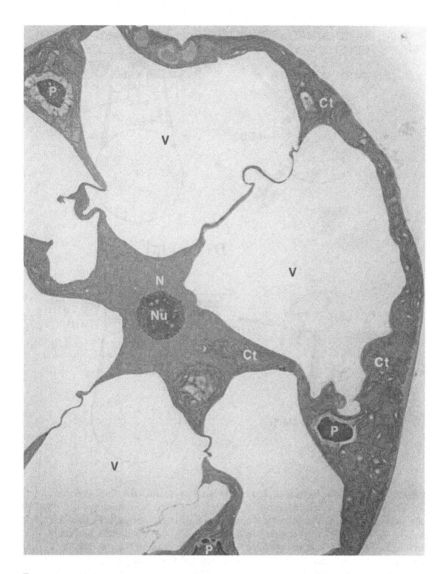

B

Caption to Figure 5.2 *(cont.)*
periphery of the cell, but also extends into each of the cytoplasmic strands, with
some lobes extending far enough to contact the nucleus; in contrast to the somatic
chloroplast, it contains numerous pyrenoids.

breaks the apparent spherical symmetry of a *Volvox* gonidium. More details
of the structure and function of the BA in dividing embryos will be considered later.

The BA of a *V. carteri* somatic cell contains components similar to those

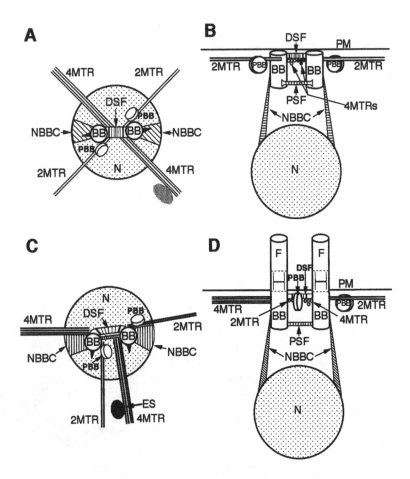

Figure 5.3 Schematic representation of the basal apparatus (BA) in a gonidium (**A, B**) and a somatic cell (**C, D**) of *V. carteri*, as viewed from the top (A, C) and one side (B, D). Like the BA of a *Chlamydomonas* interphase cell (Figures 4.2 and 4.4A), the BA of a *V. carteri* gonidium consists of two half-BAs that are arranged with 180° rotational symmetry and that are connected across the midline by symmetrical proximal and distal striated fibers, etc. In A, arrowheads on BBs indicate the direction in which flagella would be expected to beat, if they were present (which they are not), and the stippled ellipse indicates approximately where an eyespot would be expected to be located if it were present (which it is not). The 180° rotational symmetry of the BA is maintained throughout cleavage and thus is inherited by all cells of the embryo. As somatic cells differentiate in the postembryonic period, however, the two halves of the BA rotate with respect to one another by nearly 90°, so that in a mature somatic cell the two BBs are facing in nearly the same direction, and the flagella beat with effective strokes that are very nearly parallel (arrowheads). This rotation is accompanied by a severe distortion of the proximal striated fiber, which remains attached to the same BB subunits after the rotation. Note also that, in distinction to those of *Chlamydomonas*, the long axes of the BBs are parallel in the mature, flagellated cell.

of a gonidium or a *Chlamydomonas* cell, but they are arranged very differently: in a wholly asymmetric manner (Figure 5.3C,D). A similar type of asymmetric BA arrangement has now been observed in the motile cells of many rather distantly related species of colonial and multicellular green flagellates (Hobbs 1971; Hoops 1981, 1984, 1993; Hoops & Floyd 1983; Greuel & Floyd 1985; Taylor et al. 1985; Hoops, Long & Hilde 1994). In each of those species it now appears that during cleavage the BA has 180° rotational symmetry, but between the end of cleavage and the time that flagella become functional, the two halves of each BA rotate relative to one another, so that they come to lie nearly parallel to one another in the mature flagellated cells. As a result of such rotation, the flagella of such organisms beat in the same direction, in two separate but nearly parallel planes.[8]

5.3.3 An asymmetric BA in somatic cells is essential for motility

The functional significance of this reorganization of the BA that occurs during maturation of somatic cells is abundantly clear. With the cells arranged around the surface of a sphere, as they are in *Volvox*, if the flagella were oriented so that they beat in opposite directions (as those of *Chlamydomonas* do), flagellar activity would accomplish little more than to push water toward the surface of the spheroid. But with the BA rearranged so that the two flagella of each cell beat in the same direction, they should be able to work together to move the spheroid if, but only if, all other cells in the spheroid are arranged so that their flagella are beating in this same direction. And so they are. As a result of the rotational symmetry of the BA in the dividing embryo, the somatic cells of the adult are all arranged with rotational symmetry around the anterior–posterior axis of the spheroid, and the BBs of each cell are oriented in such a way that their flagella beat toward the posterior of the spheroid (Figure 5.4). (Actually, the beat direction is apparently directed slightly to the right of the anterior–posterior axis of the spheroid, so that the organism rotates to the left as it swims forward; hence the name given to it by Linnaeus: *Volvox*, "the fierce roller.") As logical as it seems that in multicellular flagellates such a rotation of the two halves of each somatic-cell BA should occur at the end of embryogenesis, the cytological mechanism by which such a rotation occurs is entirely mysterious.

---

[8] As a result of this arrangement, when individual somatic cells of *Volvox* are mechanically released from the spheroid, they tumble in place and are unable to swim directionally.

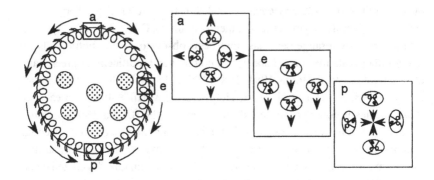

Figure 5.4 Orientation of cells within a *V. carteri* spheroid. Left: Somatic cells are oriented so that all flagella beat with their effective strokes directed toward the posterior of the spheroid (the end containing the gonidia). Right: The *en face* representations of the orientations of the BA seen in four neighboring cells at the anterior (**a**), equatorial (**e**) and posterior (**p**) regions of the spheroid (boxed in the diagram to the left). Each ellipse is meant to represent the orientation of the BBs, MTRs, and eyespots within a cell (see Figure 5.3C), and the arrows indicate the directions of the effective flagellar strokes of those cells. In the anterior, the cells are oriented so that flagella beat away from one another; along the sides, the cells are oriented so that their flagella beat in parallel, and at the posterior pole the cells are oriented so that their flagella beat toward one another. (Adapted from Hoops 1993).

5.3.4 A complex ECM holds the organism together

As in all other volvocaceans, the cells of a *V. carteri* spheroid are held in fixed relationships to one another by a complex extracellular matrix (ECM) that is transparent in the living organism. The ECM is arranged as a series of compartments that are bounded by discrete fibrous layers and filled with relatively amorphous "mucilage" (Figure 5.5). Different species of *Volvox* differ sufficiently from one another and from other volvocaceans in the details of ECM organization (Kirk, Birchem & King 1986) that ECM organization is commonly used as an important taxonomic criterion (Smith 1944; Nozaki & Itoh 1994). Nevertheless, there are many common themes. The most noteworthy of these is the ubiquitous presence in the volvocine algae of an outer "crystalline," "tripartite," or "triplet" layer, so called because it exhibits a regular crystal lattice when viewed in the electron microscope *en face* (Roberts 1974), and a characteristic dark-light-dark appearance when viewed in sections of osmium-stained specimens (Kirk, Birchem & King 1986).

The hypothesis that the complex ECM of *Volvox* evolved from the simple cell wall of a unicell resembling *Chlamydomonas reinhardtii* was reinforced

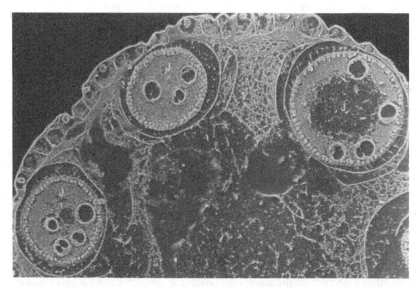

Figure 5.5 Organization of some of the fibrous elements of the ECM in a *V. carteri* spheroid. Parental spheroids containing expanding juveniles (J) were fixed, embedded, and thick-sectioned; then the embedding material was extracted, and the sections were coated with metal and viewed in the scanning electron microscope. Note the distinction between the coherent fibrous layers (such as the walls of the "compartments" surrounding all of the parental somatic cells, and the two layers surrounding each of the juveniles) and the more loosely organized meshwork that fills most of the internal spaces. The ECM is discussed in more detail in Section 5.4.4, and its major structural components are diagrammed in Figure 5.24.

when it was observed that the crystalline layers of *C. reinhardtii* and *V. aureus* are extremely similar to one another in lattice structure and distinctly different from those of many other species of *Chlamydomonas* (Roberts 1974). Further evidence that *C. reinhardtii* is much more closely related to *Volvox* than it is to certain other chlamydomonads was provided by the demonstration that *C. reinhardtii* and *V. carteri*, when stripped of their own crystalline layers, can nucleate the assembly of each other's crystalline-layer glycoproteins into a normal crystal lattice, whereas *C. reinhardtii* and *C. eugametos* are incapable of such cross-nucleation (Adair et al. 1987). Detailed analysis of certain of the genes encoding ECM glycoproteins has reinforced such views and led to the proposal that further analysis of ECM genes and proteins may provide an important tool for reconstructing the phylogenetic history of the group (Woessner & Goodenough 1994).

5.3.5 *The asexual reproductive cycle is simple and rapid*

Under standardized culture conditions, including a daily cycle of about 16 hours of light and 8 hours of darkness (16L:8D), development in *V. carteri* is synchronous, and one asexual life cycle is completed in precisely 48 hours (Figure 5.6). The cycle will be briefly summarized here without documentation, and then each phase will be discussed in more detail later.

When mature, each gonidium initiates *embryogenesis*, the first phase of which is *cleavage*, in which 11 or 12 rapid rounds of cell division (some of them visibly asymmetric) generate all of the cells that will be present in an adult of the next asexual generation. At the end of cleavage, however, the embryo is inside out with respect to the adult conformation: The gonidia are on the outside, and the flagellar ends of the somatic cells are on the interior. That awkward arrangement is quickly corrected, however, as the embryo turns itself right side out in the second phase of embryogenesis – a gastrulation-like morphogenetic rearrangement known as *inversion* that is shared with other members of the family Volvocaceae. By the end of inversion, the parental spheroid contains a cohort of juvenile spheroids, each of which resembles a miniature adult. Initially, however, the two presumptive cell types of the juvenile differ little except in size. There then follows a phase of active *cytodifferentiation*, in which the somatic cells devote themselves to assembling the organelles required for effective motility, phototaxis, and chemotaxis and elaborating most of the ECM, while the gonidia grow rapidly. Cytodifferentiation is coupled with *expansion*, in which both adult and juvenile spheroids increase in size (without further cell division, and with only modest cell growth) by the deposition of increasing amounts of ECM. About halfway through their expansion, the juveniles undergo *hatching*, by digesting individual birth canals in the parental ECM, thereby becoming free-swimming young adults. Then the parental and progeny generations go their separate ways: While the somatic cells of the parental "hulk" or "ghost" undergo *programmed cell death*, the newly released young adult spheroids continue to expand while their gonidia undergo *maturation* and subsequently initiate a new round of embryogenesis, thereby completing the cycle.

Note that although each gonidium goes around this cycle only once, the somatic cells go around twice, first as young somatic cells and then as aging parental cells.

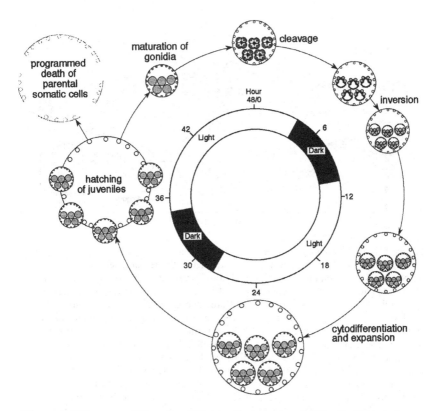

Figure 5.6 The asexual life cycle of *V. carteri*, as synchronized by the light–dark cycle indicated on the inner circle. Near the end of one illumination period the gonidia begin a series of rapid divisions that are completed early in the dark period, whereupon the embryos undergo inversion to assume the adult configuration. At that point the juveniles contain all of the cells that will be present in the adult, but these cells will not begin visible cytodifferentiation until the lights come back on (which is the basis of synchronizing effect of light). As the juvenile cells differentiate, both the parental and juvenile spheroids will expand by deposition of ECM. Shortly after the beginning of the next light period, the juveniles will digest their way out of the parent and become free-swimming adults, whereupon the parental somatic cells will undergo programmed death and dissolution. Meanwhile, the gonidia will complete their maturation and initiate a new round of embryogenesis. Note that the gonidia go around the cycle only once before they divide, but the somatic cells go around twice, first as juvenile somatic cells and then as parental ones.

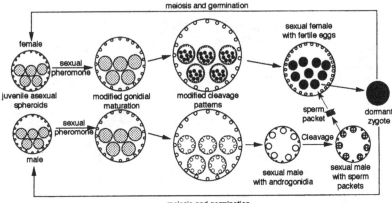

Figure 5.7 The sexual cycle of *V. carteri*. In the asexual cycle, males and females are indistinguishable, but when they are exposed to the sexual pheromone at the proper time, their gonidia initiate an additional round of embryogenesis in which cleavage patterns are modified. Embryos of induced females produce eggs instead of gonidia (although these eggs can redifferentiate as gonidia if they are not fertilized). Embryos of induced males produce a modified type of gonidia, called androgonidia, that later undergo an additional round of cleavage to form sperm packets. These sperm packets swim as multicellular units, and if they happen to encounter a sexual female, they may fertilize the eggs to form dormant zygotes, which constitute the only diploid phase of the life cycle and are resistant to desiccation, freezing, and other environmental insults. Under appropriate conditions, the zygotes can be reactivated and can execute meiosis to form one viable haploid germling and three polar bodies. Each germling will divide to form a small spheroid of one genetic sex or the other that will reenter the asexual pathway; its progeny will continue to reproduce asexually until they are once again exposed to the sexual pheromone.

5.3.6 The sexual cycle begins with a modified asexual cycle

Volvox carteri is "heterothallic," which is to say that it has genetically distinct male and female strains.[9] In the asexual phase, spheroids of the two mating types are morphologically indistinguishable. When they are exposed to the sex-inducing pheromone, their gonidia do not transform directly into

[9] There is a surprising diversity of patterns of sexual development among the approximately 18 species of the genus *Volvox* (Smith 1944; Darden 1966; Starr 1968). In *V. carteri* and a few other species (*V. gigas*, *V. obversus*, *V. perglobator*), all isolates that have been studied and described are heterothallic. There are also some species (e.g., *V. barberi*, *V. globator*, *V. powersii*) in which all isolates thus far studied are homothallic, that is, in which eggs and sperm can be produced by members of a single clone. Homothallic forms are in some cases dioecious (producing separate male and female spheroids) and in others monoecious (producing both eggs and sperm within a single spheroid). In many species, however (e.g., *V. africanus V. aureus, V. dissipatrix, V. rousseletii, V. spermatosphaera, V. tertius*), some isolates are

gametes; rather, they initiate a new round of asexual embryogenesis in which temporal and spatial patterns of asymmetric division are modified, resulting in juveniles that contain more germ cells than are present in asexual juveniles (Figure 5.7). In induced females, these germ cells develop directly into fertilizable eggs. However, the germ cells of induced males, called "androgonidia," must undergo a second round of cleavage and differentiation, as a result of which each androgonidium forms a packet of 64 or 128 sperm. Shortly after it is released from its parent, a male spheroid releases its sperm packets, which then swim as multicellular units in search of fertile females. When a sperm packet contacts a fertile female, it dissociates into individual sperm that invade the female spheroid and move about, searching for eggs ripe for fertilization.

Fertilized eggs transform into dormant zygospores surrounded by specialized coats that make them resistant to drought, freezing, and other environmental insults. After a few weeks of maturation they will undergo meiosis and germinate if washed and returned to normal culture medium. The single haploid germling that emerges from each zygote soon divides to form a small asexual spheroid that inaugurates a new period of rapid asexual proliferation.

Further details of the asexual and sexual cycles will now be considered.

5.4 Asexual morphogenesis and cytodifferentiation

Volvox carteri gonidia can be viewed as potentially immortal stem cells, each of which divides mitotically to produce more gonidia plus a cohort of mortal somatic cells. One of the most appealing features of *V. carteri* as a model developmental system is that under optimum conditions its morphogenetic program is sufficiently regular to ensure that in each generation these two entirely different cell types will be produced in quite predictable numbers and locations. The source of this regularity is the stereotyped manner in which the gonidium and the resulting embryo divide (Starr 1969, 1970a; Green & Kirk 1981, 1982). Thus, the details of cleavage are worthy of careful consideration.

heterothallic and others are homothallic. *V. africanus* is particularly interesting in this respect: Starr (1971) described *V. africanus* isolates from separate continents that were heterothallic (Australia), dioecious homothallic (North America), monoecious homothallic (Africa), and homothallic, with the regular production of both monoecious spheroids and pure males (Asia). *V. aureus* adds yet another dimension to this diversity in reproductive pattern: In this species, both heterothallic and homothallic dioecious strains have been described (Darden 1966), but there are also strains in which formation of gametes is bypassed, and gonidia exposed to the sex-inducing substance of the species develop directly into "parthenospores" (resistant resting spores resembling the zygospores that are usually formed following fusion of sperm and eggs) (Darden & Sayers 1969).

5.4.1 Cleavage: creating two cell lineages from one

Throughout most of its growth and maturation, a *V. carteri* gonidium is spherical, has a central nucleus, is highly vacuolated (Figure 5.2A), and is intimately connected to the glycoprotein vesicle that surrounds it (Figure 5.8). The onset of embryogenesis is signaled by simultaneous changes in all four of these features.

5.4.1.1 Incipient cleavage is signaled by gonidial condensation. The first visible indication that a gonidium is about to divide is that it withdraws from the surrounding vesicle and begins to flatten on its anterior and posterior surfaces (Figure 5.9). From that time until the time of hatching, the gonidium/embryo/juvenile[10] will remain surrounded by the vesicle, but not in intimate contact with it. The fact that the cell is suddenly able to withdraw from the vesicle indicates that the intimate connections between the vesicle and the plasmalemma that previously existed (Figure 5. 8) have been broken, but the way that this occurs remains to be elucidated.

Gonidial condensation is accompanied by two substantial cytological modifications. The more obvious change, visible at modest magnification, is that the vacuoles that characterized the cell during growth and maturation collapse, and the cell changes from a light-green, translucent sphere to a dark-green, opaque discoid with a distinct pit at its anterior end (Figures 5.9B and 5.11a). The second change, which can be fully appreciated only at much higher magnifications, is that the cytoplasm is thoroughly rearranged, so that the gonidium takes on a polarized organization more nearly resembling that of a somatic cell: The chloroplast that previously occupied the entire cell periphery is pushed aside at the anterior end of the cell as the nucleus moves forward to take a position immediately below the BA; meanwhile, the cytoplasm that previously surrounded the nucleus is also pushed aside at the anterior and accumulates in a bowl-shaped zone lying between the nucleus and the chloroplast (Green et al. 1981).

The emptying of the numerous vacuoles present in a mature gonidium greatly decreases its volume, permitting it to change shape. But the details of the shape change are attributable to cytoskeletal activities, particularly contraction of the centrin-rich nucleus-BB connectors (NBBCs). In a mature but uncondensed gonidium, the NBBCs are so long and fine that they are

---

[10] By current convention, the gonidium becomes an embryo when it begins to divide, the embryo becomes a juvenile when it completes the process of inversion, the juvenile becomes a young adult when it hatches out of the parental spheroid, the young adult becomes a parental spheroid when its gonidia cleave, and the parental spheroid becomes a "ghost" or "hulk" when its juvenile spheroids hatch.

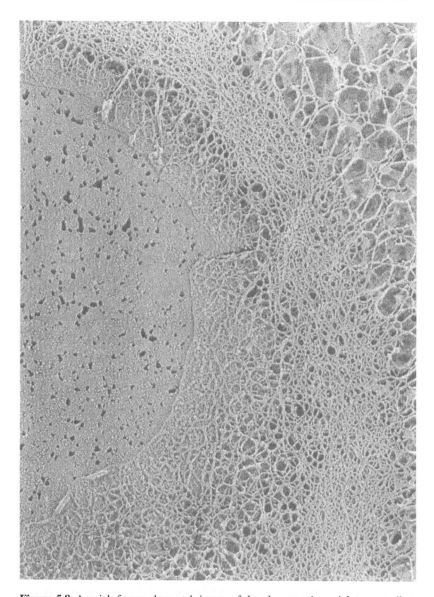

Figure 5.8 A quick-freeze, deep-etch image of the glycoprotein vesicle surrounding a *V. carteri* gonidium. The surface exposed at the left center is the inner leaflet of the plasma membrane. The surface surrounding the latter is the outer face of the plasma membrane, to which the many fibers of the vesicle appear to be firmly attached. Just before cleavage begins, these attachments are all broken, and the gonidium withdraws from the vesicle. Thereafter, the gonidium and its progeny cells remain free of any detectable extracellular coatings until after the end of embryogenesis, when a new round of ECM deposition begins. (Micrograph courtesy of U. W. Goodenough and J. E. Heuser.)

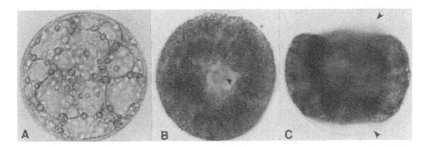

Figure 5.9 Gonidial condensation. A mature precleavage gonidium (**A**) is spherical and highly vacuolate. Incipient cleavage is signaled when the gonidium withdraws from its vesicle, empties its vacuoles, and condenses, taking on a much denser appearance. Viewed from the anterior (**B**), the gonidium is seen to have a central "pit" where the nucleus (arrowhead) has been drawn toward the cell surface by contraction of the NBBCs. In a side view (**C**) it is clear that the cell has contracted more on its anterior (upper) than its posterior (lower) surface. Note the gonidial vesicle (arrowheads).

scarcely distinguishable above the background autofluorescence by immunocytology (not shown). However, in the condensed gonidium, they have become contracted into a conspicuous "centrin plaque" that lies immediately below the BBs and above the nucleus (see Figure 5.13A' later in this chapter). It is inferred that this contraction of the centrin fibers plays a major role in the cytological rearrangements that accompany gonidial condensation. During condensation, the separation between the nucleus and BA decreases from about 30 μm to less than 1 μm. But this relative movement is not all one-way: As the nucleus moves up from the center of the cell toward the surface, the plasmalemma and the BA are drawn down, away from the surrounding vesicle, creating the pit referred to in the preceding paragraph.[11] Following condensation, the gonidium is prepared for cleavage.

When a 16L:8D illumination cycle is used to synchronize development (Figure 5.6), cleavage typically is initiated about three hours before the end of a light period and then completed in the dark.[12] Each cleavage division

---

[11] A similar, albeit somewhat more modest, distortion of the cell occurs at the onset of cell division in strains of *C. reinhardtii* that lack the constraint of an adherent cell wall (Doonan & Grief 1987).

[12] The onset of division in *V. carteri* appears to involve an internal timer that is reset during the dark period. If the lights are turned out a couple of hours before cleavage would normally begin, gonidia will not initiate cleavage in the dark. Nor will they do so within the first few hours after the lights come on the following day; rather, they wait until very nearly the normal time of day to begin dividing (our unpublished observations). An added note: The citation "our unpublished observations" is used in several places in Chapters 5–7 to refer to observations that have been made so repeatedly by various members of our research group over

takes about 35 minutes on average, and the entire cleavage period occupies six to seven hours. Division is never perfectly synchronous within the culture as a whole, at least under any conditions that we have been able to establish. It is not unusual to find, however, that in a standard culture containing approximately 10^4 adults in 300 mL of aerated medium, about 90% of the embryos are in the same division cycle at any one time, and very few are more than one division cycle ahead of or behind the mode. Thus, typically no more than one to two hours separate the first gonidia to cleave from the last. Within healthy individual parental spheroids, synchrony is generally higher, with most sibling embryos initiating equivalent divisions within a matter of minutes of one another. Moreover, within a single embryo, divisions are completely synchronous: If one cell in a fixed embryo is observed to have been caught in the middle of anaphase (which lasts less than five minutes), all others will be found to be in anaphase also, with their chromosomes located at essentially the same distance from the spindle poles.

5.4.1.2 Asymmetric divisions set apart prospective somatic and gonidial lineages. The cleavage pattern to be described next is illustrated in Figures 5.10 and 5.11. The first two cleavage planes are both meridional, but the second is not quite perpendicular to the first, and thus it divides the embryo into four quadrants, only two of which abut at the posterior pole. At this point the embryo, which began cleaving as a flattened discoid, initiates cell shape changes that will transform it into a hollow sphere. Each of the first four cells bulges at its anterior, clockwise side to form a lobe that curls in a counterclockwise direction as it also extends anteriorly. While this elongation is in progress, an oblique third cleavage furrow divides each of these cells into an anterior and a posterior blastomere. Soon a fourth round of divisions, which are also oblique, will divide the embryo into four tiers of cells that overlap extensively in the anterior–posterior direction; it is from the anteriormost two of these tiers that gonidia will be derived (Starr 1970a; Green & Kirk 1982). By this time, continuation of the cellular extension that began before the third division will have brought the anterior blastomeres from all four quadrants into contact near the anterior pole and converted the embryo into a hollow sphere. The regions in which these cells have now established contacts via cellular movement (rather than via a shared cleavage furrow) constitute a swastika-shaped set of intersecting slits known as the "phialopore" that will play an important

the past 20 years that it is no longer possible to recall who may have made those observations first, or who might properly be included in more specific citations.

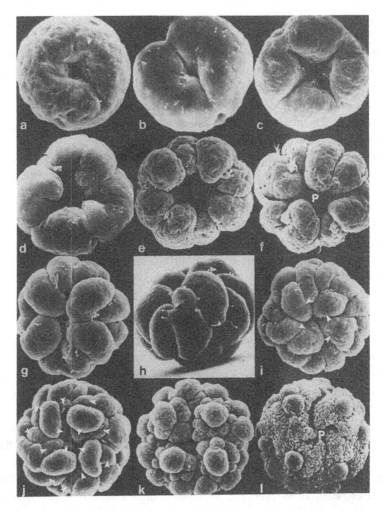

Figure 5.10 SEM analysis of cleavage in *V. carteri* (cf. Figure 5.11). Cleavage is initiated at a pit on the anterior surface of the condensed and flattened gonidium (**a**). The first two cleavage furrows occur at a right angle to one another, and frequently the second is initiated before the first is completed (**b, c**). Cellular extensions (described in the text) begin shortly after the second division (**d**) and continue during the third and fourth divisions (**e-g**); these convert the embryo into a hollow (blastula-like) sphere, but leave a pair of intersecting slits called the phialopore (P) at the anterior end of the embryo. Although the phialopore becomes less obvious as cleavage proceeds, it persists (see **l**) and will play an important role after cleavage. The third and fourth cleavage furrows are oblique, producing a 16-cell embryo (**g, h**) composed of four overlapping tiers of cells. The first five divisions are symmetrical, resulting in an embryo consisting of 32 cells of similar sizes and shapes (**i**). However, the sixth division is asymmetric in the anteriormost 16 cells and results in the production of 16 cell pairs of unequal sizes (**j**). Arrowheads in part i indicate some of the asymmetrically positioned incipient cleavage furrows, and arrowheads in part j connect large and small sister cells. The large cells produced in division six divide asymmetrically at least two more times (**k**), but then withdraw from the division cycle. Meanwhile, the small cells

role later, during inversion of the embryo. Like the first four, the fifth cleavage division is symmetric and divides the embryo into 32 crescent-shaped cells that are all similar in size.[13]

The sixth division is also symmetric in the two posterior tiers of cells. Normally, however, it is asymmetric in the 16 anterior cells, and divides each of them into a pair of large and small sister cells (Figures 5.10 and 5.11). The large cells so created are gonidial initials that will go on to produce one gonidium each, whereas all the remaining cells in the embryo will produce only somatic cells. The gonidial initials divide asymmetrically two or three more times, always cutting off a small somatic initial (as they did in the first asymmetric division) toward the anterior pole. Then the gonidial initials cease dividing, while all somatic initials continue dividing synchronously and sym-metrically until they have completed a total of 11 or 12 divisions.

Embryos that cleave a total of 12 times will have about 4,000 somatic initials, whereas those that cleave only 11 times will have half as many. A relatively healthy culture will often have a mixture of adults of these two sizes (Green & Kirk 1981). The more nearly optimum the culture conditions are, and the larger the average size of the precleavage gonidia as a result, the

---

[13] Superficially, the blastomeres of *Volvox* embryos appear to be arranged in even more of a spiral than those of "spiralian" animal (e.g., molluscan and annelid) embryos. However, contrary to statements in some early descriptions of *Volvox* cleavage (Goroshankin 1875; Overton 1889; Klein 1890), this is not because *Volvox* exhibits spiral cleavage in the sense that the term is used in animal embryology. This distinction was first pointed out quite clearly more than 75 years ago (Delsman 1919). In spiralian animal embryos, successive division planes are tipped in opposite directions (first clockwise and then counterclockwise) with respect to the animal–vegetal axis of the egg. In *Volvox*, in contrast, all division planes after the second are tipped in the same direction, so that the more anterior cell lies slightly clock-wise of its more posterior sister (although each successive division tends to be somewhat less oblique – more nearly equatorial – than the division before it) (Delsman 1919; Janet 1923a; Gerisch 1959; Starr 1970a; Green & Kirk 1981; Kirk & Harper 1986). Almost certainly this uniform orientation of cleavage planes is a result of the stereotyped behavior of the volvo-calean BA during cell division (see Sections 4.4 and 5.4.1.4), which in turn is critical for assuring that all cells in the adult will have their BBs oriented in the same way with respect to the anterior–posterior axis of the organism, and thus that all flagella will be capable of beating in a coordinated manner.

Caption to Figure 5.10 *(cont.)*
produced by asymmetric division of anterior blastomeres, as well as all cells produced by symmetric divsion of the posterior blastomeres, continue dividing symmetrically until they have completed a total of 11 or 12 divisions. At that point, the fully cleaved embryo (**I**) contains all of the cells that will be present in the adult, but it is inside out: The presumptive gonidia are on the outside, and the flagellar ends of the presumptive somatic cells are directed toward the lumen of the embryo. (From Kirk, Viamontes et al. 1982, in *Developmental Order: Its Origin and Regulation*, ed. S. Subtelny & P.B. Green, copyright © 1982 Alan R. Liss, Inc., reprinted by permission of Wiley-Liss, Inc., a subsidiary of John Wiley & Sons, Inc.)

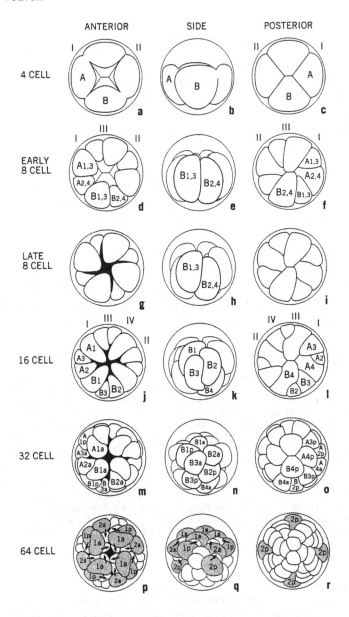

Figure 5.11 Diagrammatic representation of the first six cleavage divisions of *V. carteri* (cf. Figure 5.10). Because the two second-cleavage furrows do not quite meet at the posterior pole, the four-cell embryo of *V. carteri* (and every other volvocacean that has been examined) consists of two "B" blastomeres that are in contact at their posterior ends and two "A" blastomeres that are not. Subsequently, however, all four blastomeres cleave similarly, so the pattern can be visualized by following the behavior of either an A or a B quadrant. Here, to facilitate such visualization, one quadrant of each type is labeled with numbers and letters

larger the fraction of embryos that will divide 12 times. This is the most subtle example of a general rule: As culture conditions deviate from the optimum (e.g. because of suboptimum illumination, limiting concentrations of essential minerals, or overcrowding of the culture), the growth rate of *V. carteri* gonidia declines, and as it does, the cleavage pattern deviates progressively from the ideal one described earlier. The deviations from the idealized cleavage pattern that are observed under suboptimum conditions are as interesting and potentially as instructive – with respect to underlying patterning mechanisms – as the ideal pattern itself, and hence they are worthy of a few moments of consideration.

At only modestly suboptimum conditions, the major deviation observed is that certain of the anterior blastomeres fail to divide asymmetrically; as a result, spheroids with fewer than 16 gonidia are produced. Careful study has shown, however, that in adults with fewer than 16 gonidia the positions from which gonidia are missing are distinctly nonrandom, indicating that the failure to divide asymmetrically also occurs in a nonrandom pattern (Gilles & Jaenicke 1982). The 16 cells of the 32-cell embryo that divide asymmetrically in the ideal case are located in four tiers that have been called 1a, 1p, 2a, and 2p (Figure 5. 11) (Green & Kirk 1981, 1982). In a survey of 235 adults containing 12 to 15 gonidia it was observed that 664 of 673 "missing" gonidia (~99%) could be attributed to failure of one or more tier-2a cells to divide asymmetrically (Gilles & Jaenicke 1982). Similarly, it was found that nearly all of the additional missing gonidia in individuals with 8–11 cells could be attributed to failure of certain 1a cells to divide asymmetrically.

Gilles and Jaenicke postulated that the reason for such regularity was that when growth conditions were less than ideal, certain tier-2 cells of the 16-cell embryo would undergo asymmetric division prematurely, thereby producing gonidial initials in the 2p position, but dooming their tier-2a descendants to become symmetrically dividing somatic-cell initials (Gilles & Jaenicke 1982). They further postulated that under conditions that were sufficiently limiting to cause all tier-2 cells to divide that way, one or more tier-1 cells might also divide that way. That hypothesis was intellectually appealing because it provided such an elegantly simple explanation for the pattern of variation that was observed. However, those authors produced no

Caption to Figure 5.11 *(cont.)*
according to the conventions used by Green & Kirk (1981, 1982). Black areas in **g**, **j**, **m**, and **p** indicate the location and extent of the phialopore. stippling is used in **p** to accentuate the gonidial initials produced by the asymmetric sixth division, and (as in Figure 5.10j) arrows are used to connect pairs of large and small sister cells. For further details see Figure 5.10. (Adapted from Green & Kirk 1981, on the basis of Green & Kirk 1982.)

evidence that within a single embryo some cells would actually execute their first asymmetric division in cycle 5, whereas others would not do so until cycle 6. Starr, in his original descriptions of cleavage in *V. carteri*, had in fact reported a different basis for "missing" gonidia: "under less than optimal conditions, only 10 to 15 of these 16 cells will undergo the differentiating division *at the 32-cell stage*" (Starr 1969, emphasis added). He has since reexamined the matter carefully and has never observed the kind of intraembryo variation in the timing of the first asymmetric division that was postulated by Gilles and Jaenicke (R. C. Starr personal communication). Moreover, in our scanning electron microscope (SEM) examinations of hundreds of cleaving embryos drawn from a population in which adults with gonidial numbers ranging from 12 to 16 were regularly observed (Green & Kirk 1981, 1982; Green 1982), we never saw a single 32-cell embryo in which some blastomeres had divided asymmetrically at the previous division (K. J. Green and D. L. Kirk, unpublished observations). Thus, even though it makes the regularity of the missing-gonidia pattern more difficult to rationalize, we are driven to the conclusion that it is *spatial*, not *temporal*, variation in the asymmetric division pattern that accounts for the production of individuals with 9–16 gonidia.[14]

Observations in the preceding paragraph notwithstanding, there are conditions under which asymmetric division occurs at cycle 5 instead of cycle 6. Under significantly suboptimum conditions, the onset of cleavage can be delayed by as much as one or two days. During this delay period, gonidia usually continue to grow, but at distinctly suboptimal rates, and often they do not achieve full size before they begin to cleave. In such cases the embryos initiate asymmetric division at cycle 5, producing a maximum of 8 gonidia

---

[14] A complete model for gonidial specification will also need to account for individuals with more than the ideal number of 16 gonidia. Gilles and Jaenicke (1982) reported that they observed individuals with 17 or more gonidia only very rarely and that these rare cases could be accounted for by "twinning," in which a gonidial initial divided symmetrically to produce two smaller-than-normal gonidia that lay near each other in both the embryo and adult. However, as previously reported (Kirk & Harper 1986), under our culture conditions both the mean number of gonidia and the number of spheroids with more than 16 gonidia are higher than the numbers reported by Gilles and Jaenicke. As this book was being written, I surveyed samples of the standard female strain (HK 10) that had been obtained at various times from the University of Texas Culture Collection of Algae and maintained separately for varying numbers of years in our laboratory, and I found that the following generalizations applied to all of them: (1) The modal class of spheroids (>50%) have 16 gonidia each. (2) The mean number of gonidia per spheroid varies from culture to culture between 15.0 and 15.6. (3) Spheroids with fewer than 16 gonidia outnumber those with more than 16. (4) Spheroids with 17–20 gonidia are regularly seen (~10% of total), and those with 21–24 are occasionally seen (~1% of total). (5) In adults with 17–20 gonidia, evidence of twinning is occasionally seen, but is less common than cases in which extra gonidia are present in locations suggesting that they probably were derived from asymmetric division of tier-3a cells at cleavage cycle 6, and the rare adults with more than 20 gonidia usually have gonidia in locations suggesting that they were probably derived by asymmetric division of tier-4a cells.

(often fewer), and they also complete fewer total divisions, resulting in small progeny with subnormal numbers of somatic cells and gonidia.

Once somatic initials have completed their last cleavage division, no matter how many divisions there may have been, all of the cells that will be present in an adult of the ensuing generation have been produced. By that time, about the only visible difference between presumptive gonidia and somatic cells is in size: As a result of the combined effects of asymmetric division and different numbers of division cycles completed, by the end of cleavage the gonidial initials are about 32 times the volume (more than three times the diameter) of the somatic initials. But this difference in size is far from trivial, as we will now see.

5.4.1.3 The sizes of cells at the end of cleavage determine their fate. In the first decade of modern *Volvox* research, two hypotheses were proposed to explain the dichotomous differentiation of germ and soma in *V. carteri*. The first hypothesis was based on the changes in gonidial number and/or position that were observed following ultraviolet (UV) irradiation or centrifugation of precleavage gonidia of *V. carteri* f. *weismannia*, and it proposed that gonidial specification was mediated by a UV-sensitive, particulate morphogenetic determinant analogous to the germ plasm of certain animal embryos (Kochert & Yates 1970; Kochert 1975). That hypothesis implied that the visible, quantitatively asymmetric divisions of the embryo were accompanied by an invisible, qualitatively asymmetric division that segregated a specialized kind of cytoplasm – germ plasm – to the larger cells. The alternative hypothesis (which was based on the elevated numbers of gonidia and reduced numbers of somatic cells that were observed in *V. carteri* f. *nagariensis* mutants that ceased dividing prematurely) was that it was the sizes of cells at the end of cleavage, not the kind of cytoplasm that they contained, that would determine whether they would become mortal somatic cells or immortal germ cells (Pall 1975). More specifically, Pall postulated that any cell that was larger than about 8 μm in diameter at the end of cleavage would become a gonidium.

A few years ago, we set out to test the opposed predictions of those two hypotheses, with a rather thinly disguised predilection for the germ–plasm hypothesis (which, of course, could be traced back to August Weismann, who first proposed that *Volvox* was an appropriate model for exploring the ontogeny and phylogeny of germ–soma specification). However, we found no evidence that gonidial potential was restricted to any region of the embryo or was associated with any qualitatively unique type of cytoplasm (nor has anyone else, in any direct way). Quite to the contrary, all of the studies we performed yielded findings that were not only qualitatively but also quantitatively consistent with Pall's hypothesis, namely, that any cell that is more than 8 μm in diameter at the end of cleavage – no matter where or how it is produced – differentiates as a gonidium (Kirk, Ransick et al. 1993). Three types of experimental manipulations of wild-type embryos supported this conclusion: (1) When blastomeres were separated at the 16-cell stage and cultured individually, asymmetric division patterns were modified, and posterior blastomeres (which normally produce only somatic cells) as well as anterior blastomeres (which normally produce two gonidia each) produced an average of one gonidium each. (2) When heat shock was used to interrupt cleavage after the asymmetric division had occurred (and thus after any putative gonidial determinants should already have been segregated to the gonidial initials), presumptive somatic initials that ceased cleaving before they had fallen below the threshold size of about 8μm developed as gonidia, whereas sibling cells that divided often enough to pass that threshold developed as somatic cells. (3) When large cells were produced microsurgically in the posterior hemisphere of the embryo (an area that normally makes only somatic cells), these large cells invariably developed as gonidia. Additional studies performed with various mutants (discussed in the next chapter) all reinforced these conclusions (Kirk, Ransick et al. 1993).[15]

Although a great many studies over the past century have led to the firm conclusion that many different types of prokaryotic and eukaryotic cells have mechanisms to assess their sizes and modify their behaviors accordingly, as reviewed by Kirk, Ransick et al. (1993), there have been very few cases in which either the mechanism for assessing size or the mechanism for transducing the measured size into a cytological response has been elucidated. *Volvox* is among those cases in which these mechanisms remain wrapped in mystery. Nevertheless, the foregoing types of studies intensify interest in understanding the cytological mechanisms underlying asymmetric division and the generation of large and small cells.

5.4.1.4   The cell-division apparatus of V. carteri *resembles that of* C. reinhardtii. The nuclear and cytoplasmic events of cell division in *V. carteri* are coordinated by cytoskeletal elements quite similar to those described in Chapter 4 for *C. reinhardtii* (compare the diagrams in Figures 4.4 and 4.5 to the immunofluorescence images in Figures 5.12–5.14). Throughout the entire cleavage period, each *V. carteri* nucleus or spindle pole is held in close

---

[15] Because Pall's studies and ours were all performed with *V. carteri* f. *nagariensis*, and Kochert's were performed with *V. carteri* f. *weismannia*, it remains possible that the mechanisms of gonidial specification are different in these two formas. This possibility appears far less outlandish when it is realized that the mechanisms of gonidial specification have been shown to be different in *V. carteri* f. *nagariensis* and *V. obversus* (Ransick 1991, 1993; Kirk, Ransick et al. 1993) and that recent rRNA sequence data suggest that *V. obversus* may be more closely related to *V. carteri* f. *nagariensis* than is *V. carteri* f. *weismannia* (Figure 2.11)!

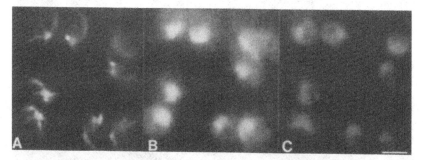

Figure 5.12 Immunofluorescent visualization of aspects of the cell-division apparatus in the posteriormost eight blastomeres of a 16-cell *V. carteri* embryo stained to reveal the locations of acetylated α-tubulin (**A**), centrin (**B**), and DNA (**C**). The appearance and proximity of the nuclei in **C** identify this as telophase of division four (i.e., very early in the 16-cell stage). Anti-acetylated α-tubulin stains specifically the microtubular rootlets (MTRs) and basal bodies (BBs). These structures lie closest to the focal plane in the pair of cells at the lower left, in which one can see the cross formed by the four MTRs, as well as the BBs (the pair of bright spots at the intersection of the MTRs). The more heavily stained MTRs are the 4MTRs (see Figures 4.4 and 5.3), which already predict the planes in which the next set of divisions will occur. In **B** it can be seen that in each cell the centrin (the principal component of the NBBCs) is concentrated as a "plaque" of intensely labeled material on the surface of each nucleus, immediately below the BBs. Note the rotational symmetry of the division pattern: Each newly-formed cell pair (and thus the cleavage plane that divides it) is rotated 90° clockwise relative to the cell pair lying counterclockwise of it. (This rotational symmetry of adjacent division planes is believed to account for the rotational symmetry of the adult that is diagrammed in Figure 5.4.) (From Kirk, Kaufman et al. 1991.)

association with a BB (with its characteristic array of MTRs) via contracted centrin fibers (Figures 5.12 and 5.13), so that the karyokinetic activities of the nuclei are closely coordinated in space and time with the cytokinetic activities of the cytoplasm (Kirk, Kaufman et al. 1991).

It appears that in *V. carteri*, as in *Chlamydomonas*, the plane of the next division is predicted fairly precisely by the location of the 4MTRs in the preceding interphase (Figure 5.13). Moreover, it is undoubtedly significant that in *V. carteri*, as in *C. reinhardtii*, 4MTRs elongate and bend sharply at the midline during formation and elongation of the spindle, so that in each case their BB-distal ends remain next to the presumptive cleavage furrow, while their BB-proximal ends move with the spindle poles, thereby keeping each BB pair precisely oriented with respect to the division plane (Figure 5.14). This appears to have an important consequence in terms of the overall organization of the embryo (and later the adult): Because of the precise and highly stereotyped patterns in which BBs are moved apart and rotated in each mitotic cycle, sister blastomeres (and later the cells of the adult) all come to

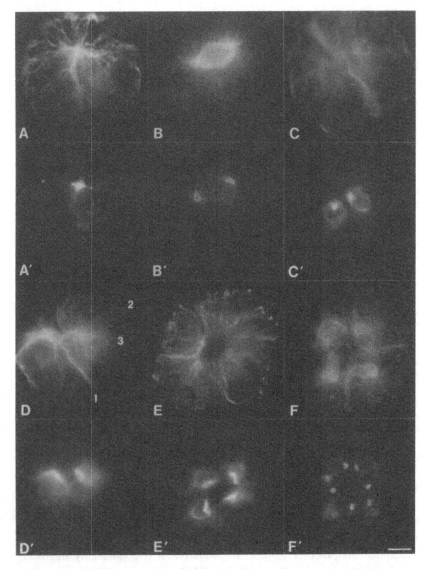

Figure 5.13 Immunofluorescent analysis of early cleavage in *V. carteri* embryos double-stained to reveal the locations of β-tubulin (**A, B, C**, etc.) and centrin (**A'**, **B', C'**, etc.). In distinction to the anti-α-tubulin used in Figure 5.12, the anti-β-tubulin used here stains all microtubules (MTs). **A** and **A'**: First prophase. In A, the MTRs are seen as a bright cross, from which many 2° MTs radiate out under the plasmalemma. In A', the NBBCs (revealed by anti-centrin staining) have contracted to form a dense plaque just below the center of the MTR cross. **B** and **B'**: First metaphase. In this focal plane, the mitotic spindle is clearly seen, but the MTRs are not; note the centrin plaques (contracted NBBCs) at each of the spindle poles. **C** and **C'**: First telophase. In the nuclear focal plane shown here, the cleavage MTs appear as an out-of-focus, diffuse line bisecting the cell, and portions of the MTRs and some 2° MTs (although out of focus also) can be discerned on the surfaces of

be arranged with rotational symmetry around the anterior–posterior axis of the embryo (Figures 5.12 and 5.14; see also Figure 5.4). This, in turn, has important consequences for the motility of the spheroid, as mentioned in Section 5.3.3, and as will be mentioned again when we consider the phototactic behavior of the adult spheroid in Section 5.4.6.

Thus, some important cytological features of the early, symmetric divisions of the *V. carteri* embryo are beginning to be understood, largely by extrapolation and modification of insights that have been gained by the study of cell division in *Chlamydomonas* and other related unicells. Sadly, however, there is as yet no information available concerning which of these features become modified, or how, at the specific times and places that asymmetric division is to occur in the *V. carteri* embryo. In light of the conclusion that in both *Chlamydomonas* and *Volvox* the location of the 4MTR in the preceding interphase appears to predict the plane of the next cleavage division, an obvious working hypothesis is that some subtle shift in the way that the BA is positioned in anterior blastomeres at telophase of division cycle 5 is the critical step that causes the sixth division to be asymmetrical in these cells. Testing of this hypothesis will, however, be a daunting task.

5.4.1.5 Incomplete cytokinesis generates a cytoplasmic-bridge system that links all cells. The predictable spatial relationships among neighboring cells that are established by the stereotyped behavior of the division apparatus at each division are maintained throughout the rest of embryogenesis by a system of cytoplasmic bridges that link all cells in the embryo into a coherent syncytium (Figure 5.15). This system of bridges arises because cell division in *V. carteri* (as in all other volvocaceans, but in distinction to division in *Chlamydomonas*) is incomplete (Metzner 1945a; Dolzmann & Dolzmann

Caption to Figure 5.13 *(cont.)*
the forming daughter cells. Note that by telophase the NBBCs (centrin plaques) have moved back toward the cleavage furrow, apparently dragging the daughter nuclei with them. **D** and **D'**: Late first telophase/early second prophase. The brightly staining lines in D constitute the MTRs, with their images reinforced by the fluorescence of the proximal ends of the 2° MTs that they have organized. Line 1 indicates the original location of the "parental" 4MTRs, and marks the location of the first furrow. Line 2 represents the newly forming 4MTRs of the daughters and predicts the plane in which the second cleavage furrows will form. Line 3 represents the newly forming 2MTRs of the daughter cells and predicts the plane in which the third cleavage furrows will form. Note that the centrin plaques in D' have begun to elongate over the surface of the nuclei and will soon split in two, to take up positions at the spindle poles, as in B'. **E** and **E'**: Late second telophase/ early third prophase; similar to D and D', but with everything doubled. **F** and **F'**: Third metaphase. Again, note NBBCs (centrin) at the poles of each spindle and, as in Figure 5.12, the rotational symmetry of the cleavage pattern. (From Kirk, Kaufman et al. 1991.)

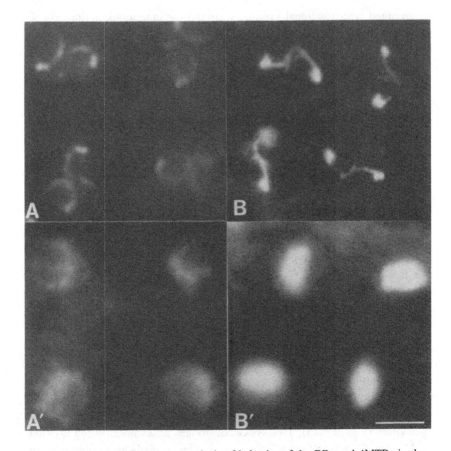

Figure 5.14 Immunofluorescent analysis of behavior of the BBs and 4MTRs in the fifth cleavage cycle, in the four posteriormost cells of a 16-cell embryo double-stained to reveal the locations of acetylated α-tubulin (**A, B**) and DNA (**A', B'**) (cf. Figure 4.4). **A** and **A'**: Prophase. As the mitotic spindle (not shown) begins to form and the chromatin condenses, the BBs (bright spots at the distal end of each arc) have moved apart to remain at the poles of the spindle. As the BBs have separated, the 2MTRs have shortened and virtually disappeared, but the 4MTRs have remained attached to the BBs, have elongated, and have bent sharply, so that their BB-proximal ends now lie nearly perpendicular to their distal ends (which remain in their original locations). **B** and **B'**: Metaphase. The BBs remain at the spindle poles; the 4MTRs (B) have overlapped one another extensively immediately above the metaphase plate. Again note the rotational symmetry of the division pattern. (From Kirk, Kaufman et al. 1991.)

1964; Bisalputra & Stein 1966; Ikushima & Maruyama 1968; Deason et al. 1969; Pickett-Heaps 1970, 1975; Deason & Darden 1971; Marchant 1976, 1977; Gottlieb & Goldstein 1977; Fulton 1978b; Birchem & Kochert 1979b; Green & Kirk, 1981; Green et al. 1981; Hoops 1981). Hundreds of bridges

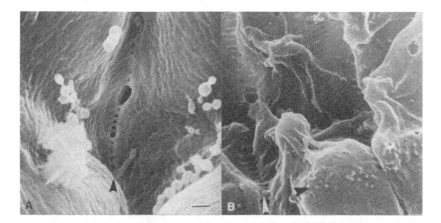

Figure 5.15 Cytoplasmic bridges in cleaving *V. carteri* embryos. **A**: Higher-magnification view of the embryo in Figure 5.10a, showing a row of cytoplasmic bridges (arrowhead) being formed very early in the first cleavage division. About 500 such bridges are formed in the first furrow; these persist and are subdivided between daughter cells, while new bridges are formed (at about the same density) in each subsequent division furrow. **B**: A fragment of a midcleavage embryo. Prior to fixation, this embryo was treated with a hypertonic solution to cause the cells to shrink and pull apart slightly, revealing more clearly the (now-stretched) bridges between neighboring cells (white arrowhead). On the cell surfaces in the foreground that were exposed by splitting the embryo in half after fixation and critical-point drying, the broken bridges are seen as a band of rather uniformly spaced bumps and pits (black arrowhead). Note that the bands of bridges girdling adjacent cells are perfectly aligned. (From Green and Kirk, *The Journal of Cell Biology*, 1981, **91**, 743–55, by copyright permission of The Rockefeller University Press.)

are formed in each *V. carteri* division cycle; these bridges are then conserved and divided between sister cells even as additional bridges are being formed in the furrow between them. By the end of cleavage the total number of bridges in the embryo will have risen to nearly 10^5, and the average cell will have about 25 bridges linking it to its neighbors (Green & Kirk 1981).

Cytoplasmic bridges apparently arise as the result of incomplete fusion of Golgi-derived vesicles in one region of the cleavage furrow (Green et al. 1981). It has long been known that cytokinesis in certain green algae, including *Chlamydomonas*, is more complex mechanistically than in either plants or animals and combines features seen in dividing cells within both of those kingdoms (Johnson & Porter 1968; Pickett-Heaps 1975). Cytokinesis in the volvocine algae involves an ingressive furrow (analogous to that of animal cells) that forms early in mitosis at the anterior end of the cell (as described in Section 4.4.3), and that subsequently extends in a lateral and posterior direction as it ingresses. But in the midregion of the dividing cell it also bears a certain resemblance to cell plate formation in plants, in that it

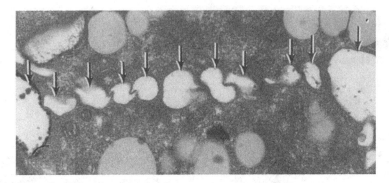

Figure 5.16 Vesicles (arrows) aligned in the presumptive cleavage furrow, in the region where cytoplasmic bridges will be formed. (From Green et al., *The Journal of Cell Biology*, 1981, **91**, 756–69, by copyright permission of The Rockefeller University Press.)

involves accumulation (and presumably fusion) of cytoplasmic vesicles along the prospective division plane (Bisalputra & Stein 1966; Johnson & Porter 1968; Fulton 1978b; Birchem & Kochert 1979b; Green et al. 1981) (Figure 5.16). It has been postulated that it is at sites where fusion of such vesicles fails to occur that cytoplasmic bridges form (Green et al. 1981; Green 1982). It is clear that in some (if not in all) cases, cytokinesis in the bridge region precedes cytokinesis at either the anterior or posterior ends of the cells (Green et al. 1981).

The bridges of the cleaving embryo are highly regular in size (~200 nm in diameter) and spacing (~500 nm center-to-center distance). They are equally regular in location and ultrastructure. Each new set of bridges is formed in a zone toward the anterior end of the furrow (Figure 5.17A) and in a very regular array that is aligned with similar arrays formed in both previous and future division furrows (Figure 5.15B). Thus, each cell comes to be girdled by a highly regular band of bridges near its anterior end, and the bridge bands of all cells are aligned into a continuous network that runs throughout the embryo. Each bridge is decorated internally with two kinds of cortical specializations: An amorphous electron-dense material that coats the inner surface of each bridge is thickest in the center of the bridge (Figure 5.17B), and associated with this material is a series of concentric (or helical?) cortical striations that not only ring the bridge proper but also extend out under the adjacent plasmalemma to eventually contact equivalent striations surrounding adjacent bridges (Figure 5.17C,D). Extremely similar cortical striations have been observed in the bridges of all other volvocacean embryos that have been examined with adequate resolution (Pickett-Heaps 1970; Marchant 1977; Birchem & Kochert 1979b); it has been proposed that they play a structural role, reinforcing the bridges and/or linking them to one another (Marchant 1977; Green et al. 1981).

The syncytium that is established by the cytoplasmic bridges that link neighboring cells is interrupted only at the edges of the phialopore, that is, in the regions where cells abut one another not across a common cleavage furrow but where they (or the cells from which they are derived) were brought into contact when anterior blastomeres extended toward the anterior pole of the embryo between the 4- and 16-cell stages (Figure 5.11). As we will see in Section 5.4.2, both this bridge-free phialopore region and the cytoplasmic-bridge network that pervades the rest of the embryo will play important roles in the next step of embryogenesis.

5.4.1.6 Uninterrupted biosynthetic and cytoskeletal activities are required during cleavage. Several reports have indicated (not surprisingly) that cleavage in *V. carteri* is adversely affected by a variety of pharmacological agents that interfere with various biosynthetic activities, cytoskeletal rearrangements, and so forth (Wenzl & Sumper 1979; Ireland & Hawkins 1980, 1981; Kurn & Duskin 1980, 1982; Green 1982; Weinheimer 1983; Desnitski 1985c, 1986, 1987, 1990; Kirk & Harper 1986). These are most logically discussed in parallel with the order observed in the gene action system, starting in the nucleus and moving toward the cytoplasm and cell surface.

Cleavage in *V. carteri* is wholly unaffected by concentrations of aminopterin (an inhibitor of deoxythymidylate synthesis) that will cause a rapid cessation of cleavage in *V. aureus* (Desnitski 1986); this has been interpreted to mean that species such as *V. carteri*, which develop large gonidia and then cleave rapidly (see Table 2.2), enter cleavage with abundant pools of DNA precursors, whereas species such as *V. aureus*, in which cleavage begins while gonidia are very small, and embryos then grow between divisions, must continuously synthesize DNA precursors in order to continue cleaving.

Not surprisingly, however, a variety of inhibitors of DNA synthesis will block cleavage in *V. carteri*, generally in a nonreversible fashion (Ireland & Hawkins 1980; Weinheimer 1983; Kirk & Harper 1986; our unpublished observations), indicating that DNA replication is required during cleavage in *Volvox*, as in all mitotic systems. However, as we will see in Section 7.1.1, it is possible that the relationship between nuclear DNA synthesis and mitosis is not as simple and direct in the volvocaceans as it is in the "typical" eukaryotic cell, and at least some of these algae may enter cell division having already undergone enough rounds of endoreduplication of nuclear DNA that they will not have to complete a full doubling of DNA in every

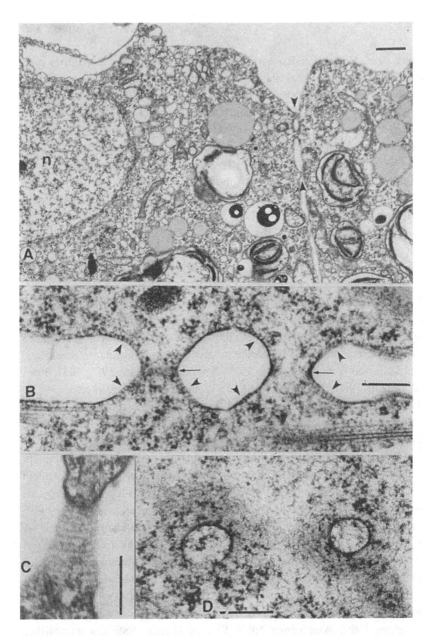

Figure 5.17 Ultrastructure of the cytoplasmic bridges. **A:** Bridges in the cleaving embryo (arrowheads) are located toward the anterior ends of the cells, at about the same level as the nuclei (n). **B:** At higher magnification, these bridges are seen to possess two types of cortical specializations. In the central region of each bridge, the plasmalemma is coated with electron-dense material; when cut in transverse section (right arrow), this material appears to be amorphous, but when cut *en face*, it appears vaguely fibrillar (left arrow). More peripherally, the inner surface of the plasma membrane in the bridge region bears a regular series of striations

division cycle. Be that as it may, it is abundantly clear that continuous DNA synthesis is required for cleavage to proceed.

Despite the extreme rapidity of the division cycle in *V. carteri* (~35 minutes under standard culture conditions) and the fact that cells of the cleaving embryo probably spend most of their time in either M or S phase, synthesis of RNA (including rRNA) appears to continue at a relatively brisk pace during cleavage (Yates & Kochert 1976). Changes are observed in the relative abundance of a variety of cell-type-specific RNAs at various stages of cleavage (Tam & Kirk 1991a,b; Tam et al. 1991). One RNA species of particular interest is G167, which is one of the most abundant polyadenylated RNAs in very early embryos; the nature of this gene and its product will be discussed in Chapter 7.

The immediacy of nuclear control over embryogenesis is indicated by the fact that actinomycin D, at concentrations well below what is required to totally inhibit incorporation of label into RNA, inhibits cleavage in *V. carteri* f. *weismannia* (Weinheimer 1983). When applied to precleavage gonidia, this level of actinomycin blocked cleavage initiation, but when cleaving embryos were exposed to the drug, they were able to complete a few divisions before arresting. Of particular interest is the report that when 32-cell embryos were exposed to this low dose of actinomycin, they cleaved up to three additional times, but they did not execute any asymmetric divisions. That observation has not been followed up, but it deserves to be. Considerably higher concentrations of actinomycin D are required to block cleavage in forma *nagariensis*, and no specific effects of the antibiotic on asymmetric division in this forma have yet been reported (Desnitski 1987).

Protein synthesis also continues apace during cleavage, and reproducible stage-specific differences in the labeling patterns of major polypeptides are seen (Kirk & Kirk 1983). In both formas of *V. carteri*, exposure to cycloheximide at concentrations below what is required to completely block incorporation of label into protein causes a virtually immediate cessation of cleavage (Weinheimer 1983; Desnitski 1990).

One very interesting protein that has recently been shown to be synthesized during cleavage is called algal-CAM, because it appears to be the first algal cell adhesion molecule (CAM) to have been characterized (Huber &

Caption to Figure 5.17 *(cont.)*
(arrowheads). Where sections pass through the bridge region tangentially (**C**) or *en face* (**D**) it is seen that these cortical striations actually constitute a series of concentric rings that extend outward under the plasma membrane adjacent to each bridge. Bars: in A, 1 μm, in B–D, 0.25 μm. (From Green et al., *The Journal of Cell Biology*, 1981, **91**, 756–69, by copyright permission of The Rockefeller University Press.)

Sumper 1994). Algal-CAM combines an *N*-terminal domain resembling higher-plant extensins (cell-wall proteins) with a pair of domains that resemble *Drosophila* fasciclin (a molecule involved in axon guidance). By indirect immunofluorescence, algal-CAM was shown to be present in the four-cell embryo in the contact zones between sister blastomeres, toward the posterior (chloroplast) ends of the cleavage furrows. When monoclonal anti-algal-CAM is added to live embryos at this stage, it causes the cells to separate at their posterior ends, giving the embryo a doughnut-like appearance (because they apparently remain coupled at their anterior ends by their cytoplasmic bridges). Continued development in the presence of this antibody causes major disruptions to the cleavage pattern, a severe reduction in the number of gonidial initials produced, and a reduction in the total number of divisions (Huber & Sumper 1994). Further details about the precise location and roles of this interesting molecule are eagerly awaited.

Algal-CAM was the first defined cell-surface molecule to be implicated in cellular adhesion in volvocacean embryos. Many years earlier, however, a number of studies had shown that interference with cell-surface glycoproteins of cleaving embryos in any of a number of ways could cause a variety of morphological abnormalities. Juveniles derived from *V. carteri* embryos that had been treated with subtilisin, or a mixture of glycosidases, or the lectin conconavalin A (Con A), or borate (a sugar-complexing agent) had gonidia in subnormal numbers and abnormal locations (Wenzl & Sumper 1979). Those findings strongly implied that such treatments had disturbed the pattern of cleavage in some way, but direct observations of cleavage in such embryos were not reported. Others observed that in *V. tertius*, Con A at high concentrations (50–100 µg/mL) would completely block cleavage or, when applied later, the subsequent process of inversion (Ireland & Hawkins 1981). In parallel with such studies, it was reported that tunicamycin (which inhibits a step required for the addition of the mannose-rich *N*-glycans to which Con A binds) had effects on *V. carteri* morphogenesis as severe as or more severe than those of Con A (Kurn & Duskin 1980, 1982). But once again, direct observations of cleavage in treated embryos were not reported. A clue to one effect of Con A came from the observation that in embryos exposed to Con A right after cleavage had been completed, regions were seen in which flagellar membranes had fused, resulting in multiple axonemes lying within a common membrane (Kurn 1981).

Jaenicke and Gilles (1982) reported that fluorescent Con A preferentially stains the anterior surfaces of the early embryo. In accord with this, Green (1982) reported that at a low concentration (10 µg/mL) Con A selectively blocks cleavage-furrow progression at the anterior ends of the embryonic cells in the next division cycle, without, however, preventing division at the chloroplast end of the cell or the formation of cytoplasmic bridges in the subnuclear region of the cell. Most surprisingly, when thin sections of these embryos were examined in the electron microscope, it was discovered that the Con A treatment had also blocked nuclear division (Figure 5.18A).[16] If the binding of Con A was reversed soon enough (by addition of the hapten, α-methyl mannoside), both karyokinesis and cytokinesis were completed, and the embryo proceeded to develop relatively normally.

Whereas Con A selectively and reversibly inhibits cytokinesis at the anterior end of the cell, at a concentration of 20–30 µg/mL, cytochalasin D (CD), an inhibitor of actin polymerization, selectively and reversibly inhibits cytokinesis at the posterior end of the cell (Figure 5.18B; Green 1982). At concentrations that do not kill the embryo outright, neither of these drugs has any effect on cytokinesis in the bridge region or on the formation of cytoplasmic bridges, supporting the contention that three different cytokinetic mechanisms are employed in each cleavage furrow (Green et al. 1981; Green 1982). Although embryos will initiate as many as three rounds of cleavage in CD at 20–30 µg/mL (dividing only at the anterior end each time), they will then arrest. Astonishingly, however, if the drug is then washed out, the embryos will begin to recover within seconds, complete all previously incomplete division furrows, and proceed with cleavage and inversion in a relatively normal fashion.

CD has other interesting effects on cleavage. When applied at a concentration as low as 5 µg/mL to an embryo that has just initiated the first cleavage furrow, the drug will completely block cleavage; the embryo will adopt an extremely dense appearance but do nothing more. If after some time the drug is washed out, however, within 30 minutes cleavage will be reinitiated and will proceed to completion (Green 1982). This concentration of CD has no perceptible effect on embryos once the first division has been completed. Why the first division is so much more sensitive to CD than the rest remains a mystery.

As noted earlier, in CD at 20–30 µg/mL, embryos fail to complete cytokinesis. If the drug is added by the 4- or 8-cell stage, they also fail to complete the cellular extensions that normally convert the flattened embryo into a hollow sphere and form the phialopore at the anterior pole Figure 5.18C). If embryos are left in CD throughout the period when "closure" of the phialo-

---

[16] At higher concentrations (~100 µg/mL), Con A causes adjacent blastomeres to fuse and can even cause all cells of a late-cleavage embryo to coalesce into a multinucleate, unicellular mass (K. J. Green and D. L. Kirk unpublished observations). Preliminary observations suggested that this effect is a result of the cytoplasmic bridges losing their integrity and "ballooning out" to the extent that the entire cleavage furrow was obliterated. Needless to say, such effects of Con A are not reversible.

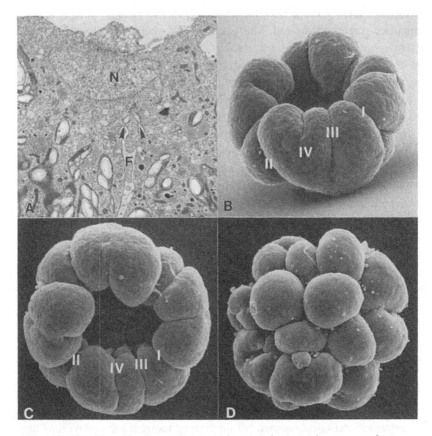

Figure 5.18 Effects of selected drugs on cleavage. **A:** An embryo exposed to concanavalin A (10 μg/mL) for one division cycle. A cleavage furrow (F) has formed in the posterior portion of the cell, and has progressed nearly to the base of the nucleus (N), and cytoplasmic bridges of relatively normal ultrastructure have been formed (arrowheads). However, neither the nucleus nor the anterior end of the cell has divided. This indicates that posterior cytokinesis and formation of cytoplasmic bridges can occur independently of karyokinesis or anterior cytokinesis. **B:** An embryo exposed to cytochalasin D (30 μg/mL) during the third and fourth division cycles; roman numerals identify the first four cleavage furrows in one quadrant of the embryo. Both the third and the fourth division furrows have been initiated normally at the anterior end of each blastomere, but no cytokinesis has occurred in posterior regions. Again, this indicates that the cytokinetic processes in the anterior and posterior regions of a blastomere are controlled independently. **C:** Anterior view of the embryo shown in B, with cleavage furrows labeled in the same way. The failure of the cytochalasin-treated cells to achieve posterior cytokinesis is associated with a failure of the cells to elongate normally and close off the phialopore (cf. Figure 5.10g). **D:** A 32-cell embryo that was exposed to Colcemid (25 μg/mL) for three division cycles. Mitosis and cytokinesis have continued on a relatively normal schedule, but cleavage planes have been randomized, producing cells of extremely abnormal sizes and shapes (cf. Figure 5.10i). (Adapted from Green 1982, with permission.)

pore is normally completed, and then washed, they will quickly reinitiate cellular extension toward the anterior pole, but they often fail to produce a normally shaped phialopore. As a consequence, once they have completed the cleavage program and inverted (both of which they appear to do relatively normally), the phialopore lips do not always come together properly, and nonspherical, misshapen adults with uncoordinated swimming behavior often result (Green 1982). This indicates that the cellular extensions that normally occur at the 4- to 16-cell stage are critical for producing a spherical, well-adapted adult organism.

Whereas CD has no discernible effect on the orientation of cleavage furrows (Figure 5.18B,C), anti-microtubule agents do, although these effects are not as readily observed as might be anticipated. Whereas colchicine at 25 μg/mL is adequate to completely block inversion of *V. carteri* (as will be discussed later), cleaving embryos exposed to the drug at 1,000 μg/mL continue to divide for a number of cycles. (Whether this difference in sensitivity is due to a stage-specific difference in permeability or in the colchicine sensitivity of the MTs is not known.) Although division continues for a time in such concentrations of colchicine (or Colcemid at 25 μg/mL, which has equivalent effects), phycoplast MTs are reduced in number, and cleavage planes are randomized; many divisions that should be symmetrical are not, and cells of greatly differing sizes and abnormal shapes are created in a chaotic pattern (Figure 5.18C) (Green 1982). This indicates, not surprisingly, that a normal MT cytoskeleton is required to establish a normal cleavage pattern.

5.4.2 Inversion: turning the best side out

Between the third and the fifth divisions, *V. carteri* embryos, like all other volvocacean embryos, develop increasing curvature, with the anterior (nuclear-BA) ends of the cells directed toward the interior.[17] A reasonable work-

---

[17] Of the organisms that have often been included in the Volvocaceae (Bourrelly 1966, 1972), *Astrephomene* (which means "not turning itself") (Pocock 1953) is the only exception to this generalization. It divides in such a way that successive division planes diverge at the BA ends of the cells (rather than converge, as they do in the rest of the volvocaceans), and thus in the fully cleaved *Astrephomene* embryo the BA ends of all cells face the exterior. Only one micrograph is available to indicate why it may be that *Astrephomene* differs from all of its colonial green-flagellate relatives in this regard; this figure (Hoops 1981) shows cytoplasmic bridges that are less numerous, longer, and much farther from the BA end of the cell than in the "bona fide" volvocaceans. If this one image is representative of the situation throughout cleavage, it could explain how embryonic cells of this genus are able to keep their BA ends pointing to the exterior at all times. In any case, because of its modified cleavage pattern, *Astrephomene* has no need to, and does not, undergo inversion. On this basis, Pocock proposed that *Astrephomene* should be placed in a monotypic family Astre-

ing hypothesis to account for this fact is that the bridge system that circles each cell at its anterior end constitutes a girdle that limits its dimensions in this region, and therefore the cells are somewhat wedge-shaped – narrower at the BA end, and broader at the chloroplast end – and naturally assume a concave arrangement. In any case, the fully cleaved *Volvox* embryo is a hollow ball in which the potential flagellar ends of all cells point inward. Obviously, if that arrangement persisted in the adult, it would make swimming rather difficult. Hence, in *V. carteri* and all other volvocaceans, cleavage is followed by the process of inversion, in which the curvature of the embryo is reversed to establish the adult configuration.[18]

Following the final cleavage division, there is a pause of up to an hour in visible activity. Then, suddenly, a wave of contraction appears to pass over the embryo, causing it to "dent" in first one region and then another. That is soon followed by a uniform contraction, during which the phialopore, which had become inconspicuous during the late stages of cleavage, widens (Figure 5.19a,g). Next, the four lips of cells bounded by the phialopore slits bend outward and backward, over the adjacent cells (Figure 5.19b,c,g,h). As the region of maximum curvature then moves progressively toward the posterior pole, so do the inverted regions (Figure 5.19d,i), until they nearly surround the uninverted posterior hemisphere (Figure 5.19e). At that point the posterior hemisphere "snaps" through the opening at the equator, and eventually the phialopore lips come together once again at what will become the posterior pole of the adult (Figure 5.19j). At that point the flagellar ends of all somatic cells are on the exterior, and the gonidia, which protruded from

phomenaceae, rather than being grouped with the Volvocaceae (Pocock 1953). Most authors have followed this suggestion. However, it has recently been suggested, on the basis of certain shared morphological traits of other sorts, that *Astrephomene* should be grouped with certain (but not all) members of the genus *Gonium* in a new family Goniaceae (Nozaki 1993; Nozaki & Itoh 1994). It seems unlikely that this suggestion will win support among algal taxonomists, particularly because those same authors have subsequently performed a molecular-phylogenetic study that nests this *Astrephomene-Gonium* clade within the Volvocaceae (Nozaki et al. 1995).

[18] For more than 200 years after *Volvox* was discovered, the phenomenon of inversion went undiscovered, not only in *Volvox* but also in all of the other volvocaceans. Then, when it was eventually discovered and documented photographically by Powers (1908), studying *V. carteri* f. *weismannia*, it was misinterpreted. Because Powers examined fixed specimens only, he got the sequence backward. He assumed, as did previous authors, that *Volvox* embryos cleaved with the cells oriented in the adult manner. Thus he interpreted inversion as a pathological reaction in which embryos that had originally been in the correct orientation turned inside out to put their gonidia on the outside and their flagella on the inside. He admitted to being baffled by the fact that although he could find these "inverted" (actually pre-inversion) *embryos* in fixed samples of all species of *Volvox* examined, he could not find juveniles or adults in this configuration, nor could he find evidence that these "pathological" individuals deteriorated. It would be more than a decade before inversion was rediscovered and correctly interpreted: as a process by which *Volvox* embryos that had been inside out at the end of cleavage turned right side out (Kuschakewitsche 1923, 1931).

the surface of the pre-inversion embryo, have been moved to the interior. With the completion of inversion, we consider the embryonic period to have ended and the postembryonic, juvenile period to have begun.

The entire process of inversion in *V. carteri* takes about 45 minutes under standard culture conditions. In the process, the anterior–posterior axis of the spheroid is reversed: As the embryo turns inside out, the anterior pole of the embryo (defined, as it is in all volvocaleans, by the initial location of the BA in the uncleaved gonidium) becomes the posterior pole of the adult (defined by the direction of swimming). Thus, although gonidia are derived by asymmetric division of anterior blastomeres in the embryo, they always end up in the posterior part of the adult spheroid.[19]

As is so often the case in the history of *Volvox* research, we turn to Mary Agard Pocock for the first insight into the kinds of activities that drive inversion. Previous workers had speculated that inversion was driven by the secretion or swelling of extracellular materials between the apical ends of the cells (Janet 1923a; Kuschakewitsch 1923, 1931; Zimmermann 1925). That was an idea that was to reemerge later, but in the absence of convincing evidence (Gottlieb & Goldstein 1979). But Pocock rejected such ideas as "wholly inadequate" and wrote that "during inversion the cells act in concert with one another, but each is at the same time altering its shape and must to some extent act independently; while some are being compressed, others are in a state of great tension, tending to separate.... since the globoid inverts as a whole, the connection between the component cells must be very intimate. It would appear that this is obtained by the presence of protoplasmic strands, which therefore play an important role in inversion ..." (Pocock 1933a).

5.4.2.1 Inversion involves dramatic changes in cell shapes and locations of cytoplasmic bridges. Kelland described in more detail than previous workers certain of the shape changes that cells in various regions of two species of *Volvox* embryos undergo during inversion, and he showed that similar shape changes occur within a minute or two in single cells or cell clusters that have been mechanically released from the embryo. From that he drew the important conclusion that such changes in cellular shape were causes, rather than effects, of inversion movements of the embryo as a whole (Kelland 1964,

---

[19] It is a matter of some interest that in all green flagellates that exhibit any degree of germ–soma differentiation, cells with reproductive potential are formed preferentially in the anterior of the embryo, and somatic cells are formed preferentially in the posterior. In volvocaceans with somatic cells (*Volvox*, *Pleodorina*, and sometimes *Eudorina*), inversion causes the somatic cells to be concentrated toward the anterior end of the adult. However, because *Astrephomene* does not invert, its somatic cells (which are always four in number) are located exclusively at the posterior pole in the adult (Pocock 1953; Nozaki 1983).

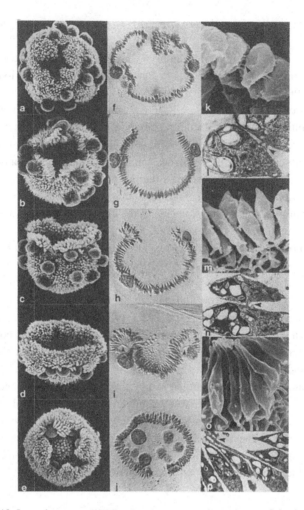

Figure 5.19 Inversion: **a–e**, SEMs of embryos at various stages of the inversion process; **f–j**, light micrographs of embryos fixed and sectioned at various stages of the inversion process; **k–p**, SEMs and TEMs of cells at selected stages of inversion. At the end of cleavage (**a, f**), the cells are pear-shaped and linked anterior to their widest points by cytoplasmic bridges (arrowheads in **k, l**), which causes the cells near the phialopore to curl inward (**f**). Inversion begins when the cells contract in the medial direction while elongating in the anterior–posterior direction, becoming spindle-shaped (**b, g, m, n**) and attached by cytoplasmic bridges at their widest points (arrowheads in **m, n**); this causes the phialopore to open widely (**b, g**). Next, negative curvature is generated at the edges of the phialopore, causing the cell sheet to bend outward in this region (**c, h**). The region of negative curvature then moves progressively farther from the phialopore (**d, i**), causing the anterior part of the embryo to curl over the posterior part. Eventually, the posterior hemisphere ''pops through'' at the equator, and cells bordering the phialopore approach each other at what has become the posterior pole of the juvenile (**e, j**). The critical step in inversion is the creation of negative curvature of the cell sheet (e.g., at arrowheads in **i**). This occurs as cells that have been spindle-shaped and linked at their widest points (**m, n**) become flask-shaped, with long,

1977). He also made the important observation that changes in cell shape (and formation of a new inversion center) were triggered immediately next to slits that he made anywhere in the embryo with a glass needle, thus focusing attention on a free edge as a site of inversion initiation.

Parallel but somewhat more detailed studies defined the sequence of shape changes in inverting *V. carteri* embryos, and led to the important additional observation that these changes in shape are accompanied by changes in the locations of cytoplasmic bridges (Viamontes & Kirk 1977). Prior to inversion, all cells are pear-shaped, with their nuclei at the inner, apical end and their chloroplasts in the outer, broad end, and at that stage the cells are linked by bridges in the vicinity of the nuclei (Figure 5.19f,k,l). As the phialopore opens, all cells will have elongated and become spindle-shaped, with distinctly pointed outer ends, and bridges will be located at a slightly subnuclear level (Figure 5.19b,g,m). (Note also that by that time each somatic cell will already have formed two short flagellar stubs, Figure 5.19m.) As the lips curl back, cells in the region of maximum curvature are seen to be flask-shaped; they have developed long, slender stalks distal to the chloroplasts and are now joined by bridges only at the very ends of these stalks (Figure 5.19h,o,p). As cells farther and farther from the phialopore undergo the spindle-to-flask transformation (with the accompanying change in bridge location), the region of maximum curvature progresses toward the posterior pole of the embryo (Figure 5.19i). As that happens, cells that previously were in the region of maximum curvature, but no longer are, relax into a simple, more or less columnar shape.

5.4.2.2 The cells move relative to the bridges, not vice versa. In confirmation of the observations of Kelland, it was shown that cell clusters released mechanically from a pre-inversion or early-inversion embryo of *V. carteri* executed a sequence of shape changes similar to those observed in the intact embryo (Viamontes & Kirk 1977; Green et al. 1981). Moreover, observation of such clusters revealed that each cell moved individually with respect to the bridge system that linked it to its neighbors – rather than remaining stationary while its bridges were translocated (Figure 5.20). The conclusion that it is the cells that move individually relative to the bridge system, rather

Caption to Figure 5.19 *(cont.)*
narrow posterior projections, and come to be linked by their cytoplasmic bridges at their outermost, narrowest points (arrowheads in **o, p**). After passing the region of maximum curvature, the cells assume simple columnar shapes (**j**). (Adapted from Kirk, Viamontes et al. 1982, in *Developmental Order: Its Origin and Regulation*, ed. S. Subtelny & P. B. Green, copyright © 1982 Alan R. Liss, Inc., adapted by permission of Wiley-Liss, Inc., a subsidiary of John Wiley & Sons, Inc.)

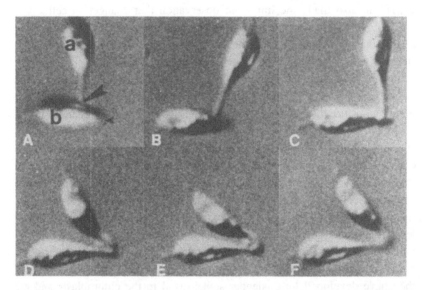

Figure 5.20 Inversion movements in a pair of isolated cells. When these cells were released by mechanical fragmentation of an embryo, they were both spindle-shaped and joined at their widest points. By the time filming began, however (**A**), cell "a" had become flask-shaped and had also moved relative to its point of attachment to "b" (large arrowhead). As this photo was taken, cell b was moving visibly in the direction indicated by the small arrowhead. In subsequent frames, cell b moves relative to the common attachment point and becomes flask-shaped (**B–F**), while cell a transforms from the flask shape characteristic of cells in the region of negative curvature to the columnar shape characteristic of cells that have passed through such a region (**C–F**). Total elapsed time, A–F, ~4 minutes. (From Green et al., *The Journal of Cell Biology*, 1981, **91**, 756–69, by copyright permission of The Rockefeller University Press.)

than the bridges that move individually relative to the cells, was reinforced by SEM analysis of bridge locations and spacings on cells undergoing the spindle-to-flask transformation in the region of greatest curvature (Green et al. 1981).

Thus, the current model to explain inversion is that it is the cytoplasmic-bridge system formed during cleavage that provides the structural framework against which the cells exert force (via their changes in shape and their movements) to execute the inversion process (Figure 5.21). The observation that all cytoplasmic bridges are regularly associated with extensive cortical specializations and that they remain in coherent, regularly spaced arrays throughout both cleavage (Figure 5.10) and inversion (Figure 5.19) appears to support the idea that they constitute a coherent structural entity – a bridge system, rather than just an array of independent bridges – that has the ability

Figure 5.21 A model of the key step in the inversion process: generation of negative curvature. The cytoplasmic bridges linking each cell to its neighbors are aligned and linked into a single, coherent bridge system (heavy line running through the embryo). The bridge system not only holds the embryo together but also constitutes the only structural element against which the cells can exert force to effect inversion. Negative curvature is first generated when cells near the phialopore individually (but successively) transform from spindle to flask shapes and simultaneously move relative to the bridges that link them to their neighbors. (From Green et al., *The Journal of Cell Biology*, 1981, **91**, 756–69, by copyright permission of The Rockefeller University Press.)

to maintain its integrity despite the considerable forces that must be exerted on it during the inversion process (Green et al. 1981).

5.4.2.3 Cell shape changes and movements are necessary and sufficient for inversion. The only treatments reported to have reversible inhibitory effects on inversion are ones that interfere with cytoskeletal and/or plasma-membrane activities. They have led to the less than startling conclusion that the cell shape changes and movements that accompany inversion are mediated by the cytoskeleton.

Throughout inversion, cells are girdled by cortical microtubules that originate in the vicinity of the BA and extend to the opposite end of the cell; these are particularly closely packed in the stalks of flask cells, to the virtual exclusion of other cytoplasmic components (Kirk, Viamontes et al. 1982). If mid-inversion embryos are subjected to treatments (such as cold or nocadazole treatment) that cause MTs to break down, all cells will collapse to simple globose shapes and inversion will immediately cease. Once the inhibitory conditions are reversed, however, all the cells will quickly resume their region-specific, pretreatment shapes, and inversion will promptly pick up precisely where it left off and will be completed normally (Viamontes et al. 1979).

If, on the other hand, pre-inversion embryos are treated with cytochalasin D (CD), cells will elongate and eventually form the long basal projections typical of flask cells. In the process, the phialopore opens widely, but no movement of cells relative to the cytoplasmic bridges occurs, and the phialopore lips never curl outward to initiate inversion. Similarly, if mid-inversion embryos are treated with CD, all cellular movements and all inversion movements will be rapidly arrested; but if the drug is washed out, inversion will resume where it was interrupted and will be completed normally (Viamontes et al. 1979).

An extremely similar result was obtained when embryos of *V. tertius* or *V. carteri* were exposed to high concentrations (100 µg/mL) of Con A: Cells elongated and became spindle-shaped, but went no further with the inversion process until the Con A was inactivated with a-methyl mannoside, whereupon inversion was completed normally (Ireland & Hawkins 1980; Kurn 1981). An antibody to an unusual glycosphingolipid membrane component that is synthesized during cleavage also blocks inversion (Wenzl & Sumper 1986b), as does antibody to algal-CAM, another surface component synthesized during cleavage (Huber & Sumper 1994).[20] An extraordinary amount of surface membrane reorganization must be required to permit a cell to move rapidly past the cortical specializations of the bridge system that links it to its neighbors during inversion. We may speculate that multivalent reagents such as Con A, or antibodies that bind to cell-surface molecules, immobilize the cell membranes sufficiently to prevent such movements.

From such inhibitor studies it was concluded that cell shape changes (mediated largely by microtubular elongation) plus movement of cells relative to the cytoplasmic-bridge system (mediated largely by a cytochalasin-sensitive motility system) are essential for inversion. Moreover, mathematical and geometric analyses have led to the conclusion that these changes are sufficient to explain inversion, because at all stages of inversion the shape of the embryo can be accounted for by the shapes of its component cells and the locations of their bridges (Viamontes et al. 1979).

Two kinds of observations support the concept that free edges play an important role in the initiation of inversion: (1) Adventitious slits introduced

---

[20] Fluphenazine, a calmodulin inhibitor, also blocks inversion, but the significance of this is not clear, because the drug has many other drastic effects on *Volvox*, such as rapid abolition of flagellar motility, flagellar fragmentation, inhibition of contractile-vacuole activity, and cell lysis (Kurn 1981).

into an embryo either surgically or as a result of certain cleavage defects act as secondary inversion centers, at which inversion movements are initiated rapidly and proceed at about the same time and in about the same manner as they do in the vicinity of a true phialopore (Sessoms & Huskey 1973; Kelland 1977; Kirk, Viamontes et al. 1982). Thus, for example, if a slit is produced at the posterior pole, as a result of either surgery or mutation, the embryo inverts from both ends and forms a "doughnut" rather than the usual spheroid (Sessoms & Huskey 1973; Kelland 1977). (2) When cells are released mechanically from any part of a pre-inversion embryo, they initiate shape changes and inversion movements almost immediately, many minutes prematurely (Kelland 1977; Viamontes & Kirk 1977; Green et al. 1981). Such observations have been used to develop the working hypothesis that the onset of inversion is controlled by an inhibitor that can be lost by diffusion at any free edge (Kirk, Viamontes et al. 1982); tests of this hypothesis have not been reported, however.

5.4.3 Cytodifferentiation: converting the germ–soma dichotomy from potential to actual

At the end of embryogenesis, the embryo consists of two classes of cells that differ visibly in size, but little else. At this stage, both the prospective gonidia and the prospective somatic cells possess short flagellar stubs, and all cells remain linked by cytoplasmic bridges and are devoid of any visible ECM. In the alternating light–dark cycle used to synchronize development, cleavage is completed and inversion occurs in the dark, and little visible change occurs in either cell type as long as the cultures remain dark.

5.4.3.1 Cell-type-specific transcription begins in advance of overt differentiation. In situ hybridization studies of embryos with cell-type-specific cDNA probes have revealed the surprising fact that shortly after somatic initials have been set apart from prospective gonidia by asymmetric division, and well before the onset of inversion, they have already begun to accumulate transcripts of "early-somatic" genes in a coordinate manner, and to a significantly higher level than have the nearby presumptive gonidia (Tam et al. 1991). That finding was unanticipated, in light of the fact that (as mentioned earlier) these two cell types are linked by numerous cytoplasmic bridges at that time, and EM examination of these bridges has never revealed any obvious barrier to diffusion in the bridges (Figures 5.17 and 5.18). The specificity of accumulation of early transcripts is not symmetrical, however: At very early stages, presumptive somatic cells accumulate "gonidia-specific" transcripts at nearly the same low rate as do the young gonidia; it is only just after cytodifferentiation has begun that these transcripts disappear from somatic cells and earn the term ''gonidia-specific.'' (This may speak to the ability of presumptive somatic cells to differentiate as gonidia under some conditions, as discussed in Section 5.4.1.3.)

Of the early-somatic genes studied to date, the only two for which functions are known are expressed well before their protein products are to be used in visible differentiation of the juvenile spheroid. The transcript of the gene encoding SSG 185 (a 185-kDa sulfated glycoprotein of the ECM that forms the walls of the compartments that surround all somatic cells in the adult) (Ertl et al. 1989) is one of the early-somatic transcripts whose accumulation patterns were described earlier (Tam et al. 1991). Of particular interest is the expression pattern of the gene encoding ISG (''inversion-specific sulfated glycoprotein''), a sulfated, hydroxyproline-rich glycoprotein that was, as its name implies, first discovered as a protein produced at the time of inversion (Wenzl & Sumper 1982). When more precise studies revealed that the protein was made only during a 10-minute period in the last half of inversion, it naturally was assumed that ISG played some important role in inversion (Schlipfenbacher et al. 1986). However, when the corresponding gene was cloned and sequenced, and antibodies were prepared and used to localize the protein product, it turned out that ISG was an ECM component localized in the outer, ''boundary'' layer of the ECM (Ertl et al. 1992). In cultures synchronized by a light–dark cycle, however, inversion occurs in the dark, but assembly of visible ECM does not take place until after the lights come back on, many hours later.[21] In short, it appears clear that during cleavage and inversion, the embryo is ''planning ahead'' and preparing for the visible stages of cytodifferentiation that will occur next. We will return to the roles of SSG 185 and ISG in Section 5.4.4.

5.4.3.2 Light triggers the onset of differentiation at the translational and visible levels. Under a standardized illumination cycle (Figure 5.6) there is little discernible change in the patterns of polypeptide synthesis for the next several hours after inversion has been completed. But at the moment that the

---

[21] The fact that synthesis of the ISG protein, as well as the corresponding transcript, was observed to occur during inversion may have been due to the fact that in at least two experiments in which it was studied, inverting spheroids for analysis were selected under a dissection microscope (Schlipfenbacher et al. 1986; Ertl et al. 1992), and we have found that such a level of illumination is more than adequate to trigger the translational effects discussed in Section 5.4.3.2 (D. L. Kirk and M. M. Kirk, unpublished observations). Alternatively, it may be that ISG is an exception to the rule that most of the messages that accumulate in the dark during and after the late stages of embryogenesis are not translated until the lights come on again.

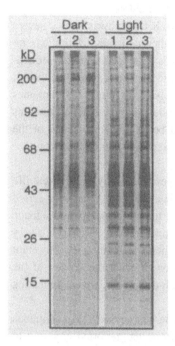 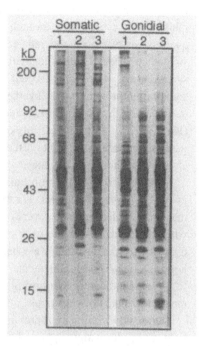

Figure 5.22 Light-induced changes in polypeptide labeling patterns. Juveniles that were near the end of the dark period in which they had completed cleavage and inversion were either kept in darkness (**Dark**) or illuminated (**Light**) while being exposed to ³⁵S sulfate for one hour. Labeling periods **1**, **2** and **3** began, respectively, one hour before, at, and one hour after the usual time of the dark-to-light transition; then the spheroids were harvested and soluble proteins were extracted for analysis by SDS-PAGE. Clearly, the resulting labeling patterns were much more strongly influenced by illumination conditions than by the age of the juveniles. (Adapted from Kirk & Kirk 1983, with permission.)

Figure 5.23 Changes in polypeptide labeling patterns in differentiating somatic cells and gonidia. Juvenile spheroids were exposed to ³⁵S sulfate for one hour at various times after the dark-to-light transition; then the two cell types were separated, and their soluble proteins were extracted for analysis by SDS-PAGE. Exposures to label occurred in the first (**1**), fifth (**2**), or ninth (**3**) hour after the dark-to-light transition. The labeling patterns of the two cell types diverged progressively with time. (Adapted from Kirk & Kirk 1983, with permission.)

lights come back on, such labeling patterns change dramatically (Figure 5.22) (Kirk & Kirk 1983). This transition has been shown to occur almost entirely at the translational level: Light triggers virtually instantaneous, massive, and reversible changes in the relative efficiencies with which various preexisting mRNA molecules are translated (Kirk & Kirk 1985). Additional important points are that (1) the action spectrum for triggering the change in protein synthesis is entirely different from that for photosynthesis (resembling the absorption spectrum of rhodopsin more than that of chlorophyll), (2) although inhibition of energy flow through the photosynthetic pathways has a severe quantitative effect on the rate of protein synthesis in the light, it has no significant effect on the qualitative differences seen between the protein-synthesis profiles of cells labeled in light and in darkness, and (3) although some of the proteins whose synthesis is markedly enhanced by light are ones destined for the chloroplast, others are cytosolic proteins, such as tubulins (Kirk & Kirk 1985). Thus, this effect cannot be explained in terms of the well-known effects of light on the energy metabolism of the chloroplasts or chloroplast photomorphogenesis.

Striking effects of light on translation can be seen at any stage of the life cycle. However, such effects are most dramatic at the end of the dark period that follows cleavage and inversion. Moreover, the major polypeptides made when the lights first come back on after the end of embryogenesis are very different from those that were being made when the lights went out eight hours earlier. This complements the conclusion drawn from studies at the RNA level indicating that during the dark period the cells are accumulating many new kinds of transcripts.

Although *V. carteri* cultures reproduce at least as rapidly in continuous light as in alternating light and darkness, no one has reported any way to synchronize development under such conditions. All of the different protocols for synchronizing *V. carteri* development that have been described in the literature involve alternating periods of light and dark (Yates & Kochert 1972; Kirk & Kirk 1976; Wenzl & Sumper 1979; Gilles, Bittner et al. 1981; Pommerville & Kochert 1981; Coggin & Kochert 1986). Only one dark period is required every 48 hours to synchronize development (Kirk & Kirk 1983; Coggin & Kochert 1986), and under such conditions (as with a more natural 24-hour light–dark cycle) cleavage and inversion are completed in the dark and cytodifferentiation begins early in the light period. Although systematic studies have not been reported, several early investigators indicated that wild populations of various species of *Volvox* appeared to be at least partially synchronized in their asexual development in a similar way (Zimmermann 1921; Shaw 1922c; Pocock 1933b; Metzner 1945a). Karsten (1918), having discovered this phenomenon in other species of algae, speculated that the significance of such a rhythm was that it permitted the algae to produce and store foodstuffs during the day that they then used for reproduction at night.[22]

---

[22] Interestingly, Desnitski (1992) has found that *V. tertius* and *V. aureus* exhibit developmental regularities with respect to the stage of the light cycle that are very different from the ones observed in *V. carteri* (see Table 2.2). In contrast to *V. carteri*, *V. tertius* and *V. aureus* initiate cleavage at the beginning of a light period and cease cleaving if the lights are turned

Our working hypothesis is that it is the massive changes in protein synthesis that are triggered in all *V. carteri* cells simultaneously at the dark-to-light transition that underlie the synchronizing effect of light. Embryos entering the dark period in less than perfect developmental synchrony may complete cleavage and inversion at slightly different times. If the lights were on at that time, each embryo should be able to initiate the next stages of development as soon as it completed inversion. But when a prolonged period of translational stasis is imposed, via darkness, it apparently provides tardy individuals an opportunity to catch up with their more advanced siblings in the accumulation of stage- and cell-type-specific transcripts. As a result, when the lights come on and translation of such transcripts is instantaneously initiated, all individuals commence cytodifferentiation in concert.

At the beginning of the light period, the young somatic cells and gonidia exhibit some differences in their polypeptide labeling patterns, but the similarities are more numerous than the differences. That soon changes. Within four hours, the two cell types will have diverged sufficiently that the differences in their protein labeling patterns, as observed by one-dimensional SDS-PAGE (which only reveals the most abundant classes of polypeptides being synthesized), will have begun to outnumber the similarities (Figure 5.23). Such divergence increases with time. In a more refined two-dimensional electrophoretic analysis (in which several hundred labeled spots were resolved), Baran estimated that nearly a quarter of all labeled polypeptides of developing gonidia and somatic cells (and a larger fraction of the most actively synthesized ones) were cell-type-specific (Baran 1984). One other particularly interesting observation made in that study was that many of the polypeptides that appeared to be synthesized only in gonidia were nevertheless abundant in somatic cells; he postulated that they were maternal products inherited by presumptive somatic cells during cleavage (Baran 1984). As we will see in Chapter 7, certain recent studies have brought that observation by Baran back into focus and have suggested that such "maternally inherited" proteins, and the genes encoding them, may play central roles in germ–soma specification.

Two visible changes begin more or less simultaneously during the first one to two hours after the lights come back on: ECM begins to accumulate within the juvenile spheroids (as discussed in more detail later), and the cytoplasmic bridges begin to break down. From that time forward, the ECM

---

out. Moreover, in both of these species cleavage proceeds much more slowly than in *V. carteri* and can be inhibited by aminopterin. He therefore postulates that the difference in the light dependence of cleavage in these three species is a result of the fact (which was discussed earlier) that *V. carteri* builds up a large pool of DNA precursors before cleavage begins, but the other two species do not, and so they require continuous photosynthetic activity to make the DNA required for cleavage (Desnitski 1992).

takes over a responsibility formerly exercised by the bridges: to hold all cells of the spheroid in fixed orientations with respect to one another. Probably the loss of cytoplasmic continuity between the two cell types that accompanies breakdown of the cytoplasmic bridges plays an important role in accelerating the divergence between the biosynthetic activities of the two cell types.

Although presumptive somatic and gonidial cells begin to differ from one another in their patterns of transcript abundance at the beginning of the light period, such differences are rapidly magnified in the first few hours after cytoplasmic bridges have broken down (Tam & Kirk 1991a). These differences in transcript abundance, which will be discussed in much more detail in Chapter 7, undoubtedly are related to the progressively changing patterns of protein synthesis referred to earlier.

5.4.3.3 The two flagella of each somatic cell have very different growth patterns. Another visible form of cytodifferentiation that begins very early in the light period is flagellar elongation, but it is well established that the two flagella of a young somatic cell do not elongate synchronously (Kochert 1968; Starr 1970a; Coggin & Kochert 1986; Nozaki 1994). From the end of inversion through the rest of the dark period, all cells (presumptive gonidia as well as presumptive somatic cells) of a *V. carteri* juvenile have two short (~2 μm) flagella. Within an hour after the lights come on, both flagella of each somatic cell will begin to elongate. However, the *cis* flagellum (the one that will be connected to the eyespot via its 4MTR) (Hoops 1993) will always elongate much faster than the *trans* flagellum: At the end of three hours, the *cis* flagellum will have grown to about 12 μm, whereas the *trans* flagellum will have reached only 4 μm in length (Coggin & Kochert 1986). For the next eight hours, no further elongation of either flagellum will occur, and so every cell will have one long and one short flagellum throughout that period. But then the *trans* flagellum will suddenly elongate very rapidly until it reaches the length of the *cis* flagellum, whereupon both flagella will elongate more slowly for many hours, finally reaching a length of about 20 μm by the time of hatching (Coggin & Kochert 1986). If these newly hatched spheroids are then deflagellated, their flagella are quickly regenerated, elongating at 10 times the rate that they elongated during development. In virtually all aspects that have been examined, flagellar regeneration in adult *V. carteri* is extremely similar to that in its much more extensively studied relative, *C. reinhardtii* (Coggin & Kochert 1986).

Differential elongation of the two flagella during development has been observed in three other species of *Volvox* and in many other volvocalean genera and species (Goldstein 1964; Harris & Starr 1969; Rayburn & Starr

1974; Nozaki 1983, 1984, 1989; Nozaki & Kuroiwa 1991).[23] Neither the functional significance nor the molecular basis of this profound difference in the elongation kinetics of the two flagella is understood. However, the existence of the phenomenon underscores an important point made in Chapter 4, namely, that in all green flagellates thus far studied the younger and older BBs of a cell (and/or the flagella that grow from these BBs) are distinguishably different in their physiological properties.

While the flagella of somatic cells elongate, the flagella of gonidia are resorbed. It has been reported by Nozaki that the gonidial flagella, before shortening, briefly undergo asymmetric elongation similar to that of somatic cells (Nozaki 1994), but we have not observed this under our culture conditions. In any case, the gonidial flagella are resorbed early in the period of cytodifferentiation, but not before formation of a pair of flagellar tunnels in the ECM above each of them. As its flagella are resorbed, each gonidium assumes the spherical shape that will characterize it until it initiates a new round of embryogenesis. Then, while the somatic cells are developing such specializations as the functional flagella and eyespots required for motility and phototaxis, the major visible change in the gonidia is growth, which is considerably greater than that of the somatic cells (Figure 3.7).

5.4.4 ECM deposition: architectural dexterity by remote control

Once cytodifferentiation begins, it is the ECM that welds the cells of the *Volvox* spheroid into a coordinated multicellular organism. The ECM accounts for more than 99% of the volume of a mature *V. carteri* spheroid, and hence elaboration of the ECM constitutes one of the major activities of the developing organism. As the spheroid enlarges, all the structural components that will be described next will enlarge proportionately and maintain their distinctive symmetries, even though some of them will be located at considerable distances from the cell bodies. Understanding how this feat is accomplished must be considered a significant intellectual challenge. The only EM study that was performed with the specific objective of studying secretory processes in *V. carteri* did identify fibrils that were presumed to be ECM precursors within Golgi cisternae and secretory vesicles located between the Golgi and the cell surface, as had others previously, in other *Volvox* cells

---

[23] Not all green flagellates exhibit such differential elongation kinetics; differences in this regard are considered by Nozaki to provide important clues to phylogenetic relationships among the group (Nozaki 1994; Nozaki & Itoh 1994).

(Bisalputra & Stein 1966; Deason & Darden 1971; Birchem 1977) (Figure 5.2A), but that study made no attempt to track such components systematically within the cell, let alone beyond the cell surface (Dauwalder et al. 1980). Further morphological studies using more powerful identification methods (such as immunogold labeling) appear to be warranted.

5.4.4.1   The V. carteri *ECM is complex and regular in its structure.* EM examination has revealed that the mature ECM is exceptionally complex, containing many region-specific morphological components (Figures 5.24 and 5.25), most of which also vary in appearance in a highly species-specific manner (Kirk, Birchem & King 1986). Therefore, we have proposed that for descriptive and comparative purposes it may be useful to subdivide the ECM into four zones. Proceeding from exterior to interior, these are the flagellar, boundary, cellular, and deep zones (Figure 5.24) (Kirk, Birchem & King 1986), which will now be described and illustrated with respect to *V. carteri* f. *nagariensis.*

The flagellar zone (FZ) includes various fibrous appendages coating the portion of the flagellar membrane external to the spheroid (FZ1), as well as the "flagellar hillock" and "flagellar tunnel" (FZ3a and FZ3b, respectively), which are specializations that surround each flagellum where it emerges from the spheroid, and the "flagellar collar" (FZ2), a set of elements that physically connect the base of the flagellum to the flagellar tunnel.

The boundary zone (BZ) is a region of the ECM that is continuous over the surface of the spheroid except in regions where it is interrupted by flagella. The BZ contains not only the "tripartite" or "crystalline" layer (BZ2) that is found in all volvocine algae, including *C. reinhardtii* (Lang 1963; Bisalputra & Stein 1966; Deason et al. 1969; Kochert & Olson 1970a; Pickett-Heaps 1970, 1975; Burr & McCracken 1973; Roberts 1974; McCracken & Barcellona 1976, 1981; Birchem 1977; Gottlieb & Goldstein 1977; Fulton 1978a,b; Birchem & Kochert 1979a; Dauwalder et al. 1980; Domozych, Stewart & Mattox 1981; Kirk, Birchem & King 1986; Adair et al. 1987; Goodenough & Heuser 1988) but also dense, fibrous layers coating BZ2 on its outer and inner surfaces (BZ1 and BZ3, respectively).

The cellular zone (CZ) is the region of the ECM internal to the BZ that exhibits specializations around individual cells. The plasmalemma of each somatic cell and gonidium is coated with a coherent meshwork of filaments (CZ1). In the case of the gonidium and the embryo or juvenile derived from it, the CZ1 layer is commonly referred to as the "vesicle" (see Section 5.4.1 and Figure 5.8). At a greater distance from each cell is a second coherent fibrous layer (CZ3) that forms a complete cubicle, or "cellular compart-

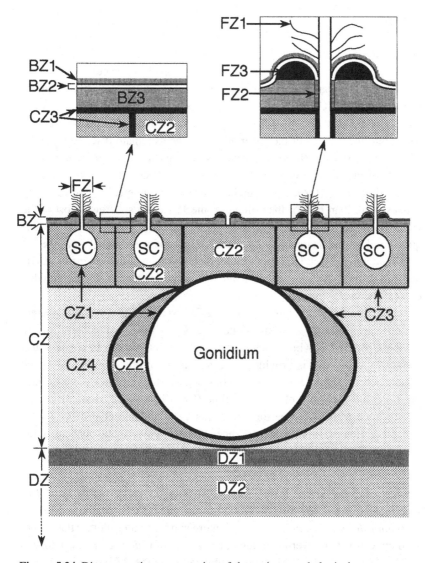

Figure 5.24 Diagrammatic representation of the major morphological components of the *V. carteri* ECM (cf. Figures 5.5 and 5.25). The nomenclature used here is that proposed by Kirk, Birchem & Kirk (1986), in which four major regions of the ECM are distinguished: the flagellar zone (FZ), the boundary zone (BZ), the cellular zone (CZ), and the deep zone (DZ). Within each of these zones, regions of distinctly different locations and/or appearances are then distinguished by numerical suffixes. For example, in the cellular zone, CZ1 identifies the fibrous layer closest to the plasma membrane of each cell, CZ3 identifies the fibrous component that surrounds each cell at a greater distance to form a "cellular compartment," CZ2 identifies the "mucilaginous" material that fills the space between CZ1 and CZ3, and CZ4 identifies the relatively amorphous layer that is internal to the somatic-cell layer (but external to the deep zone) and within which the gonidia are suspended as they undergo enlargement in preparation for division. For more details, and for the comparative anatomy of the ECM in different species of *Volvox* (which is an important consideration in *Volvox* taxonomy), see Kirk, Birchem & King (1986). (From Kirk, Birchem & King 1986.)

ment," around each cell.[24] The side walls of adjacent cubicles are fused to form a honeycomb-like arrangement within which all of the cells are located, and in each compartment of this honeycomb a relatively amorphous, mucilaginous type of ECM (CZ2) fills the space between CZ1 and CZ3.[25] As each gonidium grows, the floor of the compartment surrounding it balloons extensively into the interior of the spheroid; yet its walls remain fused to the walls of adjacent somatic-cell compartments, so that each gonidium is held in the vicinity of the somatic cells to which it is most closely related by descent.

The deep zone (DZ), which includes everything internal to the CZ, constitutes more than half of the volume of the *V. carteri* spheroid. Most of the DZ is composed of a very loose, amorphous, mucilaginous component (DZ2) quite similar in appearance to that in CZ2, but this is separated from the CZ by a moderately thick band of circumferentially aligned fine filaments (DZ1) that run around the spheroid just internal to the inner ends of the gonidial compartments.

Although the fibrous nature of many ECM components is readily apparent in EM views of thin sections (Figure 5.25C), a fuller appreciation of the complex patterns in which such components are interwoven to create a lush extracellular tapestry is provided by examination of quick-freeze, deep-etch specimens (Figures 4.7 and 5.8). A comparative study of the structure, composition, and assembly of the crystalline (BZ2) layers of *C. reinhardtii* and *V. carteri* ECM has been performed by this method (Adair et al. 1987; Goodenough & Heuser 1988; Adair & Appel 1989), but systematic analysis of the rest of the *Volvox* ECM with this method has not yet been reported.

5.4.4.2   The crystalline layer of the ECM is highly conserved in the volvocine lineage. As noted earlier (Section 5.3.4), the BZ2 (crystalline) layers of *C. reinhardtii* and *V. carteri* are much more similar to one another than either is to the equivalent layer in certain other species of *Chlamydomonas* (Roberts 1974; Adair et al. 1987). Proteins solubilized from the BZ2 layer of *C. reinhardtii* or *V. carteri* by extraction with sodium perchlorate will reassemble in vitro on the perchlorate-stripped walls of either species to reconstruct, in detail, the crystalline structure normally seen in vivo. Subsequently, that kind

---

[24] Although the cellular compartments are complete in *V. carteri* and all other members of the section Merrillosphaera, in other species of *Volvox* they may lack floors and complete walls (Kirk, Birchem & King 1986); such differences are among the features used to subdivide the genus *Volvox* into four sections (see Table 2.1).

[25] This mucilaginous substance probably consists of a mixture of polysaccharides (at least some of them sulfated) and proteoglycans (Tautvydis 1978; Crayton 1980), but studies of the mucilaginous components of *V. carteri* and other volvocine algae lag behind studies of the fibrous glycoproteins.

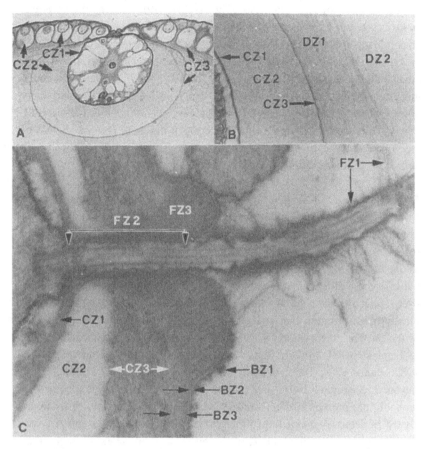

Figure 5.25 Micrographs illustrating some important ECM features diagrammed in Figure 5.24. **A**: Light micrograph illustrating the organization of cellular-zone components in the vicinity of somatic cells and a gonidium. **B**: Light micrograph illustrating the relationship of the deep zone to the cellular zone (the object at the lower left is the edge of a juvenile that has just completed embryogenesis). **C**: TEM illustrating the relationships among flagellar, boundary, and cellular zone components in the region surrounding the base of a flagellum. (Adapted from Kirk, Birchem & King 1986.)

of relationship was found to extend to *Gonium pectorale* (Adair & Snell 1990), but it could not be extended to *C. eugametos*. These relationships were reinforced by comparative analysis of "GP2," the perchlorate-soluble hydroxyproline-rich glycoprotein (HRGP) that is the major component of the crystalline lattice. *C. reinhardtii*, *V. carteri* , and *G. pectorale* all possess GP2s that have similar molecular weights, peptide maps, immunological reactivities, and EM morphologies (Roberts et al. 1985; Matsuda et al. 1987;

Goodenough & Heuser 1988; Adair & Appel 1989), whereas the *C. euga-metos* crystalline layer has a different set of perchlorate-soluble HRGPs that does not include a protein sharing any of these properties with GP2 (Adair & Appel 1989). Thus, all such studies reinforce the conclusion drawn from comparative analysis of rRNA sequences, namely, that *C. reinhardtii* is much more closely related to the Volvocaceae than it is to *C. eugametos* and many other species of *Chlamydomonas* (Buchheim et al. 1990, 1996; Larson et al. 1992). Moreover, they support the concept that the more complex ECMs of the volvocaceans evolved by elaboration of a relatively simple cell wall such as that of *C. reinhardtii*.

5.4.4.3 Analysis of deeper layers of the ECM is just beginning. The cell wall of vegetative *C. reinhardtii* cells contains about 30 different types of glycoproteins, many of which are exceptionally rich in hydroxyproline, but others of which are glycine-rich (Adair & Snell 1990). It would be quite astonishing if the morphologically more complex *Volvox* ECM did not turn out to contain at least that many kinds of glycoproteins. Unfortunately, how-ever, molecular analysis of other parts of the ECM has proved to be tech-nically more difficult than analysis of the crystalline layer. Several aspects of the protein glycosylation systems of *Volvox* have been elucidated (Bause & Jaenicke 1979; Müller, Bause & Jaenicke 1981, 1984; Kurn & Duskin 1982; Bause et al. 1983; Günther, Bause & Jaenicke 1987; Jaenicke, van Leyen & Siegmund 1991), but only a few of the many glycoproteins that probably are present in the *Volvox* ECM have been identified, let alone an-alyzed in detail. A major impediment to such studies is the extensive cross-linking that exists in the matrix, rendering most components insoluble in all conventional protein-extraction media (Kirk & Kirk 1983). Isolation of most ECM components will require methods, such as proteolytic digestion, alka-line hydrolysis, or hydrofluoric-acid deglycosylation, that can break various covalent crosslinks (Mitchell 1980; Wenzl et al. 1984). In some cases, ECM components are not even accessible to proteases except in the presence of strong denaturants (Ertl et al. 1989). Despite these difficulties, recent studies have begun to provide important insights into the composition of key portions of the *V. carteri* ECM internal to the crystalline layer.

Such studies began with a rather different purpose in mind: to examine the roles of membrane glycoproteins in *V. carteri* embryogenesis. In those studies it was observed that a variety of *V. carteri* glycoproteins were exten-sively sulfated and that many such sulfated glycoproteins (SGs), or sulfated surface glycoproteins (SSGs), appeared to be synthesized in highly stage-specific patterns during embryogenesis (Sumper & Wenzl 1980; Wenzl & Sumper 1981, 1982, 1986a; Willadsen & Sumper 1982; Wenzl et al. 1984).

For example, it was observed that SSG 140[26] was produced maximally when asexual embryos were dividing, whereas ISG was synthesized only during inversion, and female surface glycoprotein (FSG) and male surface glycoprotein (MSG) were synthesized only when sexual female and male embryos, respectively, were dividing (Wenzl & Sumper 1981, 1982).

The sulfated glycoprotein that attracted the most attention was SSG 185. The data indicated that this molecule was produced maximally when embryos (either asexual or sexual) were dividing, and that changes in both its rate of synthesis and its molecular weight occurred when the embryos were dividing asymmetrically (Wenzl & Sumper 1981, 1982). Based on these intriguing correlations, it was postulated that SSG 185 was a membrane glycoprotein of the cleaving embryo and that it was intimately involved in germ-cell specification, via a mechanism that had been proposed earlier (Sumper 1979, 1984; Wenzl & Sumper 1979, 1981, 1982; Sumper & Wenzl 1980; Willadsen & Sumper 1982). However, such a role became unlikely when it was discovered that SSG 185 is a product of the parental somatic cells, not the cleaving embryos that they surround (Kirk & Kirk 1983). This observation was confirmed, and it was established that SSG 185 is actually a highly sulfated, hydroxyproline-rich matrix glycoprotein synthesized predominantly, if not exclusively, by somatic cells (Wenzl et al. 1984). Transcripts of the gene encoding the core polypeptide of SSG 185 begin to accumulate in presumptive somatic cells shortly after these cells have been set apart by asymmetric division, and they remain abundant in these cells throughout most of the asexual life cycle; but they are not present at detectable levels in gonidia at any stage (Tam & Kirk 1991a; Tam et al. 1991).

Although SSG 185 is cross-linked into the ECM within minutes of its synthesis and is thereby rendered wholly insoluble,[27] detailed analysis of the molecule was made possible by the observation that specific fragments of it can be released by two different types of treatments of exhaustively extracted, insoluble matrices. A 145-kDa, protease-resistant glycopeptide fragment is released when such matrices are digested with subtilisin, and a 135-kDa fragment is released when either the intact ECM or the 145-kDa fragment is subjected to alkaline hydrolysis or β-elimination conditions (sodium borate in very dilute alkali) (Wenzl et al. 1984).

---

[26] When a number is included in the name of such a glycoprotein, it indicates the apparent molecular mass, as estimated from mobility in SDS-PAGE.

[27] The transient nature of the soluble form of SSG 185 has confounded attempts to quantitate its rate of synthesis by conventional methods. Small changes in the rate of insolubilization could have profound effects on the amount of labeled SSG 185 found in the soluble form after short labeling periods. Such effects may have complicated the early studies, which reported that its rate of synthesis was closely correlated with the cleavage state of the gonidia (Wenzl & Sumper 1981, 1982).

Monoclonal antibodies raised to the purified 145-kDa fragment were shown to be specific for SSG 185, and when used for immunolocalization they revealed the important fact that the insoluble form of SSG 185 is present in CZ3, the fibrous boundaries of the cellular compartments (Ertl et al. 1989). When juveniles were cultured in the presence of antibody, spheroid expansion was severely inhibited, and cellular organization within the resulting spheroids was highly aberrant. (Nevertheless, cellular differentiation was not inhibited, as evidenced by the ability of the gonidia to initiate normal embryogenesis.) The severe inhibition of overall matrix accumulation caused by the antibody may indicate that SSG 185 is required not only for assembly of the cellular-compartment boundaries but also for promoting accumulation of other portions of the ECM.

The antibody was also used to screen an expression library and to recover cDNA clones encoding the SSG 185 polypeptide backbone that were then sequenced (Ertl et al. 1989). The most unusual aspect of the deduced amino acid sequence of the polypeptide is a central domain in which 65 of 80 (and, in one region, 37 of 40) residues are prolines! This central domain is flanked by amino- and carboxy-terminal domains whose amino acid compositions are quite unexceptional.

The structure of mature SSG 185 is complex (Ertl et al. 1989). In the 145-kDa peptide isolated by protease digestion of the ECM, most or all of the proline residues of the central domain have been converted to hydroxyprolines that are O-glycosylated with di- and tri-arabinosides (and in a few cases with gulose) (Mengele & Sumper 1992), transforming the approximately 50-kDa polypeptide into an approximately 90-kDa glycoprotein. But a second type of glycoconjugate is also present: a 28-kDa sulfated polysaccharide that consists of di-arabinoside side chains projecting from a 40-residue-long polymannose backbone; both the mannose and arabinose residues are heavily sulfated and contain all of the sulfate residues (>150) of the SSG 185 molecule. EM examination indicates that the di-arabinosylated, hydroxyproline-rich central domain has the conformation of a thick, rigid rod, from the center of which the 28-kDa sulfated-polysaccharide chain projects as the leg of a T (Ertl et al. 1989).

These probably are not the only conjugates present in SSG 185, however. It is important to note two facts: (1) Subtilisin digestion releases the same 145-kDa fragment whether the substrate is the newly synthesized, still-soluble form of SSG 185 or the insoluble matrix-bound form. (2) Alkaline digestion or β-elimination conditions release the same 135-kDa fragment whether the substrate is the newly synthesized, soluble form of SSG 185 or the 145-kDa fragment produced by subtilisin digestion or the undigested ECM (Wenzl et al. 1984). These facts suggest that the subtilisin-sensitive component of SSG

185 is not the core polypeptide of SSG 185 itself, but a second ("linking") peptide that is attached to it by an alkali-sensitive glycosidic bond. Both alkaline hydrolysis and the β-elimination reaction typically release carbohydrate adducts that had been linked to peptides at threonine or serine residues. Neither of the SSG 185 glycoconjugates discussed earlier is of this type. Moreover, further analysis has indicated that both the 28-kDa sulfated side chain and the hydroxyproline arabinosides remain with the 135-kDa fragment after alkaline hydrolysis (Wenzl et al. 1984). Thus, one may postulate the existence of a third kind of carbohydrate adduct in SSG 185 that is interposed between the core peptide and the linking peptide and is attached to the former by an alkali-sensitive bond. Together, this additional carbohydrate and the linking peptide would appear to have an apparent molecular mass of at least 50 kDa (the change in apparent mass of SSG 185 following alkaline hydrolysis), of which 40 kDa is released by subtilisin digestion. Given that the newly synthesized, soluble form of SSG 185 detected on SDS-PAGE gels has the same protease and alkali sensitivities as the mature product (Wenzl et al. 1984), it seems that it must contain all of these components at the time it is released from the cell, and thus all that remains is for it to be cross-linked into the compartment wall via its linking glycopeptide.

What might be the nature of the cross-linkage that converts SSG 185 to an insoluble form in the CZ3 region of the ECM? The only candidate discovered so far is a phosphodiester bridge between two arabinose residues that has been shown to be present in SSG 185 (Holst et al. 1989). Noting that such bridges have been implicated in the cross-linking of polysaccharides to peptidoglycans in the walls of Gram-positive bacteria, those authors proposed that they may function similarly in *V. carteri*. However, because at least 10 such bridges are present and intact in the 135-kDa fragment of SSG 185, it is not clear to what extent they play a role in intermolecular, as opposed to intramolecular, cross-bridging. Further study will be required to clarify this.

A second sulfated HRGP once thought to be a cell-surface glycoprotein has now been characterized and shown to be another important structural component of the ECM. As its name implies, ISG (inversion-specific sulfated glycoprotein) was detected as a protein (apparent molecular mass ~200 kDa) synthesized only during inversion, in both asexual and sexual spheroids (Wenzl & Sumper 1982). As noted earlier, when it was shown that (in distinction to SSG 185) ISG is truly a product of the embryos themselves, and is synthesized at perceptible rates for only a few minutes during the last half of inversion, it was naturally assumed that ISG might play some important role in the inversion process (Schlipfenbacher et al. 1986). However, when an antibody to purified ISG was subsequently used for immunolocalization, the protein was found in the boundary zone of the ECM, with particularly strong staining being seen in the flagellar-tunnel regions (Ertl et al. 1992). Moreover, when a synthetic decapeptide corresponding to the carboxy terminus of ISG was added to cultures of pre-inversion embryos, it had no effect on inversion, but it caused drastic abnormalities in ECM deposition: Juveniles developing in the presence of modest amounts of this peptide developed as cell suspensions rather than as coherent spheroids (Ertl et al. 1992). This strongly suggests that ISG is not just another component of the ECM, but a component that may play some crucial role in establishing normal organization of the rest of the ECM. We will return to that theme shortly.

The amino acid sequence of ISG (deduced from genomic and cDNA nucleotide sequences, and partially confirmed by peptide sequences) consists of two distinctly different domains: an amino-terminal half of relatively conventional amino acid sequence[28] and a carboxyl half that is extremely rich in proline (>50%) and serine plus threonine (~20%) and bears a strong resemblance to extensins – the major HRGPs of higher-plant cell walls (Ertl et al. 1992). Peptide sequencing has revealed that virtually all of the prolines in the amino half of the molecule are hydroxylated. Whereas the characteristic peptide motif in extensins is $SerHyp_4$, in ISG there are 21 $SerHyp_x$ motifs, in which x ranges from 3 to 7, in addition to several other variations on a $Ser_n Hyp_m$ theme. Virtually all of the hydroxy amino acids in this region appear to be glycosylated, with the dominant sugars being arabinose and galactose (plus lesser amounts of xylose). The one potential N-glycosylation site in the molecule (located in the carboxyl half) is also utilized, and the native glycoprotein is approximately 70% carbohydrate. Thus, deglycosylation reduces the apparent molecular mass of ISG from about 200 kDa to about 60 kDa. EM examination reveals that the ISG molecule consists of a globular domain attached to a rigid rod with dimensions consistent with those predicted from sequence data, on the assumption that the glycosylation holds the carboxyl half of the polypeptide in a rigid polyproline II helix (Ertl et al. 1992). As is the case with other volvocalean HRGPs that are composed of rod-like and globular domains (Goodenough & Heuser 1988), ISG molecules

---

[28] It is a matter of some interest that the amino half of ISG bears a strong resemblance (52% similarity over 228 amino acid residues) to the equivalent region of a putative HRGP of the *C. reinhardtii* cell wall, whereas the hydroxyproline-rich carboxy halves of the two molecules exhibit no significant sequence homologies (Woessner et al. 1994). In contrast, the carboxy halves of the proteins encoded by this *C. reinhardtii* gene and a *C. eugametos* gene are 72% similar in amino acid sequence; but these two proteins exhibit no similarities in their amino halves (Woessner et al. 1994). The occurrence of such relationships among the few volvocine ECM genes that have been sequenced to date has led to the hypothesis that diversification of such genes has been accelerated by shuffling of exons encoding globular (hydroxyproline-poor) and rod-like (hydroxyproline-rich) domains (Woessner & Goodenough 1994; Woessner et al. 1994).

exhibit a propensity to associate via their globular domains to form multimers (Ertl et al. 1992).

5.4.4.4   Initiation of ECM assembly: a possible scenario. As mentioned earlier, when a synthetic decapeptide corresponding to the carboxy terminus of ISG was added to cultures of pre-inversion embryos, it caused them to develop as cell suspensions rather than coherent spheroids (Ertl et al. 1992). However, when this same peptide was added to the culture medium after ECM assembly had been under way for a few hours, no effects were seen. This suggests that ISG may have a critical role to play in the very early stages of ECM assembly and that once this step has been executed the assembly process becomes much less sensitive to perturbations. This proposed early role is consistent, of course, with the observation that ISG is apparently the first ECM molecule to be synthesized: The RNA encoding ISG is both transcribed and translated in abundance during the last half of inversion, before any other known ECM component.

As noted in Section 5.4.3.2, one of the first events to occur in the postembryonic stage is the breakdown of cytoplasmic bridges. If the spheroid developing from the embryo is to be able to swim directionally and find its place in the sunlight as an adult, it is essential that before the bridges are broken there must already be in place an alternative method of maintaining the precise orientations of all cells that were so carefully established during cleavage (Section 5.4.1). The validity of this statement has been clearly documented by study of mutants: *V. carteri* adults whose cells lack the normal rotational symmetry about the anterior–posterior axis of the spheroid cannot swim directionally, even though each of them has a normal intracellular organization and flagellar beat pattern (Huskey 1979).

Of the three types of mutants examined in the study just cited, the most instructive for present considerations was a temperature-sensitive flagellaless mutant (*flgC11*), because it revealed the important role played by flagella in maintaining the adaptive orientation of cells within the spheroid. When *flgC11* was allowed to develop at the permissive temperature, cells of the adult were oriented as in the wild type, and the mutant spheroids were able to swim directionally and execute normal phototaxis, no matter how many times their flagella were caused to resorb and regenerate by shifting them between permissive and restrictive temperatures. However, when the same strain was held at the restrictive temperature during inversion and early postembryonic development, and then shifted to the permissive temperature, spheroids developed flagella that appeared to beat normally, but the spheroids were incapable of any coordinated swimming behavior: They just oscillated in place rather than swimming forward. Closer examination showed that in such spheroids each cell had normal internal organization in terms of the relationship between flagella and eyespots, but the cells were oriented randomly with respect to the anterior–posterior axis of the spheroid (Huskey 1979). As a result, flagellar beat directions were also oriented randomly, and thus flagella of different cells were working antagonistically rather than in a coordinated manner. Not surprisingly, this abnormality could not be relieved by using temperature shifts to cause the flagella to resorb and then regenerate. Observation of nine nonconditional flagellaless mutants revealed that every one of them also had cells with normal internal organization, but with random orientation of the cells about the anterior–posterior axis of the spheroid (Huskey 1979). From this we can conclude that flagella must be present at the time that the ECM is first laid down in order for the cells to be locked into the correct orientation within the spheroid as a whole.

Based on Huskey's observations, a reasonable working hypothesis is as follows: A very important early stage in ECM construction, prior to cytoplasmic-bridge breakdown, involves deposition of boundary-zone matrix components around the bases of the flagella, and also out over the rest of the surface of the spheroid. If properly executed, this has the effect of permanently trapping all flagella (and hence the cells to which they are attached) in a fixed orientation within the surface layer, thereby preserving the cellular orientations that had been established by cleavage and maintained thus far by the cytoplasmic bridges.

What might be the ECM component that executes this essential function? Ertl et al. (1992) interpreted their immunofluorescence images as indicating that ISG is part of the tripartite, crystalline layer, BZ2. However, the particularly intense staining that they observed in the flagellar-tunnel region, relative to the rest of the surface, seems to be more consistent with a BZ3 localization for ISG. BZ3 is much thicker in the flagellar-tunnel region (where it constitutes the swelling around each flagellar base that is known as the flagellar "hillock") than it is in the rest of the surface; but the same is not true of BZ2 (Figures 5.24 and 5.25C). I postulate, therefore, that it is ISG, the first ECM component to be synthesized, that serves the function proposed in the preceding paragraph: to form an initial surface layer of ECM that includes the flagellar hillocks that will hold the flagellar bases (and hence the cells) in place while the bridges break down and while additional ECM components are accumulated.

If ISG is indeed located in BZ3, a second very important role for it in ECM organization can be postulated. The BZ3 layer of *V. carteri* occupies a location analogous to that occupied by the so-called W2 layer in the *Chlamydomonas* cell wall (Goodenough & Heuser 1985). Studies of *Chlamydo-*

monas mutants lacking parts of the inner cell wall suggest that the W2 layer nucleates assembly of the crystalline layer (Adair et al. 1987). It is quite likely that BZ3 (hence ISG?) plays a similar role in *V. carteri*.

Finally, it seems quite probable that the boundary zone (BZ3 plus the BZ2 layer it is postulated to organize) plays another important role in ECM organization: By forming a surface coat over the spheroid, the boundary zone probably restricts the diffusion of other ECM components that are secreted by the cells later on, preventing them from diffusing out into the medium before they can undergo assembly into the various internal components of the matrix. The combination of these three proposed roles for ISG, if shown to be valid, would go a long way toward explaining the observation that the peptide analogue of ISG has such a drastic effect on the organization of all parts of the ECM. By interfering with proper self-assembly of ISG into a coherent structure, it may perturb the critical first step in ECM construction that is an essential preliminary to all of the other steps that normally follow.

Many aspects of this proposed scenario are subject to test by systematic studies at the EM level. Foremost among such studies, of course, might be EM-level immunogold localization of ISG. It is hoped that such studies will be forthcoming.

5.4.5 Hatching: breaking up the old family homestead

About halfway through the period of expansion, *V. carteri* juveniles normally hatch out of their parental spheroids and become free-swimming young adults (Figure 5.7).[29] When an immobilized parental spheroid is observed during the birth process, the following sequence of events is seen (S.E. McRae and D. L. Kirk unpublished time-lapse cinemicrographic observations): The first sign that hatching is imminent comes when the parental somatic cells located closest to the center of an underlying juvenile spheroid begin to rock within their cellular compartments. These oscillations, which are feeble at first, become more vigorous until, one after another, the cells wrench free of the spheroid and tumble away. This process spreads laterally until all of the cells lying above the juvenile have tumbled off – but it spreads no farther. At about that time, the juvenile, which has been rotating restlessly

---

[29] Hatching frequently occurs much less synchronously within a culture than does embryogenesis, and we have repeatedly observed cultures in which hatching was delayed from its usual time by more than a day (for unknown reasons), but in which the gonidia nevertheless cleaved at the usual time, and with the usual synchrony, despite the fact that their parents were still located within the "grandparental" spheroids. Similar observations regarding the somewhat whimsical timing of hatching have been reported by others (Starr 1970a), but they have never really been explained.

beneath the somatic cell layer, typically becomes frozen in place momentarily (presumably as it gets caught in an opening whose diameter is slightly smaller than its own); then after rolling outward ever so slowly, it suddenly bursts free of its invisible restraints and spins rapidly out of view. Within an hour or two after the process begins, the parent has become a mere "ghost": a spheroid with gaping holes equal in number and diameter to the juveniles that were within it earlier, and with the rest of its surface dotted with somatic cells that are already beginning to fade as their chlorophyll content diminishes.

When a ghost from which all juveniles have recently escaped is fixed, stained, and examined microscopically, it is seen that the enzymatic process that formed the birth canals for the young has been highly selective: the "roof" of each ECM compartment that once held a juvenile has been digested away, but its "floor" and "walls" appear unscathed. Similarly, all of the ECM that once held the somatic cells positioned above a juvenile has vanished, but the ECM surrounding the somatic cells flanking the hole remains largely intact. When such observations are repeated at intervals, it is seen that the ghosts swim about for many hours with no perceptible change in the sizes of the holes that perforate their surfaces (N. King and D. L. Kirk unpublished observations).

The hatching of *Volvox* juveniles is analogous to the more extensively studied process by which young *Chlamydomonas* cells recently formed by multiple fission within the mother-cell wall escape from that wall. For some time it has been known that the latter process is mediated by secreted enzymes known as "autolysins" (Schlösser 1976, 1984). More recently these enzymes have come to be called vegetative lysins or V-lysins, to distinguish them from the very different gamete lysins or G-lysins that gametes use to shed their walls in preparation for sexual fusion (Jaenicke & Waffenschmidt 1981; Jaenicke, Kuhne et al. 1987; Matsuda et al. 1987; Buchanan & Snell 1988). The *Chlamydomonas* V-lysins exhibit both stage and species specificity with respect to their substrates: They attack only the wall derived from a mother cell that has initiated cell division, and they do so with sufficient species specificity that cross-sensitivity to V-lysins can be used as a taxonomic criterion to subdivide the genus into species groups that reflect relationships deduced from other criteria (Schlösser 1976, 1984; Ettl & Schlösser 1992). The G-lysins, in contrast, will degrade the walls of all the life-cycle stages except the zygospore, but the species specificities of G-lysins have not been examined as systematically as those of the V-lysins. It is of considerable interest to note that although the *C. reinhardtii* V-lysin has no effect on the walls of most other species of *Chlamydomonas*, it is able to degrade the ECM of several species of *Gonium* and *Astrephomene* (but not that of "higher" members of the family Volvocaceae) (Matsuda et al. 1987). Once again, this implies a close phylogenetic relationship between the *C. reinhardtii* cell wall and the ECM of the colonial volvocaleans.

Enzymes analogous to both the V-lysins and G-lysins of *Chlamydomonas* have been isolated from *V. carteri*, namely, a hatching lysin or H-lysin and a sperm lysin or S-lysin, respectively (Jaenicke & Waffenschmidt 1979, 1981; Waffenschmidt et al. 1990). These enzymes have certain basic similarities: Both are Ca^{2+}-activated serine proteases that have rather generalized proteolytic activities, although they (like the *C. reinhardtii* lysins) (Jaenicke, Kuhne et al. 1987) have a clear preference for hydoxyproline-rich substrates such as denatured collagen or gelatin (Waffenschmidt et al. 1990). In parallel with their *Chlamydomonas* analogues, the H-lysin attacks only the ECM of parental spheroids that are approaching the time of juvenile release, whereas the S-lysin will attack the ECM of spheroids at any stage of the life cycle (Waffenschmidt, et al. 1990). In further parallel with the *C. reinhardtii* V-lysin, the *V. carteri* H-lysin exhibits high, but incomplete, species specificity: It will degrade the ECM of a few species of *Volvox*, but is without any effect on other species of *Volvox* or on *Chlamydomonas* (Jaenicke & Waffenschmidt 1981).

H-lysin activity can be detected only in extracts of *V. carteri* spheroids before hatching; but as hatching begins, it accumulates in the culture medium, reaching its maximum level about when hatching is completed. Curiously, this enzymatic activity can be detected and assayed only by using pre-hatching adults that have been killed by heating, freezing, or formaldehyde fixation (from which it will release juveniles within 10–20 minutes); it is without any perceptible effect on live spheroids! This curious fact may somehow be related to the limited range of action exhibited during the natural hatching process, as described earlier. It was suggested that it may reflect the ability of live cells to protect their enveloping ECM against attack by moderate concentrations of H-lysin by producing inhibitors or by sequestering activators of the enzyme (Jaenicke & Waffenschmidt 1981).

Those studies, interesting as they are, raise as many questions about the hatching process as they answer: How is the ECM of the pre-hatching adult modified to make it susceptible to H-lysin, whereas that of the juvenile (or even a slightly younger adult) is not? By which cells is the H-lysin made and secreted? By the juvenile itself? Or by the overlying somatic cells, as the result of some inductive signal from the juvenile? If the enzyme is made by the juvenile, why is only the roof of its cellular compartment dissolved, while the rest of the chamber is left intact? How do parental somatic cells adjacent to those released during hatching escape having their ECM degraded? And why do they do so? (Given that these cells are already well on the way to programmed death, what selective advantage is there, if any, to thus restricting the range of action of the enzyme?)

5.4.6 Phototaxis: 4,000 oars responding to the commands of a single coxswain

At the moment that a young spheroid has hatched out, it must assume responsibility for a critical function previously executed for it by its parent: It must find a spot where the sunlight is bright enough that it will not limit the spheroid's rate of photosynthesis, growth, and reproduction, but not so bright that it will cause the spheroid's chloroplasts to accumulate excess photooxidants and self-destruct. As we have seen in Chapter 4, this capacity is clearly essential for survival and reproductive success in the environments that *Volvox* and its relatives inhabit.

Inherent in all discussions of phototaxis in *Volvox*, of course, is the recognition that as a spheroid swims forward it rotates about its anterior–posterior axis (in a counterclockwise direction as viewed from behind the spheroid), making one complete rotation about every two seconds (Sakeguchi & Iwasa 1979). Thus, when a spheroid is illuminated from one side, individual cells experience continuously changing light intensities as they are rotated cyclically toward and away from the light source. Two hypotheses have been proposed to account for the rotational component of *Volvox* motility: (1) The two flagella beat in different planes, with the effective stroke of one directed toward the posterior and that of the other directed somewhat to the right (again, as viewed from behind the spheroid). (2) The two flagella beat in the same plane, but with the effective strokes of both directed slightly to the right, rather than directly toward the posterior (Huskey 1979). Recent ultrastructural and high-speed cinemicrographic studies favor the latter hypothesis: Adjacent flagella are observed to beat in essentially the same plane, and the orientation of the flagellar axonemes suggests that the plane of these beats probably is directed very slightly toward the right (Hoops 1993).

Given the fact that in *Volvox* there are thousands of motile cells whose activities must be coordinated to perform directional swimming, it should come as no surprise to find that phototaxis in *Volvox* is rather more complex than in *Chlamydomonas*. As discussed in Section 4.1, the phototactic behavior of *Chlamydomonas* is mediated by differences in the frequencies or strengths with which the two flagella of a single cell beat in response to a particular level of photostimulation. In *Volvox*, in contrast, there is no evidence that the two flagella of a given cell ever beat in different manners. Instead, the phototactic behavior of a *Volvox* spheroid is believed to be mediated by cells in different parts of the spheroid beating their flagella with different frequencies or force, because they are experiencing different degrees of photostimulation at any given instant.

Although *Volvox* phototaxis has been studied for more than a century, and although the basic aspects of the system are now fairly well understood, there are many aspects that remain quite mystifying. The qualitative features of positive and negative phototaxis are quite invariant; that is to say, the spatial patterns in which flagellar activities in different parts of the spheroid are regulated and coordinated to achieve movement toward or away from the light appear to be similar under all conditions that have been examined. However, many quantitative aspects of these responses vary enormously, even within a single spheroid at different times – to say nothing about the variations that exist among different natural populations or different species. The phototactic threshold (the minimum light intensity required to get any phototactic response), the reversal threshold (the intensity at which organisms switch from positive to negative phototaxis), the speed of swimming in both modes, the photo-optimum (the illumination level at which spheroids accumulate within a light gradient), and many other quantitative parameters vary extremely widely with the species, age, and developmental and physiological status of the spheroid, its recent (and not so recent) illumination history, the temperature, the pH and ionic composition of the medium, the spectral quality of the light, the physical parameters of the observation chamber, the direction of the light with respect to earth's gravitational field, and many other factors. As just one example, the phototactic threshold was reported to rise by more than three orders of magnitude after spheroids that had been kept in the dark for a few hours were exposed to direct sunlight for "a few moments" (Mast 1907). Discussion of all of these complexities is well beyond the scope of this text, however, particularly because many of these parameters have not been systematically examined, and none of them can be said to be well understood in terms of either their underlying cellular/molecular mechanisms or their ecological significance.[30] Here we will be concerned primarily with a review of studies conducted to determine the basic mechanism(s) by which *Volvox* orients with respect to a source of light and then moves toward or away from that source.

Although the phototactic behavior of *Volvox* was noted by nearly everyone who studied *Volvox* during the first two centuries after van Leeuwenhoek

---

[30] The following studies, not all of which are cited in the text, will provide a starting place for the reader interested in learning more about these complexities: Oltmanns (1892, 1917), Holmes (1903), Mast (1907, 1917, 1919, 1926, 1927), Huth (1970), Hand & Haupt (1971), Schletz (1976), Sakeguchi & Tawada (1977), Sakeguchi (1979), and Sakeguchi & Iwasa (1979).

discovered it, the first attempt to study the phenomenon systematically was made by Oltmanns (1892). He examined the behavior of *Volvox* in a light gradient that he produced by illuminating an otherwise-opaque, long, narrow chamber through a side panel that consisted of a long, hollow glass wedge filled with a suspension of India ink in gelatin. He found that spheroids would accumulate in a specific region of the vessel at any given intensity of illumination. Then as he raised or lowered the external light intensity, they would move (with no regard for the direction from which the light rays were impinging on the vessel) to the region where the internal light intensity was now similar to what it had been in the region where they had accumulated earlier. For example, in dim external light they accumulated near the thin end of the wedge, but as the external light intensity was raised, they moved to a zone farther from the end, whether that required moving toward or away from the external source of light. Based on those observations, he rejected the hypothesis (derived from earlier studies of other types of organisms) that *Volvox* phototaxis involved movements that were oriented with respect to the direction of the light rays falling on the spheroids (i.e., toward or away from the light source). He postulated instead that they move about somewhat more randomly until they find a spot where the intensity of illumination is favorable, and then settle down, much as some cold-blooded animals are known to move about rather randomly until they locate an area that is neither too cool nor too warm (Oltmanns 1892).

A few years later, Holmes (1903) set out to test Oltmanns's conclusion that *Volvox* does not orient its movements with respect to the direction of the light, and he proceeded in about the most direct way possible: by observing the movements of individual spheroids in a long, narrow glass trough that was illuminated from one end by an arc lamp. He concluded that Oltmanns's central conclusion was flat wrong, basing his opinion on the following three types of observations: (1) Spheroids would swim toward the light for more than 30 cm without deviating by as much as 5 mm to either side of a straight line. (2) If the position of the light was suddenly changed, the spheroids would quickly turn and swim in equally straight lines toward the light in its new position. (3) If the light was above a certain intensity, the spheroids would swim away from it in an equally straight line. He then performed simple but important quantitative studies, and obtained data that any satisfactory theory of phototaxis would have to explain. He placed the trough over a piece of paper that was ruled in 1-cm divisions and carefully noted the time it took a spheroid to pass over each 1-cm interval under various conditions. He began by placing the spheroids at such a distance from the light that they could barely detect its presence and begin to move toward it. His own summary of the results can scarcely be improved upon:

It was found that, as the *Volvox* traveled toward the light, their movement was at first slow, their orientation not precise, and their course crooked. Gradually their path became straighter, the orientation to the light rays more exact, and their speed more rapid. After traveling over a few spaces, however, their speed became remarkably uniform until they reached the end of the trough, where they would remain. If the light is so intense that one end of the trough is above the optimum intensity of illumination, the speed of the *Volvox* is decreased as it approaches the optimum, where it finally stops. In going away from very intense light, *Volvox* moves at a nearly uniform rate until within a few centimeters of the optimum, when the speed begins to diminish. There is thus a lessening of speed as the optimum is reached from either direction. The distance over which there is a marked increase or decrease of speed is considerably less, however, than the space over which the speed is nearly uniform [Holmes 1903].

Holmes realized that the easiest way to account for the fact that a spheroid illuminated from one side turns toward the light was to postulate that flagella on the side facing the light beat less forcefully than those on the opposite side. He explained this with an analogy that has been adopted by others down to the present: the so-called rowboat analogy: "If we imagine a machine in the form of a *Volvox* colony and provided on all sides with small moveable paddles so adjusted that when they came into regions of diminished light as the machine rolled through the water their effective beat would be increased, it is clear that such a machine might orient itself to the direction of the rays and move toward the source of the illumination . . ." (Holmes 1903). To rephrase his analogy, just as a scull turns to the right when rowers on the left side stroke more strongly than those on the right, a *Volvox* spheroid will turn toward the light if the flagella on the shaded side beat more strongly than those on the illuminated side.

Holmes correctly concluded that the reason that *Volvox* could swim away from intense light had to be that above some threshold intensity the effect of light on flagellar beating was being reversed. The fact that perplexed and confused him, however, was that during most of the time that *Volvox* was swimming either toward or away from a light source, its speed was constant. He mistakenly assumed that if the light intensity regulated flagellar beating, then the speed of swimming should necessarily change continuously as the organism moved either toward or away from the light – but he had already proved that such was not the case. In view of the fact that he correctly used the concept of a threshold to explain the reversal from positive to negative phototaxis, it is rather surprising that he did not realize that a similar threshold concept could have been used to explain the uniform velocity observed when the organism was swimming in either mode. Having come to the very edge of getting it all right, Holmes deviated in his course: He rejected his initial concept that phototactic behavior might be explained in terms of the differing *intensities* of light falling on cells in different parts of the spheroid, and he substituted a more complex theory according to which it was the differing *angles* at which light rays were striking eyespots in different parts of the spheroid that determined how forcefully flagella in different parts of the spheroid would beat. Unfortunately, however, that theory was doomed from the outset by virtue of the fact that he based it on the erroneous assertion of Overton (1889) that all *Volvox* cells had their eyespots positioned so that they faced the anterior of the spheroid. Had he looked for himself, he would have found that the reverse is much closer to the truth: In all species examined, the eyespots of cells near the anterior pole face toward the sides, but on the rest of the spheroid, eyespots face toward the posterior of the spheroid (Figure 5.5) (Mast 1907; Pocock 1933b, 1938; Metzner 1945a; Hoops 1993).

A few years later, Mast devised a simple method to test the two hypotheses discussed by Holmes, namely, whether it was intensity differences or angular differences in the light rays reaching cells on opposite sides of the spheroid that caused *Volvox* to turn and swim toward a light source of moderate intensity (Mast 1907). He used two light sources that were oriented either in parallel or at an angle and were either equal or different in intensity. When parallel beams of different intensities were used, the spheroids swam toward the brighter light, with an angular bias that was proportional to the intensity difference between the two sources. When lights of equal intensities impinged on the spheroids from different angles, the organisms swam toward a point midway between the light sources, but when both the angles and the intensities of the light sources were different, the spheroids swam in a line that was biased toward the brighter light, in proportion to the intensity difference. He went on to show that over a wide range of absolute light intensities an intensity difference of about 5% between two light sources was adequate to bias the movement of *Volvox* toward the brighter light. Using microsurgical manipulations, he was also able to show that even in species such as *V. globator*, in which adult cells remain interconnected by cytoplasmic bridges, communication between cells could be ruled out as making any contribution to phototactic coordination of the cells. After many years of detailed study, however, he concluded that (although he was never able to observe this) the flagella on the shaded and illuminated sides of a spheroid must beat in different directions to cause the spheroid to turn (much as a single rower can turn a boat sharply by stroking one oar in the usual direction and the other in the reverse direction) (Mast 1926).

Some 30 years later, Gerisch (1959) reexamined the issue and deduced that the turning behavior of *Volvox* probably was to be accounted for not by a change in beat direction but by a cessation of beat, a "stop response," of flagella on the illuminated side of the spheroid. Subsequently, Huth (1970) observed such flagellar stop responses on the illuminated sides of free-swimming *Volvox* spheroids, and he developed a detailed theory in which this stop response was the main basis for orientation of spheroids with respect to light. The following year, Hand and Haupt (1971) confirmed Huth's observations and showed that the probability that cells would halt flagellar beating following a flash of light of any particular intensity or direction was graded along the spheroid from anterior to posterior, with anterior cells being more sensitive. In the process, Hand and Haupt developed a simple method for visualizing and quantifying flagellar activities in all parts of the spheroid at once: They merely introduced a suspension of small latex beads into the culture and used photographic images of the movements of those beads on the periphery of the spheroid in various regions to measure the turbulence in the water that was being created by flagellar beating in those areas.

Subsequently, Sakeguchi and co-workers used the latex-bead method to show that in both *V. aureus* and *V. carteri* the response involved in positive phototaxis is somewhat more complex than previous workers had realized: It not only involves a stop response that is executed by the cells that are experiencing a sudden increase in light intensity but also involves elevated flagellar activity (supernormal frequency and/or beat strength) by cells on the other side of the spheroid that are experiencing a sudden decrease in light intensity. Moreover, they showed that under conditions in which spheroids exhibit negative phototaxis, these responses are just the opposite: The stop response is stimulated by a decrease in light intensity, and elevation of flagellar activity occurs in response to an increase in light intensity (Sakeguchi & Tawada 1977; Sakeguchi 1979; Sakeguchi & Iwasa 1979).

We can piece together the findings summarized in the preceding paragraphs to form the following picture: When *Volvox* is illuminated from one side with light of moderate intensity, as cells rotate into the light they briefly stop beating their flagella, while at the same time cells that are rotating out of the light beat their flagella more strongly. This pair of reactions causes the anterior end of the spheroid to swing around in the direction of the light. As the spheroid swims directly toward the light, flagella of anterior cells are relatively inactive, while those toward the posterior pole are beating actively. Because cells in any particular part of the spheroid now receive the same stimulation throughout all parts of each rotation cycle, they all beat their flagella with constant force. As the spheroid swims closer to the light, and the light intensity slowly rises, there is not a proportional decrease in swimming speed, because of the graded light sensitivity of cells along the anterior–posterior axis, plus the fact that all cells except those near the anterior pole have their eyespots directed toward the posterior, away from the light. Only as the light intensity approaches some rather high level do cells in more posterior regions exhibit a stop response. At that point the spheroid has reached its "optimum zone" and will only mill about slowly, rather than swimming directionally. If the light intensity suddenly rises above the optimum level, however (e.g., if the sun comes out from behind the clouds), the effects of light on flagellar beating will immediately be reversed: Cells will beat strongly when they are facing such a bright light and stop beating as they rotate away from it. Thus, under these conditions, the spheroid will turn and swim away from the light until it once again approaches a region where the light intensity is in its optimum range, whereupon it once again will slow down and "tread water."

As mentioned earlier, the studies of Mast and others have shown that even in those species in which cytoplasmic bridges between cells persist in the adult, communication via these channels is not a requirement for achieving coordination of phototactic responses among cells in different parts of the spheroid. On the other hand, as clearly shown by the studies of Huskey with respect to the *flgC11* mutant (reviewed in Section 5.4.4.4), coordinated swimming behavior is wholly dependent on the regularity with which cells are oriented with respect to the anterior–posterior axis of the spheroid (Huskey 1979).

5.4.7 Programmed somatic-cell suicide: the last full measure of devotion

Two days after a cohort of *V. carteri* gonidia have arisen during cleavage, they will initiate a new round of embryogenesis and thereby cease to exist as individual entities (Figure 5.6). The somatic cells that were formed at the same time, however, will be only about halfway through with their respon sibilities; for nearly two more days they will remain responsible for keeping the embryos and juveniles in the optimum part of the euphotic zone, where they can photosynthesize, grow, and mature. When these juveniles hatch out and swim away, it might seem that the parental somatic cells would have outlived their usefulness, but in fact they have one last contribution to make to the family: They rapidly undergo total self-destruction and dissolution, and in the process return all of their stored nutrients to the pond that now fosters a new generation of their clonal relatives. Thus, in death, the somatic cells, which were denied the opportunity to produce progeny, promote perpetuation of their own genotype by indirect means.

Three related studies of *V. carteri* f. *weismannia* have provided the only details of we have of the demise and death of *Volvox* somatic cells (Hagen & Kochert 1980; Pommerville & Kochert 1981, 1982). Under culture conditions such that the interval between successive cycles of embryogenesis was about three days, somatic cells completed their motile functions and bade their juvenile passengers farewell in their sixth day of life. At that point, dye-exclusion tests indicated that all of these parental somatic cells were still alive, but clearly they were already moribund. Within another day, half (and within two days all) would be judged dead by all available criteria. However, signs of imminent demise had set in much earlier. By day 4 those cells had already begun to accumulate lipid droplets, a sure sign of deterioration in photoautotrophic cells. Between days 4 and 5, growth of the cells had stopped, their contents of soluble protein and chlorophyll had peaked, and they had already experienced significant declines in photosynthetic activity and in both the quantity and diversity of proteins being synthesized. By day 6, while they were still feebly executing their motile functions, the cells had increased their lipid content sevenfold, had lost 75% of their chlorophyll in the previous 24 hours, and had become visibly bleached.

The conclusion that all of those signs of deterioration and imminent death were results of metabolically active, genetically programmed cellular suicide was supported by several observations made by Pommerville and Kochert (1982). When the spheroids were placed in the dark for 24 hours at the end of day 3, while they were still vigorous, their life span was extended by 24 hours; the same treatment at the end of day 4 was only slightly less effective. In a similar vein, when protein synthesis was continuously inhibited by addition of cycloheximide at the end of day 3, the life span of the cells (as measured by dye exclusion) was extended by three days, and even if the drug was not added until the end of day 5, the life span of the cells was extended by one to two days. Furthermore, it was reported that the timetable for the decline and death of the somatic cells did not change when gonidia were removed from the spheroid at an early stage (Pommerville & Kochert 1982), indicating that those processes were not induced by signals from the juvenile spheroids nor by the hatching process.

The latter observation reveals one of two ways in which the demise of *Volvox* somatic cells differs from programmed cell death, or "apoptosis," in many animal systems that have been studied: In most of the programmed-cell-death phenomena that have been studied in animals, death is triggered (or in some cases inhibited) by signals coming from elsewhere in the organism (Gerschenson & Rotello 1992), whereas in *V. carteri* death appears to be the result of a cell-autonomous program, a basic component of the program of differentiation. Furthermore, the best-defined event in most (if not all) cases of animal-cell apoptosis is the early dissolution of the nucleus as the consequence of double-strand breaks in the DNA between adjacent nucleosomes, which results in formation of a characteristic ladder of DNA fragments that some would use as a defining feature of apoptosis (Gerschenson & Rotello 1992). If such a form of degradation of nuclear DNA occurs in *Volvox* somatic cells, it must be a very late event, because the nucleus is about the last organelle to undergo visible degradative changes in dying somatic cells (Pommerville & Kochert 1981).

Two other types of observations that have been reported suggest ways in which the cell-death phenomenon in *V. carteri* might profitably be studied further. The first is that somatic cells of sexual (egg-bearing) females were shown to have a life expectancy several days longer than that of somatic cells in their asexual siblings, but no further studies of these cells were reported (Pommerville & Kochert 1982). As we will see in the next section, the pheromone that triggers the switch from the asexual to the sexual pathway has several rapid and pronounced effects on the metabolism and biosynthetic activities of somatic cells. Until now, most discussions of these effects in the literature have centered on the ways in which they might contribute to the sexual differentiation of the gonidia. But it might be profitable to reconsider them with respect to roles they might play in delaying the onset of programmed death in somatic cells of sexually induced spheroids. The second observation is that several families of genes have been identified that are expressed only in the somatic cells of free-swimming asexual adults, after all visible features of the somatic cell phenotype are already present (Tam & Kirk 1991a; Tam et al. 1991). Sequencing of these genes and/or analysis of the patterns in which they are expressed in sexual females might yield insights into whether or not any of them encode proteins regulating the programmed death of the somatic cells.

5.5 Sexual induction and differentiation

The study by William Darden (1966) that inaugurated the modern era of *Volvox* research (as discussed in the opening paragraph of this chapter) was entitled "Sexual differentiation in *Volvox aureus*." A major contribution of that study was the demonstration that the culture medium Darden used made it possible, for the first time, to maintain axenic, asexually reproducing *Volvox* cultures indefinitely in the laboratory. However, the aspect of his research that he and others viewed as most important was the observation that asexual cultures could be induced to switch to sexual development simply by supplementing the medium with a bit of filtrate from a culture of mature sexual males. Thus, for the first time, the entire sexual cycle could be triggered and studied at will. Parallel demonstrations that male filtrates would induce sexual development were soon reported for several other species of *Volvox*, including both *V. carteri* f. *weismannia* and *V. carteri* f. *nagariensis* (Kochert 1968; Starr 1969, 1970a,b, 1971, 1972a; McCracken & Starr 1970; Vande Berg & Starr 1971; Karn et al. 1974).

The sexual-induction systems of these two *V. carteri* formas have attracted the attention of developmental biologists and biochemists alike, and to date the nature of the sexual pheromones and the developmental transitions that they trigger have been subjected to more scrutiny than any other single aspect of *Volvox* biology.[31] Moreover, there can be absolutely no doubt that the ability to consistently undergo a switch from asexual to sexual development (and thereby produce dormant zygotes that can resist both late-summer droughts and winter freezes) is a sine qua non for long-term survival of *Volvox* in temperate environments. Therefore, the sexual-induction system is obviously worthy of considered attention by anyone interested in *Volvox* biology.

5.5.1 The source(s) of the sex-inducing pheromone

Starr observed that both male and female strains of *V. carteri* could be induced to undergo sexual development simply by exposing them to a bit of medium from a male culture that had already undergone sexual development and released sperm (Starr 1970a). But that led to a conundrum: If sexual development in both sexes is triggered by a pheromone produced by sexual males, how does the sexual cycle get started in nature? Where does the "first male" come from? Two types of answers have been suggested.

As all who have worked with *V. carteri* appreciate, the process of maintaining male strains in stock cultures, let alone working with them experimentally, is plagued with difficulty because of the frequency with which "spontaneous" males appear: Unless a spontaneously sexual male is detected and removed from a stock culture before it has released sperm, it will release enough pheromone to induce sexuality in the rest of the culture; then, because sexual males produce only terminally differentiated cells – somatic cells and sperm – the culture will self-destruct.

---

[31] For example, see Starr (1970a, 1972b), Kochert & Yates (1974), Starr & Jaenicke (1974, 1989), Kochert (1975, 1978, 1981), Jaenicke (1979, 1982, 1991), Kochert & Crump (1979), Callahan & Huskey (1980), Gilles, Bittner et al. (1980, 1981), Jaenicke & Gilles (1982, 1985), Wenzl & Sumper (1982, 1986a, 1987), Gilles, Gilles & Jaenicke (1983, 1984), Weisshaar et al. (1984), Wenzl et al. (1984), Gilles, Moka et al. (1985a), Kirk & Kirk (1986), Gilles, Balshüsemann & Jaenicke (1987), Tschochner et al. (1987), Mages, Tschochner & Sumper (1988), Balshüsemann & Jaenicke (1990b), Haas & Sumper (1991), Al-Hasani & Jaenicke (1992), Jaenicke, Feldwisch et al. (1993), Sumper et al. (1993), Feldwisch et al. (1995), and Godl et al. (1995).

In his first report on *V. carteri*, Starr speculated that spontaneous males might arise as a result of certain environmental changes, such as depletion of the medium (Starr 1969). An alternative possibility soon suggested itself, however, when he isolated mutant females that underwent sexual development without exposure to pheromone (Starr 1970a). (More details of such "sexual-constitutive," or sex^c, strains will be discussed in Chapter 6.) When mutant alleles from such females were transmitted to male progeny, the latter also developed sexually without benefit of pheromone exposure (Starr 1972b). Thus, Starr speculated that it might be a constitutively sexual male arising by spontaneous mutation that initiates a sexual cycle in nature (Starr 1970a, 1972b). He went on to say, "If all spontaneous males originate by ... mutation, it would be interesting ... to contemplate the relative survival value of a sexual system that can be initiated only by spontaneous gene mutation rather than by some combination of environmental factors" (Starr 1972b).

Starr's hypothesis was not easily tested. "First males" would have to be rigorously searched for in uninduced asexual cultures and, when found, would have to be mated to wild-type females, so that progeny could be screened for a heritable sexual-constitutive trait. Because of the unpredictable occurrence and ephemeral existence of first males, the experiment would require having on hand at all times female cultures in several different stages of sexual development, so that when a spontaneous male was found, a willing and ready mate would be at hand. Callahan and Huskey performed such a difficult and tedious study (Callahan & Huskey 1980). They succeeded in mating about 16 first males and found that about half of them yielded spontaneously sexual progeny (R. J. Huskey personal communication). In their brief published discussion of that experiment, they stated that ". . . not all of the crosses contained sexual constitutive progeny. This suggests that not all of the spontaneous sexual males can be accounted for by stable mutational events" (Callahan & Huskey 1980). What they did not say, however, is that at a minimum their results appeared to support Starr's hypothesis that spontaneous mutation is an important (if not the only) way of initiating the sexual cycle in *V. carteri*.

Indeed, data collected in that experiment actually indicated that mutations conferring spontaneous sexuality occur in males at a significantly higher rate than other types of mutations. Callahan and Huskey (1980) first confirmed Starr's observation that all spontaneously sexual females had Mendelian mutations and then reported that these mutations appeared with a frequency of about 10^{-5}, and defined three sexual-constitutive loci. That yielded a spontaneous-mutation rate of about 3×10^{-6} per locus, very similar to the rate of spontaneous mutation measured by others at two unrelated loci (Kurn et al. 1978). First males, however, appeared with a frequency of 1.8×10^{-4}. Assuming that only half of those males resulted from mutations at one or another of the sexual-constitutive loci identified in females, that still would yield a mutation rate of about 3×10^{-5} per locus, or 10 times the rate observed in females. Subsequently, an equally high frequency of sexual-constitutive individuals was observed by others (Weisshaar et al. 1984). The latter study, which was conducted with great care, examining the progeny of single spheroids, yielded an overall mutation rate of 10^{-4}, or about 3×10^{-5} per locus, in both sexes.[32] Thus, as Starr had postulated, mutation clearly appears to be a viable way of initiating the sexual cycle in *V. carteri*.

In most cases in nature, the occurrence of a spontaneous mutation in one individual within a population can be expected to have extremely little impact on the population as a whole for many generations. However, mutation at one of the sexual-constitutive loci in a *V. carteri* male is a clear exception to this general rule. The reason for this is that the amount of sexual phero-mone produced by a single male is sufficient to induce about 500,000,000 other males and females in about 1,000 liters of water to switch from asexual to sexual development (Gilles, Gilles & Jaenicke 1984). And if that first round of sexual induction is not enough to influence all of the spheroids in a pond, the effect can be expected to rise exponentially as more and more males are recruited to the orgy.

Be all that as it may, mutation is not the only way that a sexual cycle can be initiated in this species. Some years ago, while we were using heat shock for an entirely different purpose (Kirk & Kirk 1985), we serendipitously stumbled over the fact that heat shock induces sexuality in *V. carteri* (Kirk & Kirk 1986). Because that effect could be reversed by all of the treatments that had previously been shown to reverse the inductive effects of purified pheromone (including treatment with an anti-pheromone antibody), and be-cause filtrates of heat-shocked cultures could be used to induce sexuality in other cultures, we concluded that heat shock causes an auto-induction, in which cells first secrete and then respond to the sexual pheromone. Whereas the only cells previously known to produce pheromone were germ cells of males in which sexuality had been triggered several days earlier, we found that somatic cells of heat-shocked asexual spheroids of both mating types released pheromone within less than two hours (Kirk & Kirk 1986).

The selective advantage of using heat shock as an alternative way of in-

---

[32] Interestingly, it was noted that many of the sexual-constitutive females isolated in that study had mutations that were unstable (Weisshaar et al. 1984). High mutation frequencies and high reversion rates are both hallmarks of transposon-induced mutations. It is interesting to contemplate the possibility that controlled transposition events may be involved in regulating the sexual cycle in *V. carteri*.

itiating the sexual cycle appears obvious. When a temporary pond in which
V. carteri has been reproducing asexually begins to dry up under the blaze
of summer sunlight (see the opening quotation for Chapter 3), only those
individuals that quickly switch to sex and produce dormant zygospores will
have their genes represented in the population the following year. Time may
be of the essence. Pocock described a pond that had a thriving *Volvox* population when it contained "plenty of water," but it was transformed into
mere "patches of mud" in only three days (Pocock 1933b). Because it would
take a minimum of three days from the time a mutation appeared in a gonidium of a male strain until that male matured to the point of releasing
pheromone (and then a minimum of three more days before spheroids responding to the pheromone would be mature enough to mate), the mutational
route to sexuality might not produce dormant zygotes fast enough to protect
the gene pool from the consequences of a sudden heat wave. But as noted
earlier, heat shock cuts the lag time for producing pheromone (and hence for
producing dormant zygotes) by nearly three days. Partial induction of sexuality is observed in *V. carteri* cultures at lower temperatures, but even the
temperatures used experimentally to achieve 100% sexuality of both males
and females in the next division cycle (42.5–45°C) (Kirk & Kirk 1986) are
temperatures that might well be experienced by *V. carteri* in shallow, sunlit
temporary ponds. Powers noted that the samples in which he finally found
sexual stages of what is now known as *V. carteri* f. *weismannia* (after several
years of failures that he attributed to searching in bodies of water that were
too deep) came from "a . . . shallow pond . . . [that was] decidedly warm to
the touch when the collection was made" (Powers 1908). Mary Pocock com-
mented, regarding other species of *Volvox* that she habitually found in shallow, temporary ponds in South Africa, that "this asexual phase may be
prolonged . . . or it may be short and soon succeeded by the sexual phase,
the duration apparently depending to a great extent on the temperature." And
later in the same paper: "Occasionally it is possible to point to unduly high
temperature as a possible cause" of sexuality, citing the observation made
by another that "changes in temperature . . . hasten the onset of the sexual
phase, but . . . low temperatures delay it" (Pocock 1933b).

Heat shock is not likely to play any role in regulating the sexual cycle in
Volvox populations, such as those of *V. globator* and *V. aureus*, that live in
permanent lakes with high thermal-buffering capacity. But the observations
sampled earlier indicate that it may well be involved in assuring the persis-
tence of the more numerous species of *Volvox*, including *V. carteri*, that
inhabit shallow temporary ponds and roadside puddles. Even in such environments, however, an alternative means of triggering the sexual cycle (such
as mutation?) clearly would be essential in order to assure formation of over-

wintering zygotes in years when the ponds dried up gradually, rather than as the result of a sudden, intense heat wave.

5.5.2 The nature of the sex-inducing pheromone

The sexual pheromone of *V. carteri*, an approximately 30-kDa glycoprotein, is one of the most potent bioeffector molecules known. It is fully effective for inducing sexual development at concentrations below 10^{-16} M (Gilles, Gilles & Jaenicke 1984), and it has been estimated that it may take as little as two molecules per gonidium to effect a full response (Pall 1973).

Detailed characterization of the pheromone was delayed and complicated by two factors: (1) The pheromone is so potent that even culture supernatants that can be diluted a millionfold and still induce sexuality contain only minuscule amounts of the glycoprotein. (2) Mass cultivation of males (the principal source of the pheromone) is extremely difficult because of the regularity with which sexual males appear spontaneously and cause the culture to self-destruct prematurely. Eventually, however, a clever way was found to circumvent the latter problem and obtain enough pure pheromone for microsequencing; that led, in turn, to recovery and sequencing of genomic and cDNA clones encoding the core polypeptide of the pheromone (Tschochner et al. 1987; Mages et al. 1988).

The deduced amino acid sequence of the pheromone (partially confirmed by peptide sequencing) consists of 208 residues, including a putative 11-residue signal sequence. The predicted size of the polypeptide is 22.4 kDa ($\sim$21 kDa if the signal sequence is deleted), which is in fair agreement with the estimates of 22.5–25 kDa for the size of the deglycosylated pheromone that have been derived from its mobility in SDS-PAGE (Tschochner et al. 1987; Balshüsemann & Jaenicke 1990a,b; Haas & Sumper 1991). Further analysis of Southern blots and genomic clones indicates that five or six tandem repeats of the pheromone-encoding gene, all encoding identical polypeptides, and differing only in intronic regions, are present in a tandem array in the wild-type genome (M. Sumper personal communication).

The intact pheromone purified from culture filtrates typically migrates in SDS-PAGE as a mixture of two to six (most commonly three) isoforms with apparent molecular masses centered around approximately 30 kDa, but differing from one another by about 1 kDa. Following complete deglycosylation, however, a single species is regularly observed, indicating that the isoforms differ in the extent of glycosylation of a common core polypeptide (Balshüsemann & Jaenicke 1990a,b; Haas & Sumper 1991; Jaenicke, Feldwisch et al. 1993). The two largest and most abundant isoforms (which have equivalent biological activities) have been subjected to carbohydrate analysis and shown to include roughly equal amounts of *N*- and *O*-linked glycans (Balshüsemann & Jaenicke 1990a). The polypeptide has six potential *N*-glycosylation sites, but peptide analysis reveals that only three of these near the center of the protein are used, and in the two isoforms analyzed, these carry similar *N*-glycans of the branched, mannose-rich type, with xylose residues attached to the peptide-proximal mannose. The differences between these two isoforms reside in the *O*-glycosidic sugars (arabinose, galactose, and/or xylose) linked to threonine residues; disaccharides and monosaccharides dominate in the larger and smaller isoforms, respectively, but these differences have no measurable impact on the biological activity. On the other hand, enzymatic removal of the *N*-glycans abolishes the inducing activity of the pheromone (Balshüsemann & Jaenicke 1990a).

Given the apparent importance of the carbohydrate moieties for the biological activity, it was somewhat surprising to learn that active pheromone is produced and secreted into the culture medium in considerable abundance when the pheromone-encoding cDNA is expressed in either yeast (Haas & Sumper 1991) or mammalian cells (Jaenicke, Feldwisch et al. 1993). In both of these expression systems, the pheromone is released to the medium as a mixture of isoforms that differ from one another (and from any of the isoforms present in *V. carteri* culture supernatants) in the extent of glycosylation. The principal isoform secreted by yeast has an apparent molecular mass approximately 5 kDa greater than that of the *V. carteri* product (although the deglycosylated polypeptides from the two sources co-migrate), which is consistent with the knowledge that yeast elongates *N*-glycans by adding as many as 150 mannose residues per chain, whereas *V. carteri* does not. The specific activity of this "hyperglycosylated" yeast product was not reported, but it was assumed to be substantial, because culture supernatants gave complete induction at a 10^9-fold dilution. However, the mammalian-culture supernatant also was active at high dilution (10^8-fold); nevertheless, the specific activity of the mammalian product was found to be only about 1% that of the native pheromone. From this we can conclude that high levels of total sex-inducing activity may mask the fact that changes in glycosylation patterns can reduce the specific activity of the pheromone dramatically.

An earlier report that the polypeptide backbone of the pheromone was synthesized well in advance, stored in an inactive form, and then glycosylated, activated, and released at the time that sperm packets broke down (Gilles, Bittner et al. 1981) apparently was erroneous, based on the use of an antiserum of imperfect specificity. A more recent study, using a highly specific antibody raised to deglycosylated pheromone (but cross-reactive with the glycosylated isoforms), produced the surprising finding that all steps in the production of active pheromone, including synthesis of the polypeptide, are executed at the very end of male development, after the sperm packets have been released and are swimming toward their targets, and just before inducing activity begins to accumulate in the medium (Balshüsemann & Jaenicke 1990b). This is all the more surprising given the "stripped-down" nature of the mature sperm (as discussed later).

Starr originally reported that "all attempts to effect cross-induction between . . . *V. carteri* f. *nagariensis* and . . . *V. carteri* f. *weismannia* have proved unsuccessful" (Starr 1970a). That statement – and hence the concept that the pheromones of these two forma were completely taxon-specific in their inducing activities – stood unchallenged for more than two decades. Recently, however, when pheromones from two different isolates of *V. carteri* f. *weismannia* were purified to homogeneity, they were both found to be capable of inducing sexuality in *V. carteri* f. *nagariensis* (Al-Hasani & Jaenicke 1992; Jaenicke, Feldwisch et al. 1993)! This relationship is unilateral, however: In agreement with earlier reports, the recent study reported that there was no effect of the *nagariensis* pheromone on either strain of forma *weismannia*. Moreover, a 10^4-fold molar excess of the *nagariensis* pheromone had no inhibitory effect when added to a test system in which a *weismannia* female was exposed to *weismannia* inducer at limiting concentration (Al-Hasani & Jaenicke 1992). This is taken to indicate that the *nagariensis* pheromone does not compete for pheromone-binding sites within the *weismannia* spheroid.

It requires a 10^5- or 10^6-fold higher molar concentration of one of the *weismannia* pheromones than of the *nagariensis* pheromone to induce sexuality in forma *nagariensis*. However, this is only 10 times more *weismannia* pheromone than is required to induce the *weismannia* strain from which the pheromone was isolated (Al-Hasani & Jaenicke 1992). Astonishingly, although both of the *weismannia* pheromones induce sexuality in forma *nagariensis*, they exhibit unilateral cross-inducibility with respect to one another: Strain 65-30(12) responds to either its own pheromone or that from strain 1B, but strain 1B responds to only its own pheromone.

In light of the newly discovered cross-inducibility, it is not too surprising to learn that the pheromones of these two formas of *V. carteri* are more similar in structure than had been suggested by earlier reports that were based on incompletely purified *weismannia* pheromone, and that indicated that the pheromones were very dissimilar in their peptide maps and were immunologically non-cross-reactive (Kochert & Yates 1974; Kochert 1981). The deglycosylated core polypeptides from all three *V. carteri* pheromones recently studied exhibit very similar mobilities in SDS-PAGE and are virtually indistinguishable in terms of their affinities for an antibody that was produced in response to the core polypeptide of the *nagariensis* pheromone as the immunogen (Al-Hasani & Jaenicke 1992). In their native, glycosylated forms, however, the pheromones exhibit much greater differences in immunological cross-reactivities and composition. Although the sugar residues of the three pheromones are qualitatively similar, they are quite different quantitatively, and presumably this is the source of their differences in biological activities. The strain that responds only to its own inducer (1B) is the one whose pheromone is most different from the *nagariensis* pheromone in terms of both sugar composition and immunological cross-reactivity.

Unilateral cross-inducibility had been discovered earlier when a new forma of *V. carteri*, forma *kawasakiensis*, was isolated a few years ago. It was shown that the *kawasakiensis* pheromone induced forma *nagariensis*, but reciprocal induction was not observed (Nozaki 1988).[33] No information is available yet regarding cross-inducibility of formas *kawasakiensis* and *weismannia* or the structure of the *kawasakiensis* pheromone.

5.5.3 Developmental consequences of exposure to the pheromone

As mentioned earlier, exposure of *V. carteri* spheroids to the sex-inducing pheromone does not cause their gonidia to transform directly into gametes; rather, it causes them to initiate a new round of asexual embryogenesis in which cleavage patterns are modified and gametes are eventually produced (Figure 5.7).

In females, the first sign that the pheromone has had an effect is that embryos fail to divide asymmetrically at the sixth division cycle, but then do so at the seventh cycle (Starr 1969, 1970a). There is also a spatial change in the asymmetric division pattern: Whereas only cells in the anterior half of the embryo typically divide asymmetrically in asexual embryos, in induced females as many as three-fourths of the cells may divide asymmetrically, resulting in production of as many as 48 germ cells. Following the completion of cleavage and the subsequent inversion (which do not differ significantly

---

[33] Nozaki (1988) also tested these two formas for interfertility. He obtained zygotes only when *kawasakiensis* females were mixed with *nagariensis* males, not in the reciprocal combination; furthermore, only 5% of the hybrid zygotes germinated, and none of the germlings were viable. We have studied Nozaki's *kawasakiensis* strains with slightly different results. Although we observed the same unilateral cross-inducibility that he did, we were able to obtain hybrid zygotes only in the combination with which he had no success: *kawasakiensis* males with *nagariensis* females. Although high levels of post-zygotic lethality were observed in these hybrids, the levels were not as high as in the reciprocal cross: Of 1,000 zygotes tested, 264 germinated, and 3 survived to establish asexually reproducing clones, for an overall survival rate of 0.3% (K. A. Stamer and D. L. Kirk unpublished data).

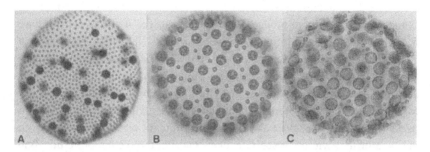

Figure 5.26 Sexual spheroids of *V. carteri*. **A**: Sexual-female spheroid containing about 48 eggs. **B**: Sexual-male spheroid containing "androgonidia" in a one-to-one ratio with somatic cells. **C**: Sexual male about 18 hours older than that in B; each androgonidium has divided six or seven times to produce a convex plate, or "packet," of 64 sperm cells.

from the equivalent processes in asexual embryos), the large cells produced by asymmetric division develop as eggs, and by the time the juvenile spheroids hatch, the eggs are fertilizable. However, if not fertilized, these eggs will enlarge and transform into gonidia that subsequently will cleave either sexually or asexually, depending on whether or not pheromone is still present. Because of this redifferentiation of the eggs, mating efficiency declines with increasing time between hatching and exposure to sperm.

Young sexual female spheroids differ in appearance from asexual spheroids by virtue of the larger number and denser appearance of their germ cells (Figure 5.26A). In detailed EM examinations of eggs and gonidia, the only consistent difference we have detected is a substantial reduction in the size (and perhaps number) of the vacuoles present in eggs, which accounts for their much denser, darker-green appearance in life (J. Hoffman and D. L. Kirk unpublished data). No substantial studies of biochemical or molecular differences between eggs and gonidia have yet been reported.

Induced male embryos exhibit yet a different pattern of asymmetric division (Starr 1969, 1970a). They divide symmetrically six to eight times before all cells in the embryo divide asymmetrically one time. Following the asymmetric division – whenever it occurs – cleavage ends, and the embryo inverts. At that point a juvenile male spheroid consists of a 1:1 ratio of small somatic cells and larger cells called "androgonidia" (Figure 5.26B). Surprisingly, these androgonidia, in contrast to asexual gonidia or eggs, develop functional flagella.[34] After about a day of growth, the androgonidia resorb

[34] The reason that this is surprising is that the male spheroids will not hatch out of the parent and become free-swimming until after these androgonidial flagella have been resorbed and sperm development has been completed; thus androgonidial flagella have no obvious adaptive value. Their formation may reflect the fact that because androgonidia will later be required to

their flagella and initiate a round of cleavage that consists of six or seven symmetric divisions and produces a concave plate of 64 or 128 cells (Figure 5.26C). A rudimentary form of inversion converts this concave plate to a convex one, whereupon the cells begin to differentiate into sperm. Spermatogenesis takes about 12 hours, and is completed while the male spheroids are still within the parental spheroid. Thus, nearly as soon as the male spheroids hatch from the parental spheroids they begin to release their sperm packets, which swim away as intact, multicellular units. We will return to the saga of the swimming sperm packets later, in the section on fertilization and zygote formation.

In general outline, cleavage of androgonidia appears to resemble that of asexual gonidia (Section 5.3), except that the starting cell is smaller, there are fewer divisions, and the divisions are all symmetrical (Deason et al. 1969; Deason & Darden 1971; Birchem 1977; Birchem & Kochert 1979b). During cleavage, however, the cells have already begun to take on some of the specialized characters that will characterize the mature sperm. Like those of most animals, *Volvox* sperm become highly specialized and efficient nuclear transportation devices, shedding cellular components that would be essential for long-term survival, while elaborating components that will prepare them for the long swim to locate a receptive egg and for the sexual fusion that will occur if their search is successful.[35]

At the light-microscopic level, one of the most obvious changes during spermatogenesis is that the cells lose the typical grass-green (chlorophyll *a*/*b*) coloration of green algae, becoming a light golden color by the time they are released as mature sperm. At the EM level, that change becomes evident as early as during androgonidial cleavage, as increasingly large portions of the chloroplast lose both stromal and thylakoid elements. By the end of spermatogenesis, the chloroplast has been reduced to a small organelle at the posterior end of the cell in which a few thylakoid lamellae surround a cluster of prominent starch granules that in turn surround the pyrenoid. Meanwhile, at the other end of the cell, mitochondria have accumulated and become tightly clustered around the bases of the flagella (Birchem & Kochert 1979a). Together, these observations suggest that during its differentiation the sperm cell is preparing to switch from photoautotrophic to a respiratory metabolism,

---

produce sperm, they cannot afford to employ the system that asexual gonidia and eggs use to repress formation of flagella.

[35] Most published descriptions of *V. carteri* spermatogenesis and fertilization have dealt with forma *weismannia* (Hutt & Kochert 1971; Birchem 1977; Birchem & Kochert 1979a,b; Coggin et al. 1979). However, our (unpublished) examinations of these phenomena in forma *nagariensis* indicate that these two forma are quite similar in this aspect of the life cycle. Hence, the descriptions given here are fusions of published and unpublished studies.

drawing the energy required for motility from starch reserves accumulated earlier.

Flagella that are fully as long as the cells, as well as prominent eyespots, are the dominant features seen at the light-microscopic level as differentiation proceeds; by the time they are mature, the sperm will have a ratio of flagellar mass to total mass far exceeding that at any other stage of the life cycle. At the EM level, one of the most obvious and striking features seen is the extent to which the cortical, MT-based cytoskeleton found in all *V. carteri* cells has become exaggerated, with MTs much more concentrated and more regularly spaced than at any other stage of the life cycle. Although the standard components of the BA (Section 4.2) all appear to be present in the sperm, additional components not seen in asexual cells are also present; it is postulated that these function to reinforce the BA during the vigorous swimming that the sperm packet will soon undertake (Birchem & Kochert 1979a).

Finally, there are two specializations of sperm that prepare them for their activities once they have reached a fertile female. First, they produce a highly modified ECM that in its abundance and structure more nearly resembles the *Chlamydomonas* cell wall than the ECM seen at other life-cycle stages. In particular, whereas at all other stages of the life cycle the "tripartite" layer of the ECM is continuous over the surface of a spheroid (providing coherence to the multicellular unit), in the sperm packet the tripartite layer surrounds each individual cell. This is undoubtedly related to the fact that as soon as a sperm packet makes effective contact with a fertile female spheroid it will break up into individual cells (Section 5.5.5). Second, a unique structure, a small, organelle-free protuberance, develops at the anterior end of each sperm, near the flagellar bases. Although quite tiny initially, this protuberance is located at the spot where, after a sperm has contacted an egg, it will extend a long "beak," or "proboscis," that will be swept back and forth over the surface of the egg just prior to fusion (Birchem & Kochert 1979a). It is difficult to escape the conclusion that this structure is analogous, if not homologous, to the "fertilization tubule" that is present in the equivalent location on mt^+ gametes of *C. reinhardtii*; that fertilization tubule (like the acrosomal process of metazoan sperm cells) is filled with actin microfilaments and makes contact with the mating structure on mt^- gametes to initiate cell fusion (Goodenough & Weiss 1975; Detmers, Carboni & Condeelis 1985).

5.5.4 Biochemical effects of the pheromone and the riddle of its mechanism of action

As stated earlier, sexual induction has attracted more attention over the past quarter-century than any other aspect of *Volvox* biology. Nevertheless, we are still without a satisfying answer to the question of paramount interest: How does the pheromone act to divert spheroids from the asexual to the sexual pathway of development? Indeed, as new data have accumulated, the cloud of mystery that surrounds the action of the pheromone often has tended to intensify rather than dissipate.

5.5.4.1 Puzzle 1: Why is the response to the pheromone so sluggish? The primary paradox that a satisfying hypothesis about the mechanism of action of the pheromone must explain is the fact that although spheroids are exquisitely sensitive to the pheromone with respect to the concentration that they require to produce a full response ($<10^{-16}$ M), they are both extremely languid and surprisingly fickle about committing themselves to sexual development in response to it. In order to assure that all progeny of the next generation will be sexual, it is necessary to expose adult spheroids to the pheromone continuously for a minimum of 8–10 hours prior to the time that their gonidia begin to cleave (Gilles, Gilles & Jaenicke 1984), and under some conditions substantially longer exposure periods are required (Callahan & Huskey 1980). But under no conditions has it been reported that the duration of exposure necessary to elicit a full response is reduced in the least by having the pheromone present in an enormous superabundance!

During this prolonged exposure period, individual gonidia appear to become committed to the sexual pathway in a stochastic manner, with some becoming committed after little more than an hour of exposure, with others (possibly even within the same spheroid) requiring a full 8–10 hours, and with the population as a whole exhibiting a half-time for commitment of about five hours (Gilles, Gilles & Jaenicke 1984). Thus, if a population of spheroids is exposed to pheromone for five hours, about half of all gonidia will cleave sexually while the rest will cleave asexually, and parental spheroids containing all possible combinations of asexual and sexual progeny will be produced.[36] Moreover, until cleavage actually begins, the "committed" state is fully reversible, and it decays with kinetics that are qualitatively and quantitatively similar to those characterizing its establishment. Thus, if young

---

[36] It is important to note, however, that although adjacent gonidia within a single spheroid may respond differentially following a marginal pheromone exposure, each individual gonidium exhibits an all-or-none response: It produces a juvenile containing either gonidia or gametes, never both.

adult spheroids are first exposed to pheromone for 10 hours or more (which by itself will commit all gonidia to sexual development) and then are washed and cultured without pheromone for the last five hours preceding cleavage, the outcome will be the same as it is when spheroids are exposed to pheromone for only that last five-hour period; that is, only half of the gonidia will cleave in the sexual pattern (Gilles, Gilles & Jaenicke 1984). These unusual kinetics have led some to assume that exposure to pheromone must cause a gradual buildup of some metabolite (or state) that must reach some threshold level before sexual commitment will occur and that this metabolite (or state) decays with a similar time course when pheromone is removed. However, it is not clear how well such a threshold effect can be used to explain the fact that some gonidia are affected by pheromone so much earlier than others that a virtually linear relationship is obtained when one plots the percentage of gonidia developing sexually versus the time of exposure to pheromone (Gilles, Gilles & Jaenicke 1984).

5.5.4.2 Puzzle 2: What is the real target of the pheromone? The second major paradox to be explained is that although the only *visible* effect of the pheromone yet discovered is that it changes the developmental behavior of the gonidia (clearly implying that gonidia are the ultimate targets of pheromone action), the only *biochemical* effects of the pheromone that have been detected so far occur outside of the gonidia. Many biochemical changes have been detected in the somatic cells and/or the ECM within minutes after pheromone addition (as will be discussed in more detail shortly), making it seem certain that these cells must possess pheromone receptors and a signal-transduction chain capable of revamping the metabolic program of the cell in response to bound pheromone. However, no equivalent biochemical changes within gonidia following pheromone addition have yet been described.[37] Thus, it becomes relevant to ask an apparently simple question: is the gonidium itself sensitive to the pheromone, or are the effects of the pheromone on the gonidium all indirect, mediated by the somatic cells and their ECM products?

Attempts to answer this presumably simple question experimentally have led to conflicting findings and interpretations. For many years it was widely accepted that gonidia did not respond to the pheromone directly, because when isolated gonidia were exposed to pheromone under the same conditions routinely used to assess its effects on intact spheroids (namely, suspension

---

[37] Actually, it was asserted without further elaboration on two occasions that biochemical changes had been detected in gonidia following pheromone exposure (Gilles, Gilles & Jaenicke 1984; Gilles, Balshüsemann & Jaenicke 1987), but no additional information on this point has ever been published.

in tubes or flasks of liquid medium to which pheromone was added) they exhibited virtually no response to the pheromone at any concentration (Jaenicke 1979; Gilles, Gilles & Jaenicke 1984; Wenzl & Sumper 1986a, 1987; Gilles, Balshüsemann & Jaenicke 1987; R. C. Starr and R. J. Huskey personal communications; our unpublished observations). However, Wenzl and Sumper (1986a, 1987) discovered that isolated gonidia would respond to the pheromone provided that two conditions were met: (1) The concentration of pheromone had to be at least 100-fold higher than that required to induce gonidia in intact spheroids. (2) The gonidia had to be exposed to pheromone on the surface of a semisolid substrate. Even at such elevated pheromone concentrations, the frequency of sexual development dropped significantly when identical gonidia were exposed to the pheromone in a thin layer of liquid medium (instead of on the surface of a semisolid substrate), and it virtually disappeared when they were exposed to it in a standard tube of culture medium. Gilles, Balshüsemann & Jaenicke (1987) confirmed these observations and reported that when the concentration of the pheromone in the medium was held constant, (1) the percentage induction that was observed declined progressively as the volume of medium was increased, (2) in any volume of medium the percentage induction was elevated if somatic cells were added to the culture, but (3) even under the latter conditions induction was reduced 10-fold if the medium was removed every two hours and replaced with fresh medium containing the same concentration of pheromone.

Those observations were interpreted at the time in two opposing ways. Wenzl and Sumper (1986a, 1987) concluded that the findings demonstrated that gonidia could respond to the pheromone directly, and they postulated that the role that the somatic cells played in the induction process was to produce ECM components that would channel the pheromone to the vicinity of the gonidia, thereby raising the effective concentration to the concentration at which the gonidia could respond. In contrast, Gilles, Balshüsemann & Jaenicke (1987) interpreted the findings as demonstrating that the induction observed with isolated gonidia was not the result of a direct action of the pheromone on the gonidia, but was mediated by an intermediate that was accumulated slowly, and passed a critical threshold concentration only if the volume of the culture was kept sufficiently small. They postulated that this intermediate was produced as a result of the pheromone acting on an extracellular signal-cascade system located in the ECM to generate a second messenger, and in the case of isolated gonidia the response was weak because the only ECM present to generate this messenger was the gonidial vesicle.[38]

---

[38] Wenzl and Sumper (1986a) obtained 30% induction of isolated gonidia that had been digested for an hour at 30°C with pronase or subtilisin, and they argued that "this treatment should

One thing on which those groups agreed was that the ECM plays an important role in the induction process, even though they disagreed about whether that role was to channel the pheromone to the gonidium or to produce a completely different kind of second messenger. Thus we turn our attention to the ECM.

5.5.4.3 Puzzle 3: Does the ECM mediate or inhibit induction? A third major paradox to be confronted is that although all the early studies and some very recent studies have led to the conclusion that the ECM plays a positive role in induction, certain findings have been interpreted as indicating an inhibitory role for an ECM component.

The evidence favoring some role for the ECM in sexual induction takes at least two forms. The first type of evidence is somewhat indirect: Added pheromone is bound to, and concentrated in, the ECM. After adding increasing numbers of spheroids to a limiting amount of pheromone solution and measuring the rate at which the percentage sexual induction declined with increasing spheroid numbers, Gilles, Gilles & Jaenicke (1984) calculated that in that concentration range the pheromone was concentrated 300-fold within the spheroids. When they then fractionated spheroids and used cellular and extracellular fractions representing equal numbers of spheroids to compete with a small number of live test spheroids for a limiting dose of pheromone, they found that the most effective competitor was an ECM fraction from the deep zone of the spheroid, suggesting that that was where the pheromone was most efficiently bound, rather than by the cells. (However, it should be noted that the second most effective competitor for the pheromone was the gonidial fraction.) They then reported that under conditions in which isolated gonidia ordinarily would not respond to pheromone, a portion of them could be induced if they were exposed to the same quantity of inducer in the presence of this kind of matrix fraction (Gilles, Gilles & Jaenicke 1984). From that finding, they concluded that the binding of the pheromone to this matrix fraction was biologically significant.

The second type of evidence suggesting an important role for the ECM in sexual induction is the observation that all of the earliest biochemical

degrade any remaining extracellular matrix material." However, our observation is that the gonidial vesicle is much more resistant to enzymatic degradation than any other part of the ECM. We have repeatedly tried to obtain "naked" gonidia over the years, for a number of different reasons, and we have never found conditions under which treatment with pronase, subtilisin, or any other enzyme resulted in complete removal of the vesicle (as assayed by electron microscopy) without first killing the cell. On the other hand, none of the biochemical changes that have been detected elsewhere in the ECM in response to pheromone exposure have ever been shown to occur in isolated gonidia – not even gonidia that have never been exposed to proteolytic enzymes (and thus have intact vesicles).

events that have been detected after pheromone addition are ECM-related. Within the first hour, substantial changes are seen in the ^{35}S labeling patterns of four sulfated glycoproteins (SGs) of the ECM that are all somatic-cell products (Wenzl & Sumper 1982, 1986a, 1987; Wenzl et al. 1984; Sumper et al. 1993). Labeling of a novel 70-kDa component, SG 70, can be detected within 10 minutes after pheromone addition; the labeling peaks within the first hour and then declines by the third hour. Labeling of female surface glycoprotein (FSG) begins in the first half hour and continues at least until the beginning of gonidial cleavage. At about that time, incorporation of label into SG 140 (a product seen before pheromone addition) begins to decline, while labeling of SG 110 (a presumed replacement form of SG 140) rises accordingly. All of these changes in ^{35}S incorporation exhibit the same dose-response relationship with pheromone concentration as does sexual induction, which is taken to mean that they are relevant to the induction process (Wenzl & Sumper 1986a, 1987). At the time those observations were first reported, one or more of these SGs were interpreted as being "part of an amplification system accumulating the pheromone and raising the actual pheromone concentration up to 100-fold at the surface of the gonidium" (Wenzl & Sumper 1987). Since that time, however, further study of SG 70 (which undergoes the earliest and most dramatic change following pheromone addition) has led to two proposals suggesting that it is involved in a rather different type of signal amplification process. We will consider these two new proposals at some length a bit later.

Phosphorylation of ECM glycoproteins is also reported to be rapidly modified after pheromone addition: ^{32}P labeling of a 290-kDa phosphoglycoprotein, pp290, declines markedly within the first 10 minutes and then rises to about 70% of control levels; meanwhile, two other proteins (pp240 and pp120) exhibit elevated labeling (Gilles, Gilles & Jaenicke 1983). Like all of the SGs discussed earlier, these pps appear to be somatic-cell products that are localized in the nearby ECM and are virtually or completely absent from the gonidium and its surrounding vesicle (Gilles, Gilles & Jaenicke 1983; Wenzl & Sumper 1987). A very different function has been proposed for these pps than for the SGs, however: Changes in their phosphorylation following pheromone addition have been interpreted as providing evidence for the existence of an extracellular signal cascade by which a putative second messenger that is required for sexual induction is generated (Jaenicke 1991). This concept will be considered in more detail shortly.

In view of the growing support for the concept that the ECM plays some positive role – at least a facilitating, if not an essential mediating role – in sexual induction, it came as more than a bit of a shock when data were presented that were interpreted by the authors as indicating that the role played by the ECM in induction was an inhibitory one. Starr and Jaenicke (1989) observed that at one very brief stage of the asexual life cycle, all that was required to cause gonidia of wild-type *V. carteri* spheroids to undergo sexual development was to isolate them from the rest of the spheroid! That startling effect, which was at odds with everything that had been published and believed up until that time, could be produced only if the gonidia were isolated from young adult spheroids in the first hour after the lights came on to start the illumination period within which the gonidia would later begin to cleave. The same effect was observed whether the gonidia were isolated by protease digestion of the ECM, as in the published report (Starr & Jaenicke 1989), or by mechanical disruption of the spheroids (R. C. Starr personal communication). Furthermore, it was reported that sexual development would also occur if (during this same narrow window of time) the somatic cells were killed with various aldehydes (Starr & Jaenicke 1988, 1989). The fact that sexual development could be prevented if antibody to the pheromone was added to the culture medium following either of those treatments was taken to indicate that the gonidia were releasing pheromone and undergoing a "self-induction" similar to that which had earlier been shown to occur following heat shock (Kirk & Kirk 1986). Starr and Jaenicke offered the following interpretation of their novel observations:

. . . we propose that the gonidia of the female strain are secreting inducer even under standard growth conditions, but in such small quantities that neither is it detectable in the surrounding medium nor does it cause self-induction. It is possible that self-induction is prevented by the binding of this small amount of inducer to some component secreted by the somatic cells into the matrix. Thus it is only when the somatic cells are removed from the young spheroid (or killed by aldehydes) . . . that the isolated gonidium becomes self-induced by its own secretion of inducer. . . . Exogenously supplied inducer in the induction process appears to saturate the neutralizing factor in the matrix and may not necessarily be involved directly in the induction process. This could well account for the extremely small amounts of inducer needed for the induction . . . [Starr & Jaenicke 1989].

Fascinating as this hypothesis is, it does not seem to provide any explanation for the fact that if the gonidia are isolated at any time other than during the critical first hour of light, they not only do not undergo any self-induction but also become *more* sensitive to exogenous pheromone if matrix or somatic cells are added to the culture – not *less* sensitive, as the hypothesis quoted in the preceding paragraph would seem to demand (Gilles, Gilles & Jaenicke 1984; Gilles, Balshüsemann & Jaenicke 1987). Moreover, the published data do not rule out an alternative interpretation: It is possible that as cells undergo a transition from darkness to light they are sufficiently sensitive to various stresses that treatment with aldehydes, or protease digestion, or mechanical disruption of the spheroid is adequate to trigger a stress response analogous to the heat-shock response that has been shown to cause all *V. carteri* cells to release pheromone (Kirk & Kirk 1986). In this regard, it should be noted that by itself the transition from dark to light has been shown to trigger production of heat-shock proteins in *C. reinhardtii* (Kropat et al. 1995). In any case, although these observations are not to be ignored, they do not provide an adequate reason to ignore all of the other, earlier observations indicating that under most conditions the ECM promotes, rather than inhibits, the sexual-induction process. Hence we will now examine the two major classes of hypotheses regarding how the ECM may have such an induction-promoting effect.

5.5.4.4 Puzzle 4: Does cyclic AMP inhibit or mediate induction? Kochert (1981) reported that cyclic adenosine 3'-5'-monophosphate (cAMP) and phosphodiesterase (PDE) could be detected in isolated gonidia of *V. carteri* f. *weismannia*, and he postulated (by analogy with certain animal-hormone systems) that cAMP might be the intracellular second messenger that was being produced in response to pheromone binding and that then was triggering a cascade leading to sexual differentiation.

In initial tests of that hypothesis, Gilles, Gilles and Jaenicke (1984) reported: (1) that sexual induction was inhibited by isobutyl methylxanthine (IBMX, a PDE inhibitor), (2) that the inhibition was greatly intensified if cAMP was added along with the IBMX, and (3) that exogenous PDE induced sexuality in the absence of any added pheromone.[39] On the basis of those observations they suggested the opposite of what had been postulated by Kochert, specifically, that "high levels of cyclic nucleotides block induction whereas low levels induce. . . . The inducer lowers – by a mechanism as yet unknown – the level of cyclic nucleotides in the matrix." For several years the hypothesis that extracellular cAMP is an inhibitor of induction was sustained. Included among the observations that were cited as being consistent with that hypothesis were the following: (1) Concentrations of cAMP were greatly elevated in spheroids of a noninducible strain, elevated less in a sexually induced wild-type strain, and elevated still less in a sexual-constitutive strain (Gilles, Moka et al. 1985a; Jaenicke & Gilles 1985). (2) Changes in phosphorylation patterns of ECM glycoproteins were seen within minutes after pheromone addition (Gilles, Gilles & Jaenicke 1983; Jaenicke & Gilles 1985). (3)

---

[39] It has been asserted by others that induction by PDE was subsequently found to have been an artifact caused by inadvertent cross-contamination of the enzyme with pheromone (Wenzl & Sumper 1987); this assertion has never been confirmed or denied in print by the original authors.

Components of an extracellular cAMP-dependent signal-cascade system, including an adenylate cyclase, a cAMP-inhibited protein kinase, and a PDE, had all been detected, as described in preliminary form in Gilles, Gilles and Jaenicke (1984), Gilles, Moka and Jaenicke (1985) and Jaenicke and Gilles (1985).[40] (4) Preliminary data suggested that the pheromone caused cAMP concentrations to decrease initially before rising to above preinduction levels (Moka 1985).

Further analysis apparently revealed, however, that the preliminary findings cited as item 4 had been misleading and that instead of the reduction of cAMP that had been predicted for several years, the pheromone actually "triggers a rapid 8-fold increase of cAMP in vegetative *Volvox* spheroids of all developmental stages (Moka 1988). Nevertheless, this signal is not sufficient for determination of the algae which occurs 3–4 h before cleavage of the embryo. . . . This determination to form gametes correlates with a second cAMP increase, only occurring if the sexual inducer is still present in the medium" (Nass, Moka & Jaenicke 1994). As the preceding quotation indicates, improved data regarding cAMP concentrations led to an abrupt paradigm shift, so that cAMP was no longer viewed by those investigators as being an inhibitor of induction, but rather "a good candidate for being the second messenger for the sexual inducer. . . . [it is] proposed that the pheromone's action is mediated by cAMP" (Nass et al. 1994; cf. Jaenicke 1991; Feldwisch et al. 1995).

The more recent data on the time course of cAMP fluctuations in induced spheroids (Moka 1988) are actually in accord with earlier data indicating that cAMP concentrations were elevated in bulk samples of unstaged sexual spheroids (Gilles, Moka et al. 1985a; Jaenicke & Gilles 1985), and thus there can be little reason to doubt that cAMP concentrations are indeed elevated following pheromone exposure. Nevertheless, because the only effect that has thus far been reported to occur following addition of exogenous cAMP is an inhibition of sexual induction by the pheromone (Gilles, Gilles & Jaenicke 1984), prudence would seem to recommend suspending judgment with respect to what *causal* role, if any, cAMP plays in the long chain of events that transpire between the initial exposure to pheromone and the final commitment of the gonidia to initiate sexual development many hours later. Attempts to clarify this matter apparently are being continued (Feldwisch et al. 1995).

---

[40] More recently, a pair of *V. carteri* cAMP-binding proteins of uncertain function have been purified and characterized (Feldwisch et al. 1995). These proteins lack detectable protein kinase activity and are able to hydrolyze cAMP to 5'-AMP, but they have physical properties that distinguish them from the known phosphodiesterases of *Volvox*. What role, if any, these proteins play in sexual induction remains to be determined.

5.5.4.5   Puzzle 5: Does SG 70 amplify the pheromone signal, and if so, how?
As noted earlier, of the many changes in labeling patterns of ECM components that have been observed following exposure of spheroids to pheromone and $^{35}SO_4^{2-}$, among the earliest (detected within 10 minutes) is intense labeling of SG 70 (Wenzl & Sumper 1986a, 1987). Thus, this molecule has been subjected to more intensive scrutiny. In initial studies (Wenzl & Sumper 1986a), SG 70 was shown to be a product of the somatic cells that is rapidly incorporated into the ECM in such a way that it cannot be resolubilized by boiling in 1-M NaCl/3% SDS, but is released after hydrofluoric-acid deglycosylation as a 60-kDa polypeptide that co-migrates with the deglycosylation product of the soluble precursor. In contrast to other SGs (in which only sugar residues appear to be sulfated), SG 70 has half of its sulfate groups in the form of tyrosine esters, with the rest as sugar esters. In those initial studies, the authors considered and rejected a possible role for SG 70 (and other SGs) as a "second messenger" that would be the actual agent of sexual induction. Instead, they postulated that the role of SG 70 (and other SGs) was "to collect the positively charged pheromone molecules and to direct the pheromone's transport through the highly negatively charged extracellular matrix to the gonidia's cell membrane." In closing, they pointed out that "pheromone-induced synthesis of SG 70 does also occur in *Volvox* spheroids at developmental stages at which the pheromone is wholly ineffectual in eliciting sexual development (e.g., at the time of early gonidial cleavages). This fact further strengthens the hypothesis that the observed changes within the extracellular matrix do not constitute the central part of the trigger mechanism" (Wenzl & Sumper 1986a).

However, when SG 70 was studied further, and the genes encoding it were cloned and sequenced, two very different views emerged sequentially (Sumper et al. 1993; Godl et al. 1995). One of the first additional observations to be made (Sumper et al. 1993) was that after SG 70 had become embedded in the ECM in an insoluble form, some of it underwent a specific cleavage (with a half-time of ~6 hours) into two fragments of approximately 42 and 30 kDa. The products of that cleavage, like the parent SG 70 molecule, remained firmly embedded in the ECM, insoluble in boiling SDS/NaCl, but capable of being solubilized by extraction with 0.1-M EDTA in the presence of detergents. Sequencing of proteolytic peptides indicated that SG 70 actually was a mixture of at least two related molecules, a conclusion that was subsequently confirmed by cDNA cloning and sequencing. The deduced amino acid (aa) sequences of these two species of SG 70 polypeptides exhibited only 31% identity, but 73% similarity. But it was when these sequences were subjected to a computerized homology search that the real surprise emerged: The C-terminal region (~200 residues) of each of the deduced SG 70 polypeptide sequences exhibited about 70% similarity to the sequence of the sex-inducing pheromone!

Because both SG 70 molecules carried a homologue of the pheromone, they were renamed pherophorins I and II, names whose close linkage belies the fundamental functional differences between them that were to emerge from further analysis (Sumper et al. 1993). Pherophorin I transcripts are present constitutively and do not increase following pheromone treatment, whereas accumulation of pherophorin II transcripts is strongly induced within 20 minutes of pheromone exposure. Moreover, whereas pherophorin II undergoes the proteolytic cleavage described earlier, producing a 42-kDa fragment that contains the domain homologous to the 30-kDa pheromone, pherophorin I is not cleaved. It was taken as particularly significant that no induction of pherophorin II transcription was detected when three different noninducible mutants were exposed to pheromone.

At that point, Sumper and co-workers postulated a new role for SG 70 (a.k.a. pherophorin II) as follows: "This glycoprotein is secreted and deposited within the ECM and within a few hours, a 42 kD domain that is a homologue of the sex-inducing pheromone is proteolytically cleaved from pherophorin II. *If this cleavage product also has the biological property to induce a gonidium*, the massive production of pherophorin II would raise the inducing principle within the ECM by orders of magnitude. Subsequently, a classical receptor/effector mediated interaction would be sufficient to trigger the gonidium" (Sumper et al. 1993). The critical phrase in that proposition is the one I have italicized for emphasis. Those authors did not report attempts to induce sexuality with the native 42-kDa proteolysis product of pherophorin II (although they had isolated it in quantities adequate for biochemical analysis), but they did report that a peptide of equivalent sequence produced in *E. coli* was without inducing activity, which they speculated might be due to differences in posttranslational modifications.

A more recent study (Godl et al. 1995) has added several new dimensions to this fascinating story. A third member of this family of ECM proteins, pherophorin III, was discovered and found to be 87% identical to pherophorin II in deduced aa sequences. Like pherophorin II, pherophorin III undergoes specific proteolytic cleavage in the ECM; but in distinction to pherophorin II – and like pherophorin I – pherophorin III is expressed constitutively, rather than being pheromone-induced. It was also established in that study that at least 10 genes encode polypeptides of the pherophorin II type! The ones that were characterized turned out to be 96–99% identical in nucleotide sequences in the coding region, as well as 91–95% identical in their (~800-bp) upstream regions. Estimates of the numbers of genes encoding pherophorins I and III were not reported.

Three new observations relating to the possible roles of the pherophorins in the sexual-induction process were reported by Godl et al. (1995): (1) A mutant that develops sexually in the absence of added pheromone also expresses the pherophorin II gene under these conditions (although the abundance of the transcript is increased following pheromone addition).[41] (2) Conditions that suppress sexual induction (incubation in the dark for the last six hours of a seven-hour exposure to pheromone) also suppress proteolytic cleavage of pherophorin II in the ECM. (3) A version of the 42-kDa C-terminal fragment of pherophorin II produced in a baculovirus expression system lacks any sex-inducing activity.

In light of the third of those observations, the authors proposed yet a third way that SG 70/pherophorin II, or more specifically, its 42kDa C-terminal fragment, might play a crucial role in the sexual-induction process:

It has been demonstrated that the ECM has the property to bind the pheromone (Gilles et al., 1984). Thus, the surface of the actively swimming organism could serve as a collector for enrichment of the pheromone. The specific production of the 42-kD cleavage products derived from pherophorin II and III would then compete for the pheromone-binding sites within the ECM, thereby releasing and concentrating the pheromone in the small volume of the spheroid's interior that contains the reproductive cells. This mode of action could certainly achieve an increase of the actual pheromone concentration by order[s] of magnitude. The outcome of a simple experiment appears to support this model. Cutting a *Volvox* spheroid into two halves does not disturb the attachment of the reproductive cells to the ECM, and yet such an opened spheroid no longer responds to the pheromone with the same sensitivity as the intact organism. This strange behavior is predicted by the latter model [Godl et al. 1995].

In evaluating the possibility that the 42-kDa fragment of pherophorin II plays a central role in the induction process, either by mimicking the pheromone and acting as the operative sexual inducer in vivo, or by competing with the pheromone for binding sites within the ECM, at least one observation seems to be worthy of more consideration than it has been given: The similarity in the amino acid sequences of the pheromone and pherophorin II is not reflected in the properties of the processed and secreted molecules: SG 70/pherophorin II is a highly acidic glycoprotein containing both tyrosine and sugar sulfates (Wenzl & Sumper 1986a), whereas the pheromone is a very basic glycoprotein with a pI (isoeletric point) of 10.5 (Gilles, Gilles & Jaenicke 1984; Jaenicke & Gilles 1985) that has never been reported to contain any sulfate residues. Although it is not difficult to imagine that these two

---

[41] Interpretation of that finding would be facilitated by knowing whether the mutant used in the study is constitutively sexual because it produces the pheromone constitutively and induces itself or whether it enters the sexual pathway in some pheromone-independent matter. Such information is not currently available (R. C. Starr personal communication).

molecules might bind to one another quite strongly by electrostatic interactions, as originally proposed (Wenzl & Sumper 1986a), it is rather more difficult to visualize the nature of a binding site either on the surface of a gonidium or in the ECM that could bind both of these two very different kinds of molecules with any significant specificity. Perhaps the current situation is best summarized by the sentence that immediately follows the preceding quotation: "Clearly more work is necessary to define the function of the pherophorins" (Godl et al. 1995).

Nearly a decade ago it was stated that "the sexual induction system of *Volvox carteri* is far from being understood on a biochemical level. . . . although many modifications of the extracellular matrix have been found" (Gilles, Balshüsemann & Jaenicke 1987). Although some even more fascinating modifications of the ECM have been described in the interim, it is not yet clear that they have brought us any closer to understanding the mechanism of sexual induction. It is to be hoped that this situation will change soon.

5.5.5 Zygote formation and germination

As mysterious as the mechanism by which sexuality is induced in *V. carteri* may be, its function in the natural world is perfectly clear: to generate resistant zygotes that will protect the germ plasm of the population against summer drought and winter cold, so that it can reemerge when more favorable conditions return.

Once gametes of both mating types have been produced in response to the sexual pheromone, as described earlier, the challenge for the sperm is to locate and fertilize the immotile eggs. This appears to be a nontrivial task, because the eggs remain ensconced within the female spheroids, hidden behind a layer of somatic cells and ECM that, to the human observer, appear indistinguishable from those of asexual spheroids. Thus the sperm must distinguish between nonfertile and fertile spheroids and selectively attach to and penetrate the latter; then it still faces the challenge of locating an egg and fusing with it.

As noted earlier, the sperm are formed and released and then swim as packets of 64 or 128 biflagellate sperm. These packets swim with the flagellar end forward; their movement through the medium is both much more rapid and much more erratic than that of spheroids propelled by somatic cells. The packets are strongly phototactic, but even when swimming toward the light they follow a much more tortuous, quasi-helical path than do spheroids, and they frequently make sharp turns, zig-zagging toward the light. In the absence of effective contact with a female, the sperm can swim about for several hours before the packet begins to break down and the cells begin to degenerate.

Both Darden (1966) and Starr (1969) reported (with respect to *V. aureus*, and *V. carteri* f. *nagariensis*, respectively) that sperm packets appear to be attracted to fertile females. However, when Coggin et al. (1979) tested the hypothesis that fertile females of *V. carteri* f. *weismannia* released a sperm-packet chemoattractant, they observed only negative results. We have not examined the matter systematically, but when we have observed sperm packets of *V. carteri* f. *nagariensis* in a dish containing asexual and sexual females in equal numbers, it has appeared to us that the sperm swim quite randomly, appearing to contact spheroids by chance, rather than by directed swimming, and with about equal chances of first contacting an asexual or a sexual female (our unpublished observations).

In any case, the sequellae of contacts with asexual and sexual spheroids are normally different (Kochert 1968; Starr 1969; Hutt & Kochert 1971).[42] After contacting an asexual spheroid, a sperm packet may skim about over the surface briefly, as if exploring the surface with its flagella, but usually it swims off within a few seconds. After contacting a female spheroid, however, such exploratory movements usually are more deliberate and prolonged and appear to have a tendency to cause the sperm packet to move toward the posterior of the female, as if there were an adhesive gradient over the spheroid, with its high point at the posterior end (Figure 5.27A). Within a minute or two, however, a sperm packet hovering over a sexual female typically will cease movement, and then within another minute or two it will begin to dissociate into individual sperm cells that will commence to wiggle about (Figure 5.27B). By that time, nearby regions of the ECM of the female will have begun to break down, permitting a few somatic cells in the area to pull loose from the spheroid and tumble away into the medium. This process gives rise to a visible hole in the surface of the adult spheroid that is called the "fertilization pore" (Figure 5.27C); several days later these pores will still

---

[42] There is an exception to this rule. If sperm are added in great superabundance to a small culture, not only will they attach to, be activated by, and digest the matrix of asexual females, but they also will do the same to asexual or sexual male spheroids (and even dead spheroids) in what can best be described as a feeding frenzy. Under such conditions I have observed upwards of 50 sperm packets attached to and dissociating on the surface of a single spheroid (be it asexual, male, or female). The attachment process under such conditions appears superficially to be autocatalytic, as if the sperm packets might be releasing some factor that stimulates indiscriminate attack of the nearest spheroid by other sperm packets. The amount of sperm lysin (see Section 5.4.5) (Waffenschmidt et al. 1990) released under such conditions often is sufficient to completely degrade the ECM of all spheroids in the dish, converting the entire culture to a suspension of single cells in 10–20 minutes. We had hoped that this might be a way to produce naked gonidia, which would be useful for a variety of purposes. Unfortunately, however, it turned out that the gonidial vesicle is about the only part of the ECM that is resistant to the sperm lysin (D. L. Kirk unpublished observations).

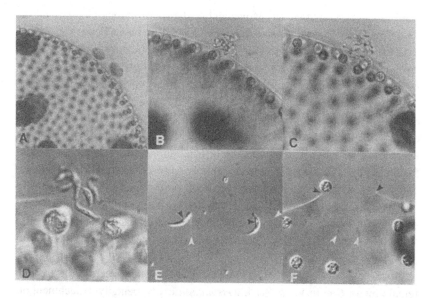

Figure 5.27 Sperm–female interaction. **A**: Two sperm packets attached to the posterior end of a sexual female (the large, dark objects are eggs). **B** and **C**: Sperm packets in the process of dissociating into individual sperm after being activated by contact with the flagella of the female. **D**: Individual sperm in the process of entering the interior of the female. **E**: Two isolated sperm; note flagella (white arrowheads) and nuclei (black arrowheads). **F**: A "fertilization pore" (between the black arrowheads) that has been left behind where the sperm digested an entry hole in the ECM of the female. One somatic cell at the edge of the pore was being released as a free cell at the time the photo was taken, and several others had been released previously, as the matrix surrounding them was digested by the sperm. Several out-of-focus sperm can be seen on the interior (white arrowheads). **A–C**, brightfield microscopy; **D–F**, Nomarski differential-interference microscopy.

be visible, so that it will be possible to count them and determine how many sperm packets attached to and attempted to fertilize each female.

Subsequent events have been described most fully for *V. carteri* f. *weismannia* (Kochert 1968; Hutt & Kochert 1971; Birchem & Kochert 1979a; Coggin et al. 1979), but events appear to be rather similar in forma *nagariensis*. As soon as the fertilization pore has begun to form (within the first 10–15 minutes after sperm–female contact) sperm can be seen penetrating the spheroid and wriggling through the liquefying ECM with a euglenoid or worm-like motility in which the cells exhibit considerable flexibility and extensibility. Whereas the flagella had been directed anteriorly and had beat in a ciliary fashion while the sperm packet was still intact, they now will be extended backward along the sides of the cell, undulating in a manner more reminiscent of the flagellar movements of metazoan sperm (Birchem & Kochert 1979a). In free liquid, these undulations cause the sperm to swim forward in a corkscrew manner. Movement of the sperm through the interior of the female spheroid appears to be random; there is no evidence for a chemoattractant released by eggs. But given the large number of gametes of both types now present within a relatively confined space, often it is only a matter of time until contact is made. As mentioned earlier, volvocacean sperm cells often extend a rather lengthy proboscis or beak from the anterior end; in *V. carteri* f. *weismannia* it is reported that when a sperm eventually contacts an egg, it sometimes sits virtually motionless while the proboscis is swept back and forth over the egg in an apparently exploratory manner. The act of cell fusion has been described in print only for *V. carteri* f. *kawasakiensis* (Nozaki 1988). In that forma, no proboscis was seen on the sperm, but as the sperm attached itself tightly to the side of the egg it was very slender, with an extended posterior "tail." Suddenly that tail was retracted, and the sperm "rapidly entered the anterior pole of the egg from its anterior to posterior ends" (Nozaki 1988).

Kochert and his colleagues have investigated several aspects of the sperm–female interaction in *V. carteri* f. *weismannia* experimentally. Attachment of the sperm packet to the spheroid and "activation" (triggering of the dissociation and pore-forming processes) both appear to be mediated by contact with the flagella of the female (Coggin et al. 1979). Sperm packets attach to and are activated by isolated flagella of a sexual female, but they do not attach to or become activated by a female that has recently been deflagellated by pH shock. However, once a female has regenerated her flagella, she is again able to bind and activate sperm packets. In light of these observations it is interesting to note that Huskey et al. (1979a) were able to mate "flagellaless"-mutant females of *V. carteri* f. *nagariensis*. Either their "flagellaless" females had short flagellar stubs that were not visible in the light microscope (they were not examined in the electron microscope) (R. J. Huskey personal communication), or else the binding and activation of sperm have different bases in these two formas.

If a *V. carteri* f. *weismannia* female is exposed to trypsin or protease VI at 50 µg/mL for five minutes, she will retain the ability to bind sperm packets and trigger their dissociation; but those sperm will not produce a fertilization pore (Coggin 1979). However, if the exposure to these enzymes is continued for 20 minutes the female will lose the ability to bind and cause the dissociation of sperm packets. Those findings are taken to indicate (1) that the enzymes responsible for the dissociation of the sperm packet and for the formation of a fertilization pore are different and (2) that the releases of these two activities are triggered separately in response to two different ligands on the flagella of the female, with the ligand triggering sperm-packet dissolution being more resistant to protease digestion than is the one triggering pore formation. Further evidence that the two enzymes are different comes from the observation that trypsin inhibitor has no effect on sperm packet-breakdown, but it blocks pore formation (Coggin et al. 1979).

It has been reported that sperm-packet breakdown occurs, but fertilization pore formation sometimes does not, if actinomycin D, cycloheximide, puromycin, or chloramphenicol is present during the sperm–female interaction (Hutt & Kochert 1971). If taken at face value, those observations would imply that in order to release the enzyme(s) that will digest the ECM of the female, the sperm must engage in continuous synthesis of RNA, as well as continuous synthesis of proteins on both cytosolic and organellar ribosomes. However, certain details of the observations suggest that considerable caution should be exercised before accepting such an interpretation: Sperm from packets that attached to a female within the first eight minutes after the male and female cultures had been mixed together were unfazed by any inhibitors that were present in the medium; it was only sperm from packets that took longer than eight minutes to locate and attach to a female that subsequently failed to form a fertilization pore. Because no data were presented by the authors concerning the successes of such "late attachers" in the absence of inhibitors, one cannot rule out the possibility that such sperm packets were physiologically impaired in some way, and that their failure to complete the fertilization process was a consequence of something other than the presence of inhibitors in the medium.

As mentioned in Section 5.4.5, a sperm lysin, or S-lysin, has been isolated from sperm packets of *V. carteri* f. *nagariensis* (Waffenschmidt et al. 1990). This enzyme is a 34-kDa serine protease with a collagenase-like preference for hydroxyproline-rich substrates; like the lytic enzymes of some animal sperm, it is strongly calcium-dependent. Although indirect, the evidence is fairly convincing that this actually is the enzyme involved in pore formation. As discussed in Section 5.4.5, the properties of this enzyme are quite different from those of the matrix-degrading enzyme called H-lysin, which had been isolated earlier from asexual spheroids by the same group (Jaenicke & Waffenschmidt 1979, 1981). As in the case of pore formation in forma *weismannia*, the S-lysin of forma *nagariensis* is inhibited by trypsin inhibitor. Moreover, although it is capable of causing complete breakdown of asexual or male spheroids when present in sufficient concentration, S-lysin preferentially digests pores in female spheroids in the neighborhood of eggs. The fact that this enzyme was released in active form, and in considerable abundance, by sonication of sperm packets that had never been exposed to female spheroids suggests either that the processes of pore formation are quite different in the two formas of *V. carteri* or that the observations discussed in the preceding paragraph are misleading.

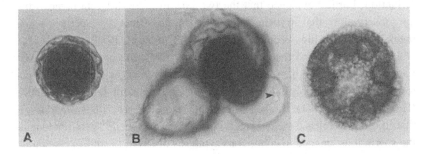

Figure 5.28 Zygote germination. **A**: Mature zygote; note the thick, crenulated cell wall. **B**: Germinating zygote. The heavy outer zygospore wall has deteriorated and split open; the dark-red, flagellated germling cell (now haploid) and the three nonviable polar bodies (arrowhead) are in the process of escaping, but are still enclosed in the thin inner wall of zygospore. **C**: Germling spheroid containing eight gonidia that was produced when a germling cell – after swimming about briefly – settled down, cleaved, and inverted in a manner similar to that of an asexual gonidium.

If an egg is not fertilized, it is capable of enlarging within a day or two, transforming into a gonidium, and cleaving to form a daughter spheroid. This transformation appears to be accelerated when the pheromone that induced egg formation is washed out (Starr 1970a), but whether the offspring of these transformed eggs will themselves be sexual or asexual will depend on whether or not pheromone was present during the transformation process.

Once fertilized, however, the egg (now a zygote, of course) takes a very different path. Within some days it transforms from a dark-green color to a bright-red color, develops a thick crenulated cell wall, and becomes dormant (Figure 5.28A). The minimum time required for the zygote to mature to the point that it will be both fully resistant to environmental insults and capable of germination when returned to favorable conditions has never been pre-cisely defined, but is probably at least two weeks. In the dormant state, mature zygotes can survive for more than a decade in a dry dish and still remain capable of germination when appropriate conditions are restored (R. C. Starr, personal communication). Alternatively, they can be maintained at −80°C indefinitely (K. A. Stamer & D. L. Kirk unpublished observations). Thus, dormant zygotes provide a means for long-term preservation of germ plasm in the laboratory as well as in nature.

When washed and transferred to fresh medium, healthy zygotes will begin to germinate within about a day. The first step in germination is meiosis, as a result of which one large viable germling and a set of three nonviable small cells resembling polar bodies will be formed (Starr 1975). The germling hatches as a red-orange biflagellate unicell that swims about for a short time before settling down to cleave and form a small germling spheroid containing four to eight gonidia (Figure 5.28B, C). Following inversion, the germling spheroid is at first distinguishable from all other spheroids in the life cycle by virtue of its retained reddish color, but within hours that is replaced by the characteristic green. Although the germling spheroid at that point resembles a small asexual spheroid, its formation obviously must involve a very different genetic program than that of a typical asexual spheroid, because mutations that will reveal themselves in all of its progeny generally are not expressed in the germling itself (R. J. Huskey, personal communication).[43] After enlarging for a day or so, however, the gonidia of the germling will cleave to produce asexual spheroids indistinguishable in type from those that will be produced in each successive round of the asexual reproductive cycle that has become reestablished.

Having surveyed the highlights of what is known – and what remains mysterious – about key aspects of asexual and sexual development in wild-type *V. carteri*, we will now turn our attention to the progress that has been made in understanding some of the genetic underpinnings of *Volvox* development.

---

[43] An alternative hypothesis to explain the germling spheroid's lack of mutant features that will be expressed in its asexual offspring is that genes whose products are required for formation of the germling spheroid are transcribed in the zygote before meiosis, and hence the germling contains mRNAs representing both parental genotypes.

6

Mutational Analysis
of the *V. carteri*
Developmental Program

> ... mutants involving the time of differentiation, the pattern of differentiation, and the nature of the differentiated reproductive cells indicate embryogenesis in *Volvox carteri* f. *nagariensis* is under the control of a number of genetic loci. ...
>
> The somatic cells of *V. carteri* show a characteristic ... loss of ability to grow and divide. ... A mutant has been isolated in which the somatic cells do not lose this ability. This mutant ... may provide us with interesting material to study controls by which such processes of cell growth and multiplication are regulated.
>
> Starr (1970a)

With those words summarizing his observations of several interesting spontaneous mutants of *V. carteri*, Richard Starr tried to direct the attention of those assembled for an annual meeting of the Society for Developmental Biology toward a promising new avenue for analyzing the genetic control of cell differentiation. But for some reason that was not an avenue destined to become quickly crowded with fellow travelers. Over the next quarter century, only nine laboratories would publish one or more studies involving the use of *V. carteri* mutants,[1] and of those, only three[2] would make any sustained

---

[1] The Blamire, Huskey, Jaenicke, Kirk, Kurn, Pall, Schmitt, Starr and Sumper laboratories.
[2] The Huskey, Kirk and Starr laboratories.

efforts to use genetics as a tool for dissecting *V. carteri* development. Despite that lukewarm response to Starr's summons, his perception that *V. carteri* had substantial promise as a developmental genetic system was soon reinforced, principally through the efforts of one laboratory.

Robert Huskey, formerly a bacteriophage geneticist, was the first to join Starr in the exploration of *V. carteri* developmental genetics. Huskey and his co-workers developed methods for chemical mutagenesis and isolation of temperature-sensitive mutants and used them to recover mutants with interesting defects in a wide range of developmental processes. Two of the 12 categories of developmental defects they described in their first paper (Sessoms & Huskey 1973) were similar to defects that Starr had described three years earlier, namely, mutants with increased numbers of gonidia, and mutants in which somatic cells redifferentiated as reproductive cells, divided, and reproduced. The rest of Huskey's mutants were new, however, and exhibited abnormalities affecting the cohesion of cells in the dividing embryo, the cleavage pattern, the establishment of organismal polarity, the inversion process, and the deposition and stability of the extracellular matrix. Subsequently Huskey's group would use such mutants to analyze several key aspects of *Volvox* development (Huskey 1979; Huskey & Griffin 1979; Huskey et al. 1979b; Callahan & Huskey 1980) and to establish a preliminary linkage map for the species (Huskey et al. 1979a).

6.1 Formal genetic analysis

Identification and recovery of *V. carteri* mutants have been facilitated by the fact that the organism is haploid in all active phases of the life cycle, and thus many mutations reveal themselves immediately. The other side of this coin, however, is that the absence of an active diploid phase in the life cycle precludes complementation analysis.

The related unicell *C. reinhardtii*, although it is also normally haploid, is amenable to two methods of assessing complementation and dominance relationships, neither of which is accessible in *Volvox*. Because *C. reinhardtii* gametes are morphologically similar to vegetative cells and fuse to form quadriflagellate cells that remain motile and continue to express many vegetative traits for several hours, complementation and/or dominance relationships often can be assessed by simply examining quadriflagellate cells and determining whether or not the mutant phenotype of one or both mating partners is cured by fusion with a cell of a different genotype. Such "dikaryon rescue" has been employed with great success to analyze, for example, the relationships among various mutations that interfere with normal flagellar assembly, motility, and length control (Lewin 1954; Starling & Randall 1971; Luck et al. 1977; Huang, Piperno et al. 1981; Lefebvre, Aselson & Tam 1995). But it is not likely that dikaryon rescue will ever become feasible with *Volvox*, because *Volvox* eggs and sperm are highly specialized cells that no longer express most vegetative features of interest.[3] Stable diploids of *C. reinhardtii* can also be obtained readily, because about 5% of all fused gamete pairs divide mitotically, rather than forming dormant zygotes (Harris 1989). A similar phenomenon has never been observed in *Volvox*.

Diploid *V. carteri* clones have been detected on two separate occasions among the progeny of crosses performed for other reasons. In one case the diploid nature of such strains was established by karyotyping (Meredith & Starr 1975; R. C. Starr personal communication); in the other, it was detected by analysis of DNA polymorphisms (Adams et al. 1990). In both cases, the diploids arose at low frequency, apparently because of a failure of occasional zygotes to execute meiosis before germination, and in both cases the diploids exhibited slow growth, a number of rather nondescript morphological abnormalities, and differentiation as males following sexual induction. Although those chance occurrences demonstrated that vegetative diploids were viable in *V. carteri*, and also indicated that the male allele at the mating-type locus was dominant to the female allele, they did not lead to a method of predictably generating diploid progeny. Huskey and Griffin began a rigorous attempt to force diploidy by crossing two strains that carried linked conditional-lethal mutations, germinating zygotes under conditions (such as the presence of spindle toxins) that they hoped would suppress meiosis, and then selecting for germlings able to grow under restrictive conditions. Unfortunately, however, that effort was not successful (R. J. Huskey personal communication).

In the absence of complementation tests, allelism in *V. carteri* is defined strictly by recombination: Mutations that exhibit less than 1% recombination are assumed to be allelic (Huskey & Griffin 1979). This value is justified by the knowledge that in *C. reinhardtii* intragenic recombination occurs at frequencies up to 1.6% (Matagne 1978; Harris 1989).

A second genetic tool that facilitates formal genetic analysis in *Chlamydomonas*, namely, tetrad analysis (Harris 1989), is also inaccessible to *V. carteri* geneticists, because each germinating *V. carteri* zygote produces only one viable meiotic product (plus three polar bodies) (Starr 1975) (see Figure 5.28B). Thus linkage in *V. carteri* must be evaluated by random spore analysis, which is a more laborious process.

---

[3] Exceptions to this generalization that *V. carteri* gametes do not express most mutant vegetative features are certain mutant traits (such as the flagellaless phenotype) that are expressed by sperm packets, but that abolish their fertility (Huskey et al. 1979a).

Despite these limitations, Huskey et al. (1979a) were able to use 105 mutants in 19 phenotypic categories to define 33 loci and place them provisionally in 14 linkage groups, thereby providing a preliminary map of the *V. carteri* genome. Fourteen also happens to be the number of chromosomes in *V. carteri* (Cave & Pocock 1951b; R. C. Starr, personal communication). But given the fact that the average number of markers per linkage group was less than three in that study, it is highly unlikely that these preliminary linkage groups actually correspond in a 1:1 manner to individual chromosomes.

The preliminary map generated by Huskey and co-workers demonstrated clearly that *V. carteri* was accessible to formal genetic analysis. However, two considerations impeded attempts to extend their map. The first was that 10 of the 14 linkage groups in the preliminary map were defined exclusively by morphological mutations, many of which lowered fecundity, particularly when efforts were made to combine them to generate multiply marked tester strains. Thus, mapping a new morphological mutation using such linkage markers turned out to be a tedious process at best. It subsequently became impossible when most of the strains used to define loci and linkage groups in the original map were lost (R. J. Huskey personal communication). Therefore, some years ago we set out to develop a new map with more useful properties. To that end, certain key markers from the original map are being combined with drug-resistance markers and DNA polymorphisms of various types that will eventually provide a much higher density of markers without reducing fertility (Kirk & Harper 1986; Harper et al. 1987; Adams et al. 1990, D. L. Kirk and co-workers unpublished observations). As the density of reliable markers is being increased, we are producing multiply marked tester strains that should eventually prove useful for more efficient mapping of morphological mutations. Some of the conventional (i.e., drug-resistance and morphological) Mendelian markers that will eventually be incorporated into this map are described in this chapter, and some of the molecular markers (DNA polymorphisms) that are now being mapped are described in the next chapter, where the present status of the genetic map is also described.

6.2   Drug-resistant mutants

V. carteri mutants resistant to the following substances have been reported: bromodeoxyuridine (BUDR) and fluphenazine (FP) (Kurn et al. 1978; Kurn & Sela 1981), chlorate and methionine sulfoximine (Huskey et al. 1979a,b), benzimidazole, canavanine, erythromycin, fluoroacetamide, hydroxyurea, isopropyl *N*-(3-chlorophenyl)carbamate (CIPC), and novobiocin (Kirk & Harper 1986). Toxicity studies have been reported for 37 additional drugs (Kirk

& Harper 1986), but strains resistant to those other agents have not been reported to date.

The frequency of appearance of BUDR-resistant clones among a wild-type population was used to estimate the spontaneous mutation rate in *V. carteri*, assuming that mutations at only a single locus conferred this phenotype; a rate of about 3×10^{-6} was obtained (Kurn et al. 1978).

FP-resistant mutants were isolated in an attempt to clarify the roles played by calmodulin (the presumed target of FP activity) in *V. carteri*. It was reported that at a concentration of 1 μg/mL, FP quickly caused wild-type *Volvox* cells to lyse, quite possibly by interfering with the activity of contractile vacuoles that is required for normal osmoregulation. At slightly lower concentrations the drug modified several processes, including uptake and metabolism of phosphate, flagellar motility, and inversion of embryos (Kurn & Sela 1981; Kurn 1982). The four FP-resistant mutants that were isolated and analyzed did not really clarify the causal relationships among these effects, because the four strains all had different pleiotropic phenotypes: The only property they all shared was survival in FP at 1 μg/mL; they differed with respect to their abilities to resist other in vivo and in vitro effects of the drug and with respect to whether or not they exhibited abnormal development in the absence of the drug (Kurn & Sela 1981; Kurn 1982). The patterns in which these different features were combined in the four mutants did not lead to any significant insight beyond the obvious one: that calmodulin may play several different roles in *V. carteri*. Because no Mendelian analysis was reported, it is not clear how many loci these four mutant strains may define. That is unfortunate, because it has been reported that in *C. reinhardtii* the cell body and flagella may contain calmodulins with quite different properties, but it is not known whether these are the products of two different genes or are two different isoforms encoded by the same gene (Van Eldik, Piperno & Watterson 1980; Schleicher, Lukas & Watterson 1984). Further analysis is being impeded because FP-resistant mutants of *C. reinhardtii* have not been recovered (Harris 1989), the original FP-resistant mutants of *V. carteri* apparently are no longer available, and our attempts to recover similar FP-resistant mutants have not succeeded (Kirk & Harper 1986).

Chlorate-resistant (CR) mutants of *V. carteri* have proved to be extremely useful for developing important molecular-genetic tools, in addition to being of some intrinsic interest. Chlorate is a nitrate analogue and a substrate for the enzyme nitrate reductase (NR) in many species. By reducing nitrate to nitrite (which is subsequently reduced to ammonia by nitrite reductase), NR makes it possible for many bacteria, fungi, algae, and higher plants to utilize nitrate as a nitrogen source; it therefore plays a pivotal role in nitrogen assimilation in nature (Guerrero, Vega & Losada 1981). When it was first discovered that many CR higher plants were deficient in NR activity, it was concluded that chlorate was not intrinsically toxic, but became toxic only when it was reduced to chlorite by NR (Aberg 1947; Liljeström & Aberg 1966). Subsequently it has become clear that in at least some organisms the situation is more complicated than that, because not all NR-deficient mutants are resistant to chlorate (Cove 1976a), and not all CR mutants lack NR activity (Nichols & Syrett 1978; Prieto & Fernandez 1993). Indeed, in one study of *C. reinhardtii* it was found that only 23 of 244 CR mutants analyzed were unable to grow on nitrate as a sole nitrogen source, and it was concluded that at least 14 different loci (seven of which appear to be unrelated to the nitrogen-assimilation pathway) can mutate to generate a CR phenotype (Prieto & Fernandez 1993). In light of such reports, it is quite astonishing how much success has been achieved in using resistance to chlorate as an efficient means of isolating NR-deficient mutants of bacteria, fungi, plants, and algae (Aberg 1947; Liljeström & Aberg 1966; Lewis & Fincham 1970; Solomonson & Vennesland 1972; Cove 1976a,b; Stouthamer 1976; Hofstra 1977; Müller & Grafe 1978; Nichols & Syrett 1978; Sosa, Ortega & Barea 1978; Tomsett & Garrett 1980; Wray 1986). *V. carteri* is no exception.

In an early study, all 20 nitrosoguanidine-induced CR *V. carteri* mutants that were analyzed genetically were able to grow on nitrite (but not nitrate) as a sole nitrogen source,[4] and none of them exhibited significant levels of NR activity after exposure to conditions that induced NR in wild-type spheroids (Huskey et al. 1979). Mendelian analysis revealed that 15 of these mutants had lesions at the *nitA* locus[5] (Huskey et al. 1979b), which has subsequently been shown to encode the NR apoenzyme (Gruber et al. 1992). The remaining five CR mutants had lesions at the *nitC* locus, which has not

---

[4] The 20 CR mutants studied for the published report were drawn from a total of 39 CR mutants that had been isolated, all of which were capable of growing on nitrite but not nitrate (D. L. Kirk unpublished observations).

[5] The conventional nomenclature currently used in *Volvox* genetics, and in this book, is as follows: Locus names are based on an italic, lowercase, three-letter abbreviation of the phenotype associated with the mutation initially used to define that locus. Where more than one locus with the same mutant phenotype has been defined genetically, the loci are distinguished by appended italic capital letters that are assigned in the order in which the loci were defined. Genotypes with respect to a specific locus are designated by appending to the locus name a superscript plus sign (wild-type), or minus sign (mutant), or an allele designation. Mutant phenotypes (as opposed to genotypes) are identified by a two- or three-letter roman abbreviation in which at least the first letter is capitalized. For example, when mutant strains are isolated on the basis of chlorate resistance, they are initially given the phenotypic designation CR; if it is determined that they are also unable to grow on nitrate, they are identified as Nit mutants; if their lesions are mapped to a specific locus, they may be identified as having the *nitA⁻* or *nitB⁻* genotype, or, more specifically, as having some particular allele at the mutant locus, such as *nitA*^CRH22. Such conventions were not always used in early descriptions of mutants, however. In those cases, the nonconventional terms used by the earlier authors are also used in this text.

been analyzed further. It has been speculated that *nitC* may correspond to the *nit2* locus of *C. reinhardtii*, a regulatory locus that controls the expression of the nitrate and nitrite permeases, NR, and nitrite reductase (Fernández et al. 1989; Galaván, Cárdenas & Fernández 1992; Schnell & Lefebvre 1993). If so, the regulatory relationships must be somewhat different in the two species, because *nitC* mutants of *Volvox* are able to grow on nitrite, whereas *nit2* mutants of *C. reinhardtii* apparently are not. A third *V. carteri* locus, *nitB*, was defined by a mutant that, although unable to grow on normal concentrations of nitrate, was chlorate-sensitive and was capable of growing on elevated concentrations of nitrate; it was assumed to have a lesion of nitrate uptake (Huskey et al. 1979b).

In a more recent study, 29 of the first 32 spontaneous or heat-shock-induced CR mutants of *V. carteri* analyzed were found to be unable to utilize nitrate (but able to utilize nitrite) as a nitrogen source, indicating that they lacked NR activity. Nearly a fourth of these CR mutants (13 of 57) were found to have insertions at the *nitA* locus large enough to be readily detected on Southern blots (Miller et al. 1993). The latter study illustrates the important role that CR mutants have played in developing two important molecular-genetic tools that will be described in more detail in the next chapter: A revertible CR mutation of the sort just described was used to identify and clone a transposable element, *Jordan* (Miller et al. 1993), that is now being used for transposon tagging of genes of developmental interest (Section 7.5). Similarly, a nonrevertible CR mutant with a lesion at the *nitA* locus was used as a recipient of cloned *nitA* DNA in studies that led to the development of a transformation system for *V. carteri* (Section 7.4) (Schiedlmeier et al. 1994).

6.3 Morphogenetic mutants

The many interesting categories of morphogenetic mutants initially described by Starr were all discovered by intermittent inspection of wild-type cultures (Starr 1970a). When Huskey began his studies, his hope was that *V. carteri* could be used as a model system for analyzing the genetic control of phototaxis (R. J. Huskey personal communication); therefore he and his associates regularly subjected mutagenized cultures to a phototactic selection scheme (Huskey et al. 1979a). Although they did not reach the original goal of recovering mutants with primary defects in phototaxis by that means, they did recover mutations affecting virtually every stage of morphogenesis, all of which shared the property of causing impaired swimming behavior. Additional morphological mutants that did not affect swimming behavior were recovered by visual screening of aliquots of a mutagenized culture in liquid medium (Huskey et al. 1979a). We have found that plating such cultures in semisolid medium (soft agarose over an agar base) greatly reduces the recovery of false positives with non-heritable abnormalities, because the small clones that are formed in agarose after two or three cycles of asexual reproduction permit one to evaluate immediately the mitotic heritability of an abnormal phenotype.

6.3.1 Cleavage mutants

In wild-type *V. carteri,* the patterns of symmetric and asymmetric divisions are highly stereotyped in both asexual and sexual *V. carteri* embryos, resulting in adult spheroids with highly predictable numbers and arrangements of germ and somatic cells. Mutants have yielded considerable information about some of the ways in which these stereotyped cleavage patterns are regulated.

6.3.1.1 Pattern-switching mutants. As described in Section 5.5.3, three quite different temporal and spatial patterns of asymmetric division are normally seen in wild-type *V. carteri* embryos under optimum culture conditions; as a result, asexual, female, and male spheroids have different, highly characteristic patterns of germ-cell abundance and distribution (see Figures 5.1 and 5.28). This regular correlation between asymmetric division pattern and germ-cell phenotype might lead one to speculate that there is a causal relationship between the pattern of asymmetric division and the type of germ cells that are subsequently produced. Mutational analysis clearly reveals that such is not the case: Mutants have been recovered that have virtually all possible combinations of these three asymmetric-division patterns and germ-cell types, resulting in production of gonidia in the sexual female or male patterns, sperm packets in the female or asexual patterns, and so forth.

Huskey and co-workers defined three multiple-gonidia (*mul*) loci at which mutation causes asexual embryos to cleave in the temporal and spatial patterns normally observed in sexually induced embryos, resulting in the formation of adults containing supernumerary gonidia that are distributed in either the egg or sperm pattern (Huskey et al. 1979a). In MulB mutants (strains with mutations at the *mulB* locus), asymmetric division occurs as in sexually induced wild-type females, but the germ cells subsequently produced are gonidia, not fertilizable eggs (Figure 6.1A). In contrast, in MulC and MulD mutants (which define two unlinked loci with similar mutant phenotypes), asexual embryos cleave like sexually induced male embryos and thus

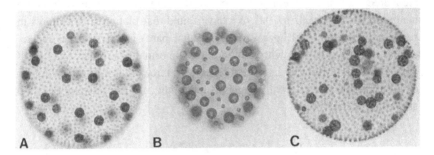

Figure 6.1 Mul mutants. **A**: MulB; gonidia produced in the pattern normally associated with egg production by a sexual female. **B**: MulC; gonidia produced in the 1:1 ratio with somatic cells normally associated with production of androgonidia by a sexual male; MulD spheroids have a similar phenotype. **C**: MulX; temporal and spatial control of asymmetric division appears to have been lost, resulting in production of cells of a variety of sizes in an unpredictable pattern. (From Kirk, Kaufman et al. 1991.)

produce gonidia in a pattern similar to that normally seen for androgonidia (Figure 6.1B). A reasonable working hypothesis is that the *mulB, C,* and *D* loci are normally involved in switching the asymmetric-division program under the control of an activated female or male mating-type locus, and that in the mutants the regulatory coupling has become defective, so that a cleavage pattern normally expressed only in one phase of the life cycle has become the default pathway. In any case, they provide evidence that the cleavage program activated following sexual induction normally is able to override the default program, whatever that may be, because none of these mutants exhibit abnormal cleavage patterns in sexually induced embryos (i.e, induced MulC/ D females and induced MulB males cleave in the wild-type female and male patterns, respectively).

Other mutations cause modifications of asymmetric-division patterns only in sexual embryos. Starr (1970a) described a mutant male (R-1) in which androgonidia, and hence sperm packets, were produced in the pattern typical for egg production in sexual females. Correspondingly, mutation at the multiple-egg (*megA*) locus causes all induced females to produce more than twice the usual number of eggs, and at the extreme, eggs are produced in the 1:1 ratio to somatic cells normally seen for androgonidia in males (Callahan & Huskey 1980). Neither of these mutations had any effect on pattern formation in either asexual spheroids or sexual spheroids of the opposite mating type.

In contrast to the foregoing, a mutation at the *mul-2* locus has been described that affects pattern formation in both asexual and sexual embryos. In Mul-2 mutants, asexual embryos, sexual male embryos, and sexual female embryos all cleave in the pattern normally observed only in sexual females (Callahan & Huskey 1980). This strain appears to have a constitutive mutation of a locus involved in establishing the sexual female cleavage pattern – a mutation that cannot be overridden, even in sexual males.

When a wild-type male or female mating-type locus is activated following exposure to the sexual pheromone, it triggers both a change in the asymmetric-division program and a change in the pathway of germ cell differentiation. The mutants just described indicate that although these two changes are normally triggered coordinately, they must be mediated by two parallel but independent pathways, because they can be dissociated by mutation. This conclusion is reinforced by consideration of certain sterile mutants, to be described later, in which the sexual pheromone activates the switch to a sexual cleavage pattern, but fails to activate the switch in germ-cell differentiation that normally leads to gamete formation.

6.3.1.2 Other types of pattern-forming mutants. Normally, the 16 cells in the anterior half of an asexual embryo first cleave asymmetrically at cleavage cycle 6 and then continue cleaving asymmetrically for two or three more cycles, producing one small somatic initial and one large gonidial initial at each division. Mutations at *mulA* do not cause any perceptible change in the asymmetric-division pattern at cycle 6, but they cause some or all of the large cells produced by this initial asymmetric division to divide symmetrically in cycles 7 and/or 8 (Starr 1970a; Huskey et al. 1979a). The result of this abnormal cleavage pattern is that MulA mutants frequently have clusters of two, three, or four gonidia in regions where individual gonidia are found in wild-type spheroids; moreover, the sizes of these gonidia decrease in proportion to the number of times that they have cleaved symmetrically after first dividing asymmetrically. Considered by itself, this phenotype suggests that *mulA* controls only the late stages of asymmetric division. However, it is not clear that this interpretation can be extended to account for pattern variations observed in sexually induced males carrying the *mulA* lesion. In a single culture of such a strain, some male spheroids possessed excess androgonidia and a deficiency of somatic cells, and other spheroids possessed just the opposite abnormality (Starr 1970a). It is conceivable that this mixture of phenotypes could arise if the final division in MulA males is symmetric, instead of asymmetric as it is in wild-type males. Because it is believed that size at the end of cleavage determines whether a cell will develop as a somatic or germ cell in *V. carteri* (Pall 1975; Kirk, Ransick et al. 1993), if MulA male embryos approaching the final division had cells that were somewhat variable in size, symmetric division in some cases might result in formation of a pair of androgonidia, and in other cases in a pair of somatic cells (as in the R-2 mutant to be described shortly). Careful examination of cleavage in MulA males should reveal whether or not this hypothesis is correct, but such an examination has not been reported.

Although careful studies of cleavage in males with a reduced androgonidia (*radA*) mutation have not been reported either, their phenotype can be rationalized somewhat more securely than can the phenotype of MulA males. In sexual RadA males, the ratio of somatic cells to gonidia is 3:1, rather than the usual 1:1 (Callahan & Huskey 1980). This is the phenotype to be expected if (instead of withdrawing from the division cycle after dividing asymmetrically as wild-type males do) RadA males execute one extra round of division that (in parallel with asexual spheroids) is symmetric in somatic initials but asymmetric in androgonidial initials.

Sexual males of the R-2 mutant described by Starr (1970a) also have reduced numbers of androgonidia and excess numbers of somatic cells, but for a different reason. In R-2 males, as in wild-type males, asymmetric division is delayed until the final division; but at that time, when some cells divide asymmetrically to produce the usual sort of androgonidium/somatic-cell pair, other cells divide symmetrically to produce a pair of somatic cells that are larger than normal (Starr 1970a). The R-2 mutation also affects the sexual female: In induced R-2 female embryos, asymmetric division occurs at the usual time, but only in cells in the anterior half of the embryo (i.e., the same region that normally cleaves asymmetrically in asexual embryos); this leads to production of more germ cells than are normally produced in asexual spheroids, but fewer than are normally produced in sexual females. Thus, the R-2 mutant further dissociates aspects of the sexual phenotype that are normally coupled: It does not affect the *temporal* change in the cleavage pattern that is normally triggered by sexual induction, but it does interfere with the *spatial* changes in the asymmetric division pattern that normally accompany sexual induction.

The five allelic *megA* mutations described by Callahan and Huskey (1980) that were mentioned earlier and the several ''multi-egg'' mutations that were described independently by Zeikus and Starr (1980) (which probably represent additional *megA* alleles) have an effect opposite to that of the R-2 mutation: They increase the number of gametes produced by sexual spheroids of both mating types. The descriptions provided by those two sets of investigators indicate that the Meg phenotype results from a one-cycle delay in the time of asymmetric division relative to sexually induced wild-type females, plus an expansion of the area in which asymmetric division occurs. In its most extreme form, this combination leads to production of eggs in the 1:1 ratio with somatic cells normally seen in sexual males. In males, a *megA* mutation is able to suppress a mutation that by itself would cause a reduction in androgonidial numbers (Callahan & Huskey 1980), and a multi-egg mutation can also cause a sexual male embryo to forgo asymmetric division completely, resulting in a spheroid in which all cells develop as androgonidia (Zeikus & Starr 1980).[6]

Mutations at an undetermined number of "*mulX*" loci cause much less regular patterns of temporal and spatial abnormalities of asymmetric division than any of those described earlier (Kirk, Kaufman et al. 1991). Indeed, in some such strains, both the timing and the locations of asymmetric division appear to be essentially random, resulting in adult spheroids with a wide range of cell sizes, arranged in wholly unpredictable patterns (Figure 6.1C). Whereas other *mul* loci appear to be involved in global patterning of asymmetric division, *mulX* loci are postulated to be involved somehow in controlling the division symmetry within individual dividing cells, such that when a *mulX* function is defective, the specification of division planes in individual cells may be at least partially uncoupled from the control of the global-patterning influences. Further study of such mutants is warranted.

Later, another class of mutations, gonidialess (*gls*), that have a profound effect on asymmetric division will be described. These mutants have lesions at one or more loci that appear to be epistatic to all of the other pattern-forming loci that have been discussed, because a *gls* mutation abolishes asymmetric division, regardless of the status of the various other pattern-forming loci. Discussion of *gls* mutations is being deferred, however, because they play such a central role in our current view of how germ–soma specification appears to be programmed in the *V. carteri* genome – which will be the last topic to be covered in this chapter.

6.3.1.3 Mutations affecting anterior–posterior polarity. Two mutants with pattern-forming defects that were interpreted as being results of aberrant anterior–posterior (A–P) polarity in the cleaving embryo were also described by Sessoms and Huskey (1973). In the "doughnut" strain, both hemispheres of the embryo cleaved asymmetrically to produce gonidia, and each hemisphere formed a phialopore at which inversion was initiated. (It was the torus-shaped adult produced when the two sets of inverting cellular lips met and fused at the equator that gave this mutant its name.) However, the somatic cells of the doughnut mutant exhibited the normal gradient of eyespot size, indicating that they had differentiated normally along the A–P axis. In contrast, in the "double-posterior" mutant (in which gonidia were also produced

---

[6] Sexual male spheroids containing only androgonidia are sometimes (but not always) produced by the wild-type male isolate of *V. carteri* forma *kawasakiensis* (Nozaki 1988), and such spheroids are a diagnostic feature of *Volvox spermatosphaera* (Smith 1944).

in both hemispheres), all adult somatic cells appeared to be of a posterior type, as judged by eyespot size. However, this mutant did not initiate inversion from both ends; it initiated inversion perfectly normally, from one end only.[7] These two mutants indicate that various aspects of A–P polarity that are associated with normal *V. carteri* development can be dissociated by mutation and therefore must be under separate genetic controls. Normal A–P polarity of an asexual individual includes, among other things, (1) restriction of asymmetric divisions and gonidial differentiation to the prospective posterior end of the spheroid, (2) formation of a phialopore and initiation of inversion at the prospective posterior end of the spheroid only, and (3) graded differentiation of somatic cells along the A–P axis of the spheroid with respect to cell size and eyespot size. The doughnut mutation appears to abrogate features 1 and 2, but not 3, whereas the double-posterior mutations appear to abrogate features 1 and 3, but not 2. This reinforces the conclusion drawn from study of the "pattern-switching" mutants described earlier that features that appear to be tightly linked in normal *Volvox* development may be mechanistically quite independent of one another.

6.3.1.4 Mutations directly affecting the cell-division process. The central role of the volvocalean basal apparatus (BA) in determining the site and plane of cell division was discussed in Sections 4.4 and 5.4. In a screen for inversion mutants, a strain called S16 was isolated that had a severely abnormal post-inversion phenotype, but it subsequently turned out to have temperature-sensitive cleavage abnormalities that were associated with severe abnormalities of BA structure and function (Mattson 1984; Kirk, Kaufman et al. 1991; Kirk, Ransick et al. 1993). Strain S16 has been used to define the cleavage A (*cleA*) locus. When grown at 24°C, S16 exhibits only modest defects in cleavage and adult morphology. But as the temperature is raised, cleavage planes become increasingly randomized, and by 37°C cytokinesis becomes so erratic that cells of widely different sizes are produced in a helter-skelter pattern (Figure 6.2). Moreover, at that temperature the normal segregation of organelles often does not occur, and thus multinucleate, anucleate, multiflagellate, and aflagellate cells are all produced (Mattson 1984; Kirk, Ransick et al. 1993). In embryos cleaving at 24°C, the S16 BBs appear to be associated with the nucleus and plasmalemma in a more or less normal manner; but in embryos cleaving at 37°C, many BBs are observed deep in the cytoplasm, with no apparent connection to either the nucleus or the plasmalemma.

---

[7] Subsequently it was found that the double-posterior strain carried two unlinked mutations, one of which caused a flagellaless phenotype and one of which caused an inversion defect, and neither of which, by itself, had any discernible effect on A–P polarity (R. J. Huskey personal commununication)

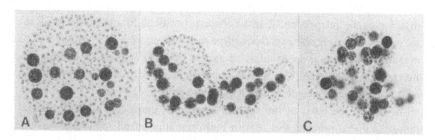

Figure 6.2 S16: A strain with a temperature-sensitive mutation of the *cleA* locus that affects the organization of the basal apparatus. **A**: Cultures maintained at 24°C exhibit relatively modest defects in the cell-division pattern. **B**: At 32°C, cleavage planes are randomized. This results in a large shift in the ratio of large to small cells and in abnormalities of intercellular relationships that cause distorted inversion patterns. These defects, in turn, result in multilobed adults containing an excess of gonidia and a deficiency of somatic cells. **C**: When cultures are shifted to 37°C during cleavage, the division pattern (and hence the adult morphology) is even more severely disturbed. Indeed, cleavage abnormalities are so severe at this temperature that continuous cultivation at 37°C is lethal for S16. (From Kirk, Ransick et al., *The Journal of Cell Biology*, **123**, 191–208, by copyright permission of The Rockefeller University Press.)

In such locations, BBs undergo abortive efforts to template flagellar development, sometimes producing as many as 15 transitional zones but no flagellar axoneme (Mattson 1984). It is anticipated that detailed studies of cell division in S16 at a range of temperatures, at which the structure of the BA will be disrupted to varying degrees, may yield additional insights into the control of various aspects of cytokinesis.

We have observed many other mutants with highly abnormal cytokinesis over the years, but have yet to undertake a systematic Mendelian or cytological analysis of other such strains. Given the complexity of the cytokinetic apparatus of *Volvox*, it is to be anticipated that mutations at many different loci might cause cytokinetic defects, but it remains to be determined at how many such loci viable mutations can be recovered and analyzed.

A very different kind of defect of the embryonic cell division program is seen in premature-cessation-of-division (Pcd) mutants (Pall 1975; Kirk, Ransick et al. 1993). As the name implies, mutation at an undetermined number of *pcd* loci causes cells to withdraw from the cleavage cycle prematurely, although in those mutants that have been examined carefully, all divisions that have taken place have appeared to be morphologically normal (Kirk, Ransick et al. 1993). Pall deduced that in the Pcd strains he studied (which have since been lost), all blastomeres within an individual embryo ceased dividing at the same time, although sibling embryos completed different numbers of divisions (Pall 1975). In contrast, in the Pcd strains we have studied, it is clear that various blastomeres within a single embryo withdraw from the division cycle independently, in an apparently stochastic manner (Kirk, Ransick et al. 1993). In the two types of mutants the consequences of premature cessation of division are similar: an excess of germ cells and a striking deficiency of somatic cells in the adult spheroids, in extreme cases resulting in a near absence of somatic cells (Kirk, Ransick et al. 1993).

It was on the basis of the superabundance of germ cells in both asexual and sexual Pcd mutants that Martin Pall originally postulated that it was the size of a cell at the end of cleavage (not the kind of cytoplasm that it contained) that would determine whether a blastomere would develop as a germ cell or a somatic cell (Pall 1975). As discussed in Section 5.4.1.3, we once expressed considerable skepticism about this hypothesis (Kirk & Harper 1986), and set out to disprove it. However, all of the studies we executed in that effort, including more detailed studies of Pcd mutants, yielded findings wholly consistent with Pall's hypothesis (Kirk, Ransick et al. 1993).

6.3.2 Inversion mutants

Among the most frequently recovered types of morphological mutants are those with inversion defects. In the first mutant screen executed by the Huskey group, fully a quarter of all mutants isolated were either noninverters (Inv) or quasi inverters (q-Inv) (Sessoms & Huskey 1973). We were able to isolate more than 60 independent Inv and/or q-Inv strains in a single screen. Some of those failed to initiate inversion, whereas others initiated inversion quite normally but then arrested at various stages before completing it (Figure 6.3) (Kirk, Viamontes et al. 1982). However, another abundant subclass of putative Inv mutants contains those that tear themselves apart to varying degrees, and in various bizarre ways, in the process of trying to invert. Strain S16, described in the preceding section, and the "fruity" mutant described earlier by Huskey's group (Sessoms & Huskey 1973) both fall into this category. Careful microscopic and/or time-lapse cinemicrographic analysis of such mutants frequently reveals that they exhibit subtle aberrations of cleavage or cytoplasmic-bridge formation that interfere with establishment of normal intercellular relationships, but that become grossly apparent as major malformations only when the embryos are exposed to the severe mechanical stresses that accompany inversion (Kirk, Viamontes et al. 1982, K. J. Green, D. M. Mattson, J. L. Bryant, Jr. & D. L. Kirk unpublished observations). A specific case in point comes from the work of Ann Sessoms, who showed that the critical period for two temperature-sensitive q-Inv mutants (*q-inv*-1 and *q-inv*-2) (Sessoms & Huskey 1973) was not during inversion, but during

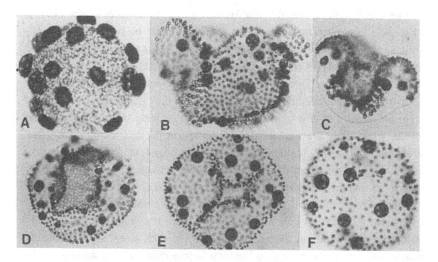

Figure 6.3 Inv mutants. **A:** "True" Inv mutants fail to ever initiate inversion. **B–E:** "quasi-inverters" (q-Inv strains) initiate inversion but then appear to be blocked at various specific stages in the process, producing a variety of bizarre adult phenotypes. **F:** A "doughnut" mutant initiates inversion from both ends, and when the lips of inverting cells meet at the equator, a torus is produced. (From Kirk, Viamontes et al. 1982, in *Developmental Order: Its Origin and Regulation*, ed. S. Subtelny & P. B. Green, copyright © 1982 Alan R. Liss, Inc., reprinted by permission of Wiley-Liss, Inc., a subsidiary of John Wiley & Sons, Inc.)

cleavage (Sessoms 1974). It appears that under restrictive conditions those *q-inv* lesions lead to formation, during cleavage, of cytoplasmic bridges that are too weak to withstand the stresses of inversion without breaking, and that such a defect ultimately results in an inability to complete inversion. In short, a sizable fraction of all strains that are identified provisionally as Inv mutants at the time of isolation turn out to be cleavage mutants when they are analyzed more carefully.

Inversion in *Volvox* exhibits certain obvious parallels to gastrulation in metazoan embryos. But Inv mutants of *V. carteri* are generally viable, whereas metazoan embryos with significant gastrulation defects generally are not. Therefore, we had originally hoped that detailed analysis of Inv mutants of *V. carteri* might yield useful insights into the way in which such a morphogenetic process is programmed genetically and executed cytologically. However, sad to say, inversion mutants of *V. carteri* have proved to be as difficult to analyze in detail as they are easy to isolate, for reasons that will now be discussed.

Genetic analysis of Inv strains has been frustrated primarily by the topological abnormalities of the sexual Inv female. Of the 19 strains with inversion defects initially isolated by the Huskey group (Sessoms & Huskey 1973), only one – a temperature-sensitive mutant that could be crossed under conditions in which it had the wild-type phenotype – was successfully subjected to Mendelian analysis (Huskey et al. 1979a; R. J. Huskey, personal communication). As described in Section 5.5, sperm packets appear to recognize sexual females and become activated only when they make effective contacts with the flagella of the female. But Inv females wear their flagella on the inside, where they are inaccessible to the sperm packets. Females of q-Inv strains have an open phialopore into which an occasional sperm packet may find its way, contact her flagella, and become activated. However, in those cases a second problem arises: After the activated sperm have separated from one another and have digested their way past the somatic-cell layer, they have great difficulty in locating the eggs of the q-Inv female in her partially inside-out configuration. In the vast majority of such cases, sperm probably wander off into the surrounding medium before ever locating an egg.

Cytological and molecular analyses of Inv and q-Inv mutants have also been frustrated by two problems: the difficulty of identifying the primary defects in such mutants by cytological examination, and the lack of molecular handles on the inversion process. However, the transposon tagging system to be described in Section 7.5 offers new hope for those interested in detailed analysis of the inversion process. Transposon-induced Inv mutants should ultimately allow an entirely fresh and productive approach to genetic and molecular analysis of this important morphogenetic process.

6.3.3 ECM mutants

Four major categories of mutations affecting the ECM have been described (Sessoms & Huskey 1973; Huskey et al. 1979a), and a fifth category has been identified (R. C. Starr, personal communication).

In dissolver (Dis) mutants, none of the major structural elements of the ECM assemble normally; thus, when the cytoplasmic bridges that linked cells during embryogenesis are broken down shortly after inversion, the juvenile spheroids dissociate into single cells (Sessoms & Huskey 1973; Huskey et al. 1979a; our unpublished observations). Following dissociation of the spheroid, the somatic cells tumble about in the medium rather aimlessly (being unable to swim in a directional manner in isolation), and the gonidia fall to the bottom of the culture vessel. Although they lack virtually all other ECM structures, most Dis strains that we have examined to date possess a normal CZ1 layer ("vesicle") around each gonidium, and probably around each somatic cell as well (our unpublished observations). In their isolated state,

Dis gonidia grow more slowly than wild-type gonidia in intact spheroids, and seldom get as large before they cleave. Nevertheless, they eventually do initiate cleavage, and they produce embryos that appear relatively normal (albeit somewhat smaller than normal).

Because it is unlikely that a single mutation could obliterate synthesis of nearly all ECM components, a reasonable working hypothesis is that most *dis* mutations affect synthesis or assembly of some critical, primary component of the ECM that serves to nucleate proper assembly of other ECM components that are produced later. Moreover, temperature-shift experiments with a temperature-sensitive (ts) Dis strain (Dis-1) revealed that the critical ts period is right at the beginning of ECM deposition (Sessoms 1974). As noted in Section 5.4.4.4, the sulfated glycoprotein ISG is a reasonable candidate to be a primary organizer of the ECM: It is the first ECM component synthesized in the juvenile (with peak synthesis occurring already during inversion), and it is located in the correct region to provide an initial scaffolding upon which much of the rest of the ECM could be constructed. Thus, the gene encoding ISG is a prime candidate to be the target of (at least some) *dis* mutations. This interpretation is reinforced by the observation that a peptide analogue of the C-terminus of ISG prevents ECM assembly and leads to a phenocopy of the Dis phenotype, if (but only if) it is added to embryos before ECM deposition has begun (Ertl et al. 1992).

From the very limited published descriptions, it might appear that the delayed-dissolver (d-Dis) phenotype has a very different basis than the Dis phenotype: A d-Dis juvenile was said to invert and expand normally, but then to fall apart before its gonidia had begun to cleave (Sessoms & Huskey 1973; Huskey et al. 1979a). From that it might be imagined that the d-Dis phenotype results from precocious and/or excess production of the hatching enzyme, H-lysin (Section 5.4.5). However, observations made during temperature-shift studies of a ts strain (d-Dis-1) lead to an alternative hypothesis. The critical ts period for d-Dis-1 is not at or just before the time that the mutant spheroid breaks down; rather, it is (as with Dis-1) at the very earliest stages of ECM deposition, a day or more before spheroid breakdown begins (Sessoms 1974). Thus, it seems possible that the defect in this strain is in the structural integrity of the ECM itself when its assembly is begun under restrictive conditions, not in the regulation of H-lysin production. Strain d-Dis-1 was successfully mated and used to define a locus on linkage group III (Huskey et al. 1979a), but no additional studies have been reported.

As their name suggests, non-expander (Exp) mutants are distinguished at the light-microscopic level by the fact that they fail, to differing degrees, to enlarge as much as wild-type spheroids. Ultrastructural examination of 13 Exp mutants revealed several types of abnormalities, many of which defied precise description (Sessoms 1974). One strain appeared to have relatively normal ECM organization, but simply subnormal amounts of it. The two Exp mutants that were most similar to wild-type organisms in expansion rates had matrix that seemed more loosely organized than normal; those strains also accumulated nondescript dense material around the flagellar bases and tended to fall apart by the time their gonidia were mature. The most abundant class of Exp mutants (10 of 13) had gross aberrations of the ECM at the ultrastructural level. Although those 10 strains differed in many details, they all tended to have two major types of defects: (1) Although various ECM components that normally are highly regular in structure (such as the spheroid boundary and the cellular-compartment boundaries; see Section 5.4.4) were present in abundance, they were highly abnormal in organization, often being present in multiple layers or disorganized heaps. (2) There was a marked deficiency of the less highly organized, mucilaginous components that usually fill all of the internal compartments of the spheroid (Sessoms 1974). In many cases, the appearance was of a balloon that had collapsed into a crumpled, nondescript mass, for lack of the internal contents that should have kept it smoothly inflated. The six Exp strains that were analyzed genetically defined four *exp* loci (Huskey et al. 1979a). An interesting additional mutant, "unequal expander," which expands normally in the anterior hemisphere but not the posterior one, defined a fifth locus (*u-exp*) (Huskey, Griffin et al. 1979a). Those findings indicate that the products of a great many genes must be involved in forming the complex ECM of the *V. carteri* spheroid. This is hardly surprising, because the relatively small fraction of all known cell-wall mutants of *C. reinhardtii* that were analyzed genetically in an early study defined 19 loci, with most loci being defined by a single allele (Hyams & Davies 1972). The large number of gene products involved in making an ECM as relatively simple as the *Chlamydomonas* wall apparently provided much grist for the evolutionary mill to work with in generating the more complex ECM of the colonial and multicellular green flagellates.

Delayed-release (Rel) mutants have a phenotype that is, in a sense, opposite that of the d-Dis strain: Their juveniles fail to hatch at the normal time (presumably as a result of defects in production or secretion of the H-lysin), and sometimes as many as three or four generations of juveniles will develop within a parental Rel spheroid. Needless to say, within such cramped quarters, development tends to become progressively more abnormal with each generation. Although Rel mutants are observed with moderate frequency in a visual mutant screen, only one has been analyzed genetically (Huskey et al. 1979a).

Premature-release mutants have also been isolated, but have yet to be described in print (R. C. Starr personal communication). In the asexual phase,

juveniles apparently are released before any significant expansion has occurred, and in sexual males, juveniles are released before androgonidia have
even begun to cleave (rather than being retained, as they are in the wild type,
until cleavage has long since been competed and sperm packets are mature).
These strains, more than the d-Dis strain discussed earlier, would appear to
be good candidates for strains with mutations causing premature or excessive
production of the H-lysin (Section 5.4.5).

6.3.4 Cell-orientation mutants

Huskey isolated three categories of mutants with deficient spheroid motility
that turned out to have defects in the orientation of cells with respect to
spheroid coordinates: flagellaless (Flg) mutants, eyespot-location-abnormal
(Eye) mutants, and rotation-abnormal (Rot) mutants (Huskey 1979).

   Of the 10 Flg strains Huskey described, the only three for which crosses
were reported defined three unlinked genes: *flgA, flgB,* and *flgC.* Six Flg
strains had complete flagellar channels in the ECM, but no flagella that pro-
jected from those channels far enough to be detected by either light microscopy or scanning electron microscopy. In all six of those strains lacking
visible flagella, the internal organization of each cell appeared to be normal,
but the cells were oriented randomly with respect to one another and to the
A–P axis of the spheroid. In contrast, all three Flg strains that had short
flagella protruding visibly from their flagellar channels had the highly regular,
wild-type orientation of cells; that is, each cell was oriented so that a line
passing through its flagellar channels was approximately perpendicular to the
A–P axis of the spheroid, and its eyespot was always posterior to that line
(cf. Section 5.3). The difference between these two categories of Flg mutants
led to the proposition that proper orientation of cells with respect to one
another and to the A–P axis of the spheroid requires that functional flagella
be present at the time that matrix deposition begins. As discussed in Section
5.4.4.4, the tenth strain (*flgC11*), a ts Flg mutant, strongly reinforced that
proposition: when *flgC11* was held at the permissive temperature during inversion and matrix deposition, cells in the adult were oriented normally, but
when sibling cultures were held at the restrictive temperature during this
critical period (so that flagella were lacking during early stages of ECM
deposition), cellular orientation was random (Huskey 1979). In neither case
could the cellular orientation be modified later by repeated cycles of flagellar
resorption and regeneration.

   Whereas the foregoing observations indicate that the presence of functional flagella during the initial stages of ECM deposition is *necessary* to

achieve normal orientation of cells within the spheroid, examination of Eye mutants has demonstrated that it is not *sufficient* (Huskey 1979): Four independent Eye mutants were recovered, and all had the same phenotype. They had beating flagella of normal length that were able to agitate the water, but the spheroids were unable to swim forward or execute phototaxis. Upon closer examination, Eye mutants were found to share two features with the first category of Flg mutants described earlier: The internal organization of each cell was normal, but the orientation of cells with respect to one another was random. Three of the Eye mutants were analyzed genetically and found to define at least two *eye* loci; thus we can conclude that at least two functions, in addition to the presence of beating flagella, are required to ensure that cells remain properly oriented at the time that ECM assembly begins – and thus to ensure that the adult will be able to swim in a directional manner and execute phototaxis.

Only one Rot strain was described (Huskey 1979). In contrast to wild-type spheroids, when Rot spheroids were viewed from the posterior end they were seen to rotate in a clockwise direction, rather than a counterclockwise direction. Most such spheroids simply rotated in place, but a few were capable of weak phototaxis – but they swam toward the light backward (i.e., with the gonidia-free end trailing, rather than leading)! Upon closer examination, the cells of those spheroids were all observed to be oriented in the same manner with respect to the A–P axis of the spheroid, but each was backward, that is, its eyespot was on the side of the cell facing the anterior pole, rather than the posterior pole, of the spheroid. It is unfortunate that this strain has been lost, because further analysis of the developmental basis for this bizarre arrangement might have proved quite illuminating.

Huskey did not achieve his original goal of dissecting phototaxis genetically by isolating mutants with primary defects in the phototactic machinery. Nonetheless, the cell-orientation mutants that he isolated and studied made an important contribution to our understanding of the mechanisms of *Volvox* phototaxis by solidifying two important inferences: (1) that the predictable orientation of cells in a *Volvox* spheroid with respect to one another and to the A–P axis of the spheroid is a sine qua non for these cells to be able to coordinate their motile activities, and (2) that the presence of functional flagella at the time of initial matrix deposition is necessary (albeit not sufficient) to achieve this kind of adaptive cellular orientation.

6.4 Mutations affecting sexual reproduction

In addition to the pattern mutations that affect the numbers and distributions
of gametes in sexual spheroids (Section 6.3.1), four major categories of mutations with more drastic effects on the sexual reproduction of *V. carteri* have
been described.

6.4.1 Constitutive sexuality

Starr (1970a) reported isolating several female clones that produced sexual
progeny in the absence of exogenous sexual pheromone and in the absence
of any detectable sexual pheromone that they themselves had produced. He
then demonstrated that one of these strains had a mutation tightly linked to
the mating type locus. Callahan and Huskey (1980) expanded on these observations, and defined three unlinked sexual-constitutive (*sex^c*) loci.

A mutation at the *sex^c A* locus caused both males and females to develop
sexually in the absence of the sexual pheromone (Callahan & Huskey 1980).
Moreover, by itself, that mutation was lethal for members of either sex that
did not find a mate and undergo sexual fusion. Sexuality is normally a dead
end for males that do not find a mate, because both the somatic cells and the
sperm of a male spheroid are terminally differentiated, postmitotic cells.
However, sexually induced wild-type females can reproduce asexually in the
absence of a mate, because eggs that are not fertilized eventually redifferentiate as gonidia, enlarge, and divide to form new spheroids (Section 5.5.3).
However, eggs produced by females carrying the *sex^c A* mutation were unable
to redifferentiate as gonidia. Hence, that mutation could only be carried on
the background in which it was recovered: as a suppressor of a sterile mutation known as *sicA* (sickle). The properties of both the SicA mutant and
the Sex^c A/SicA double mutant will be discussed later.

A mutation at the *sex^c B* locus also caused both females and males to
develop sexually in the absence of pheromone, but it did not preclude continued asexual reproduction by either sex. In Sex^c B females, all embryos
produced eggs in every generation, but these eggs, like those of wild-type
females, were capable of redifferentiating as gonidia if unfertilized. In Sex^c B
males, embryos produced a mixture of androgonidia (which went on to form
sperm packets) and small gonidia that eventually cleaved to form new embryos with a similar mixture of androgonidia and gonidia. Interestingly, exogenous sexual pheromone had no effect on the ratio of androgonidia to
gonidia that were formed by Sex^c B males (Callahan & Huskey 1980).

A mutation at the *sex^c C* locus caused females carrying it to produce a mixture of asexual and sexual offspring in the absence of pheromone. Eggs of sexual spheroids could redifferentiate as gonidia and cleave, whereupon they (like the gonidia of asexual Sex^{c}C spheroids) produced a mixture of sexual and asexual progeny. In contrast to Sex^{c}B males, Sex^{c}C females do produce a higher proportion of sexual progeny (up to 100%) when pheromone is added to the medium. The effect of the sex^cC mutation on males is unknown, because the gene is so tightly linked to mating type that no Sex^{c}C male recombinants were found among more than 6,000 progeny (Callahan & Huskey 1980).

It was postulated that the "70-7" mutation studied earlier by Starr (1970a) – which also was tightly linked to mating type – is an allele of sex^cC; however, the lack of recombinants precluded a test for allelism. Assuming that they are allelic, the 70-7 mutation appears to be the stronger allele, since 70-7 females produced exclusively sexual spheroids under standard culture conditions, whereas the Sex^{c}C mutant of Callahan and Huskey never produced 100% sexual progeny in the absence (and rarely did so in the presence) of exogenous pheromone.

Because all of their Sex^c mutants had slightly different phenotypes, Callahan and Huskey (1980) were able to examine epistatic relationships among them, with the following results: In Sex^{c}A/Sex^{c}B double mutants, just as in Sex^{c}B single mutants, both females and males were able to reproduce asexually. That was interpreted to mean that sex^cB was completely epistatic to sex^cA, and thus that these two genes must act via the same pathway. In contrast, in Sex^{c}A/Sex^{c}C females a combination of the two mutant phenotypes was seen: As in Sex^{c}A, all progeny spheroids were 100% sexual, but as in Sex^{c}C, the eggs were capable of redifferentiating as gonidia. That was taken to mean that sex^cA and sex^cC act on the sexual process via different regulatory pathways. The same conclusion was drawn with respect to sex^cB and sex^cC , when it was observed that the Sex^{c}B/Sex^{c}C double mutant combined features that were present in the two single mutants.

As noted earlier, females with a sex^cC mutation develop differently in the presence and absence of the sexual inducer, but males with a sex^cB mutation do not. Those observations, along with other evidence that neither Sex^{c}A nor Sex^{c}B mutants are responsive to the inducer, led Callahan and Huskey to propose that the sex^cA and sex^cB loci encode functions required for responding to the inducer, but that the sex^cC function is part of a different, inducer-independent pathway by which sexuality might be triggered directly by some unknown set of environmental conditions. They further speculated that this inducer-independent pathway might provide the means by which "first males" are produced to get the sexual cycle started when environmental conditions warrant it. Appealing as this idea seems, however, no one has yet identified any set of environmental conditions that will start the sexual cycle in wild-type *V. carteri* in the absence of pheromone production (see Section 5.5).

6.4.2 Reduced fertility and sterility

The first *V. carteri* mutant with reduced fertility, strain 70-3, was described by Starr (1970a). When exposed to even 10^4 times as much pheromone as is required by the wild-type strain for complete sexual induction, 70-3 responded by making somewhat more gonidia than normal, but never produced more than 2% sexual females under otherwise standard culture conditions (Starr 1970a). The basic defect in strain 70-3 segregates among progeny as a simple Mendelian trait caused by a mutation at a so-called "noninducibility" locus (Starr 1972b; Zeikus & Starr 1980). However, the severity of the phenotype caused by this mutant allele is subject to extensive modulation (from almost total noninducibility to nearly normal inducibility) by a complex set of genetic and environmental factors. Alleles at many different loci appear to act as partial suppressors of the noninducible phenotype, and the effects of the various suppressor alleles are additive. The only suppressors identified that had any detectable phenotypic effect in the absence of the noninducibility allele were the multi-egg (*megA*) alleles discussed in Section 6.3.1.2, all of which were among the strongest suppressors of noninducibility that were detected, but that nevertheless could not completely suppress the noninducible phenotype in the absence of suppressor alleles at other loci (Zeikus & Starr 1980).

Under standard culture conditions, the severity of the noninducible phenotype (measured as the percentage of the progeny that were sexual following exposure to excess pheromone) was relatively uniform among replicates of any given genotype, but all such strains produced more sexual progeny when exposed to pheromone in the presence of an elevated level of either CO_2 or light. That led the authors to test the possibility that the noninducibility allele might influence sexual induction very indirectly, through a photosynthetic defect. However, no defects in photosynthetic activity could be detected in the mutants. Moreover, sublethal concentrations of the photosynthetic inhibitor 3-(3,4 dichlorophenyl) dimethyl urea had no effect on the inducibility of wild-type strains – beyond a generalized retardation of development. Nor did either elevated CO_2 or light enhance sexual induction when wild-type strains were exposed to limiting concentrations of the pheromone. Thus, these CO_2 and light effects on expression of the noninducibility mutation appear to be mediated by some unknown, nonphotosynthetic pathway (Zeikus & Starr 1980).

Fortunately, most mutations that affect fertility do not appear to be that complex.

One of the most extreme sterility mutations described to date is *sicA* (Callahan & Huskey 1980). In the presence of the sexual pheromone, SicA gonidia become sickle-shaped and refuse to cleave as long as the pheromone is present. But when the pheromone is washed out, these gonidia resume their normal shape and cleave to form exclusively asexual progeny. Interestingly, the SicA phenotype is completely suppressed by the *sex^{c}A* mutation, and at the same time *sicA* partially suppresses the Sex^{c}A phenotype, as discussed in the preceding section. The *sexA* mutation cannot be propagated asexually in a wild-type genetic background because the eggs produced under the influence of this mutation do not redifferentiate as gonidia, but SicA/SexA double mutants can be propagated asexually because their unfertilized eggs do redifferentiate. It is noteworthy that the eggs of this strain are capable of being fertilized, despite the presence of the *sicA* ''sterility'' allele. The *sex^{c}B* mutation also suppresses both aspects of the SicA phenotype: the sickling of gonidia in the presence of pheromone, and the sterility. In contrast, in a SicA/Sex^{c}C double mutant both genes are expressed: both eggs and gonidia are formed in the absence of pheromone (as in Sex^{c}C), but when pheromone is added, the gonidia sickle and do not cleave (as in SicA).

The parthenogenetic-androgonidia (PanA) mutant responds to pheromone by cleaving in the sexual male pattern, but the germ cells it produces are all gonidia that cleave to form new juvenile spheroids, not androgonidia that cleave to form sperm packets; thus PanA is completely sterile (Callahan & Huskey 1980). This phenotype parallels one seen in the noninducibility strains discussed earlier: After being exposed to the sexual pheromone, many gonidia of noninducible females cleave to produce offspring with gonidia in the typical egg pattern (Starr 1970a; Zeikus & Starr 1980). Thus, once again, we see that the two responses that normally are coupled when wild-type strains are exposed to pheromone (i.e., the change in cleavage pattern and the change in germ-cell type) can be dissociated by mutation and therefore must be controlled by separate genetic functions that are normally activated together in response to the pheromone.

6.4.3 Male potency

The two standard wild-type strains of *V. carteri* f. *nagariensis* initially isolated by Starr were HK 10 (female) and HK 9 (male). The basic mechanism of sexual induction in this forma was discovered (following up on earlier studies of *V. aureus* and *V. carteri* f. *weismannia*) (Darden 1966; Kochert 1968) when diluted medium from a sexual culture of HK 9 was shown to be capable of causing asexual cultures of both HK 10 and HK 9 to switch to the sexual pathway (Starr 1969). Subsequently, however, it was found that fluid from a sexual culture of strain 69-1b, an F_1 male derived from an HK 10 × HK 9 cross, was 100 times as potent as fluid from a comparable sexual culture of its male parent, HK 9 (Starr 1970a). That indicated that although females do not normally make detectable quantities of the inducer, they must possess autosomal genes capable of influencing the potency of the inducer produced by their male offspring. Subsequently it was observed that different male progeny derived from a single cross could vary by six orders of magnitude with respect to the potency of the inducer they produced under highly standardized conditions! Further genetic analysis led to the conclusion that male potency is heritable as a multigene trait, probably involving at least four loci, and possibly as many as six loci, two of which may be linked to one another, but none of which appear to be linked to mating type (Meredith & Starr 1975). The evidence indicated that differences in potency were due at least in part to differences in the quality (as opposed to quantity) of the inducing molecules, but differences in the sizes or chromatographic behaviors of inducers from high-potency and low-potency males could not be detected (Meredith & Starr 1975). In light of the fact that the pheromone is now known to be a glycoprotein with a complex glycosylation pattern, the multigenic basis of male potency is not highly surprising. Interestingly, a male with significantly higher potency than strain 69-1b has not been reported, and hence it was from 69-1b that the pheromone was purified and characterized and from which the gene encoding the polypeptide backbone of the pheromone was cloned and sequenced, as described in Section 5.5.2.

By analysis of DNA polymorphism (using a pheromone-specific DNA probe provided by Manfred Sumper) we have established that whereas strain 69-1b (derived from Japan) has a tandem array of five or six genes encoding the pheromone that are arranged in a tandem array,[8] a wild-type male strain isolated from near Poona, India (Adams et al. 1990), has only one copy of this gene. Nevertheless, we have been unable to detect any significant difference in the potencies of the culture supernatants produced by these two strains under controlled conditions.

---

[8] Sequence analysis indicates that these genes all encode the same polypeptide sequence, but differ somewhat in noncoding regions (M. Sumper personal communication).

6.4.4   Female hypersensitivity

As noted in Section 5.5.2, the sexual pheromone of *V. carteri* is one of the most potent bioactive molecules known, capable of inducing full sexual development in wild-type spheroids at a concentration of less than 10^{-16} M. However, a multi-egg mutant, "Gone 12," was described that is 10-fold more sensitive to the sexual pheromone than are wild-type strains (Starr & Jaenicke 1989). No segregation was observed between the multi-egg and pheromone-hypersensitivity traits when Gone 12 was crossed to wild-type males. It has been postulated that Gone 12 either produces a sufficient amount of the sexual pheromone constitutively, or else produces enough less of the putative inhibitor of pheromone action (see Section 5.5.4.c), that it is poised on the verge of induction and thus needs very little additional exogenous pheromone to push it over the edge (Starr & Jaenicke 1989).

6.4.5   Gender reversal

Ultraviolet (UV) treatment of a wild-type male strain of *V. carteri* (Pal 3) resulted in the production of stable mutants that produced fertile eggs instead of sperm packets (Starr & Jaenicke 1989). This is the only mating-type switch that has ever been detected in heterothallic green algae (i.e., in algae with genetically distinct sexes; see Section 5.3.6). This type of mutant, which thus far has not been given an official name in print,[9] responds to pheromone by cleaving in the typical male pattern, but the germ cells it produces develop as eggs. These eggs are then capable of being fertilized by sperm produced by the unmutagenized brothers of the mutant individual. When the progeny derived from this incestuous coupling are exposed to pheromone, half of them develop as normal males, and half as egg-producing males (Starr & Jaenicke 1989), indicating that the transformation is the result of a single, stable mutation linked to the mating-type (*mt*) locus.

Several aspects of the sex-reversal phenomenon are worthy of further consideration: (1) The fact that a male strain can be converted into a fertile female by mutation indicates not only that the male genome contains all of the structural genes required for producing fertile eggs but also that the mating-type-male (*mt-m*) locus must contain, in latent but potentially active form, the regulatory element(s) required for activating expression of these genes.

---

[9] In conversation, Professor Starr tends to refer to such a mutant as a "she-male." He seems resistant to my suggestion that it be named Christine Jorgenson (or *cjo*), in honor of the first member of the species *Homo sapiens* known to have undergone a surgical gender change.

It is not surprising to learn that the two sexes share all of the autosomal genes required for gender-specific functions. But given the strictly heterothallic nature of all known isolates of *V. carteri*, it might well have been anticipated that even if sequences required for activating the female sexual program had once been present in the *mt-m* locus (and vice versa), they would not have been subjected to stabilizing natural selection, and therefore over the centuries they might have become nonfunctional through mutation and genetic drift. But apparently that is not the case, at least for the *mt-m* locus. (2) The phenotype of the egg-producing male (fertile eggs in the male spatial pattern) further reinforces the conclusion that two aspects of the sexual phenotype that normally are associated (i.e., a gender-specific cleavage pattern and germ-cell phenotype) are controlled independently by the *mt* locus when it has been activated by the sex-inducing pheromone. Moreover, this mutant indicates that one of these *mt* functions can mutate independently of the other, and hence they must be regulated by separate elements within the *mt* locus. (3) Starr (personal communication) has indicated that the appearance of she-males in cultures of wild-type males that have been UV irradiated is far from a rare event: He has observed "several" mutants appearing in the first generation after irradiating only a few hundred males. This high frequency suggests that such a gender switch is unlikely to be the result of some kind of rare point mutation. In *Chlamydomonas reinhardtii*, which is also heterothallic, it is now known that the *mt* loci of the two different mating types share several lengthy sequence elements, but those elements are present in different orders and orientations in the two mating types, presumably as the result of a sequence of inversions and rearrangements (Ferris & Goodenough 1994). It is tempting to speculate that the mating-type switch observed by Starr is the result of a UV-induced chromosomal rearrangement that switches the positions of male-specific and female-specific regulatory functions, thereby silencing the former and placing the latter in a potentially active state. This hypothesis is, of course, potentially subject to test.[10] (4) The phenotype of the egg-producing male resembles rather closely the phenotype of the MegA/multi-egg females described earlier (i.e., both types of mutants produce eggs in the male pattern). However, genetic evidence indicates that they have converged on this common phenotypic state in two quite different ways. In the egg-producing male, this phenotype arises from a mutant *mt-m* locus

---

[10] One obvious prediction derived from this hypothesis is that reversion to *mt-m* should occur under the same conditions and with about the same frequency as was observed for the forward mutation. Unfortunately, however, it is much easier to detect and recover a switch in the male-to-female direction than the reverse. This is because sexual development must be induced with pheromone in order to determine if a gender switch has occurred, and sexually induced males self-destruct, whereas sexually induced females can be cloned, as discussed earlier.

combined with wild-type pattern-forming loci, whereas in the MegA strains it arises from a wild-type *mt-f* locus combined with a mutant pattern-forming locus. (5) Finally, it should be pointed out that the progeny Starr produced by crossing an egg-producing male with his brothers should be genetically identical at all loci except those tightly linked to *mt*, thus constituting the first known isogenic male and female lines of *Volvox*.

6.5   Mutations disrupting the normal germ–soma dichotomy

As indicated in the opening chapters, I find *V. carteri* to be particularly attractive as a model for analyzing the genetic and evolutionary origins of dichotomous cellular differentiation and establishment of a germ–soma division of labor. Of the many phenotypic categories of mutants that have been detected in *V. carteri* (not all of which are described in this book, by any means), there are three categories that appear particularly relevant in this context, because they lead to a breakdown of the germ–soma division of labor in one or both cell types. Potentially, these mutations may provide insight into the types of functions that were added during the course of evolution to permit dichotomous differentiation and division of labor.

Before discussing these mutants, it may be worthwhile to digress momentarily to recall the principal way in which the *V. carteri* asexual life cycle differs from that of its simpler volvocalean relatives. In all unicellular and "homocytic" colonial volvocaleans there is but a single cell type at any one stage in the life history of the organism, and thus all cells must be capable of executing all vital functions. In such species, during asexual reproduction each cell first executes vegetative functions (motility, growth, etc.) while it is in the biflagellate state, and then it transforms into a reproductive cell in which the flagella are separated from their BBs while the cell rounds up, reorganizes its cytoplasm, and proceeds to initiate a rapid sequence of cleavage divisions. At the end of this division cycle (and in the case of the volvocaceans, after inversion), all cells once again differentiate as biflagellate vegetative cells. Thus, all reproductive cells are derived from vegetative cells, and all vegetative cells are capable of becoming reproductive cells. In *V. carteri*, however, these two sets of functions are parceled out to two different cell types: somatic cells and gonidia. Thus the question we wish to address: What genetic functions had to be added to the ancestral genome to cause development of two cell types, each of which executes only one half of the ancestral developmental program? Keys are to be sought from mutants that fail to establish such a complete division of labor.

6.5.1 Somatic-regenerator mutants

Once again, for the first clue to an interesting aspect of *Volvox* biology we return to the incisive observations of Richard Starr. While screening cultures for spontaneous females, he discovered something entirely different: a spheroid in which most of the somatic cells were undergoing the usual degeneration and programmed death, but in which a cluster of somatic cells in one region had enlarged and divided to form small asexual spheroids (Starr 1970a). When these tiny spheroids were isolated, they and their progeny repeated the cycle, producing somatic cells that appeared normal at first, but then resorbed their flagella and eyespots, enlarged, transformed into gonidia, and divided to produce spheroids of like phenotype.

Starr showed that this "somatic regeneration," or "fertile somatic-cell" trait, as he called it, was inherited in a simple Mendelian manner. Moreover, he observed that when somatic regenerators of either mating type were exposed to the sexual pheromone, their somatic cells were capable of executing all aspects of sexual, as well as asexual, reproduction. For example, following induction, the somatic cells of the mutant male strain were capable of producing small sexual male spheroids containing a 1:1 ratio of somatic cells and androgonidia, but then later the somatic cells of these sexual males divided to form tiny sperm packets. Similarly, when a female strain carrying this trait was exposed to the sexual pheromone, somatic cells could give rise to sexual spheroids containing eggs. Thus, although wild-type somatic cells are postmitotic and incapable of division, mutation at a single locus opens up to these cells all avenues of reproduction that exist in the species.

Huskey and Griffin subsequently extended the analysis of such somatic-regenerator (Reg) mutants (Huskey & Griffin 1979). They analyzed 39 Reg mutants in four phenotypic classes and showed that 38 alleles exhibited less than 0.1% recombination in crosses and thus defined a single locus, which they named *regA*. The 39th mutant had a lesion that mapped to an unlinked locus that they called *regB*. However, subsequent reexamination of this strain in the same laboratory showed that the so-called RegB mutant did not really have a somatic regenerator phenotype: in distinction to all other Reg mutants, it never formed normal-looking somatic cells, and it had a constellation of other defects not seen in any of the Reg mutants (Baran 1984). Thus, in retrospect it can be said that all mutants with the Reg phenotype that were studied by Huskey and Griffin mapped to the *regA* locus.[11] We have since

---

[11] Earlier they had concluded that the strain that had initially been identified as a "small regenerator" (s-Reg) (Sessoms & Huskey 1973) had been mischaracterized and was actually a constitutively sexual female; indeed, it was the strain used to define the *sex*ᶜ locus discussed in Section 6.4.1 (Huskey et al. 1979a; Callahan & Huskey 1980).

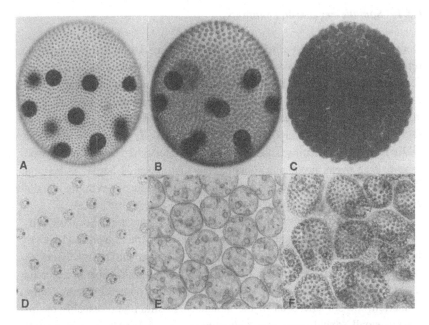

Figure 6.4 A "somatic regenerator" (Reg) mutant. **A–C**: Appearances of Reg spheroids at about 18 hour intervals after hatching. **D–F**: Higher magnification views of surface cells of the spheroids above. **A, D**: Initially, Reg spheroids appear to have a normal complement of somatic cells and gonidia, and thus they resemble wild-type spheroids. However, within a day Reg somatic cells resorb their flagella and eyespots and redifferentiate as gonidia (**B, E**) (cf. **B** with Figure 5.9A). By a day later, gonidia that were derived from somatic cells have divided to produce small spheroids (**C, F**) that will repeat the cycle. The "true" gonidia of Reg spheroids behave just like wild-type gonidia, except that the somatic cells of their progeny spheroids have the Reg phenotype. (From Kirk, Baran et al. 1987, with permission from the copyright holder, Cell Press.)

analyzed many other, independently isolated Reg strains and confirmed that they all have lesions that map to the *regA* locus also (Harper, Huson et al. 1987; Kirk, Baran et al. 1987; Adams, Stamer et al. 1990; our unpublished studies). Thus, the Reg phenotype is the only one for which the *V. carteri* genome can be considered to have been saturated with mutations.

The three phenotypic classes associated with *regA* mutations in the study of Huskey and Griffin (1979) differed with respect to the number and distribution of the somatic cells within a spheroid that regenerated. By far the largest number (33 of 38) were "total regenerators," in which all somatic cells regenerated (Figure 6.4). In three "polar regenerators," some cells at the extreme anterior end of the spheroid (and occasionally a few at the extreme posterior end) failed to regenerate. In these strains, cells at the margin between regenerating and nonregenerating regions sometimes divided while still retaining aspects of the somatic-cell phenotype. The remaining two strains were "spotty regenerators" in which approximately equal numbers of regenerating and nonregenerating cells were intermixed in a random pattern over the surface of the spheroid. This phenotypic variability raised the possibility that there might be three tightly linked genes of somewhat different functions at the *regA* locus. However, a cold-sensitive RegA mutant (RegA[150]) was used to demonstrate that all three of these phenotypes can be generated by differences in the expression of a single mutation at the *regA* locus. First it was shown that RegA[150] has its critical ts period shortly after the end of embryogenesis. Then it was shown that which of the three classes of nonconditional mutants RegA[150] adults resembled depended on when they had been shifted from the permissive to the restrictive temperature, relative to this ts period. Spheroids shifted to the restrictive temperature long before the critical period began had the "total Reg" phenotype, those shifted early in the critical period had the "polar Reg" phenotype, and those shifted late in the critical period had the "spotty Reg" phenotype (Huskey & Griffin 1979).

All such data taken together lead to the conclusion that there is but a single gene in the *V. carteri* genome – *regA* – that can mutate to give the Reg phenotype. By conventional genetic reasoning it would appear obvious that the function of the wild-type allele of *regA* is to suppress all aspects of reproductive development in somatic cells. Thus, when *regA* is mutant, somatic cells follow the ancestral pathway of development: first vegetative and then reproductive. Because no difference has yet been detected between the developmental patterns of gonidia that carry the wild-type and mutant alleles of *regA*, we conclude that *regA* is never expressed in gonidia.

An iconoclastic idea about how the *regA* gene may be regulated (turned off in gonidia and on in somatic cells) came with the serendipitous discovery that the locus exhibits extraordinary stage-specific, locus-specific, and cell-type-specific hypermutability (Kirk, Baran et al. 1987). When juveniles were exposed to UV irradiation precisely three hours after the end of inversion (shortly before studies of RegA[150] had indicated that the *regA* function is expressed), as many as 10% of all gonidia in these juveniles underwent a nonrevertible mutation to the *regA*⁻ state. Equally noteworthy was the fact that about 70% of the gonidia that did not undergo mutation to the *regA*⁻ state in response to the UV irradiation also did not continue to develop as gonidia; rather, they developed as larger-than-normal, but terminally differentiated, somatic cells. Several other important observations about this phenomenon were reported (Kirk, Baran et al. 1987): (1) No increase in mutation rate at any locus other than *regA* could be detected as a consequence of this treatment. (2) All of the mutations induced by UV irradiation arose in gonidia; none arose in somatic cells, even though *regA* mutations are expressed by somatic cells, and there were more than 100 times as many somatic cells as gonidia in the target spheroids. (3) The hypermutable period was only one hour long. (4) Agents classified as point mutagens (e.g., nitrosoguanidine or ethyl methanesulfonate) did not induce *regA* mutations at that time, although they are very effective as conventional mutagens when applied during cleavage. (5) In contrast, several agents that either induce error-prone recombination and repair or else interfere with normal recombination and repair in other systems (e.g., bleomycin, nalidixic acid, or novobiocin) induced the same types of locus-specific mutations as UV irradiation when applied at the same time. (6) Gonidia exhibited a second short period of locus-specific hypermutability after they had nearly completed their differentiation, but a few hours before they began to cleave.

Three aspects of those results struck us as particularly noteworthy as we attempted to understand the significance of this surprising phenomenon: (1) The *regA* locus becomes hypermutable in gonidia, in which it is never normally expressed, shortly before it is about to be expressed in somatic cells. (2) Most of the gonidia that do not mutate in response to UV treatment differentiate as terminally differentiated (large) somatic cells, as if UV irradiation had caused them to express the *regA* gene. (3) Neither of those responses is induced by conventional point mutagens, but only by agents that tend to cause error-prone recombination and repair. Findings 1 and 2 suggested that the unusual mutational properties of this locus might well be reflections of an unusual mechanism by which its expression normally is completely repressed in gonidia but not somatic cells, and finding 3 suggested that this regulatory mechanism might involve some form of DNA recombination or repair. The working hypothesis that evolved from these considerations is as follows:

1. For normal development to occur in *V. carteri,* the *regA* gene must exist in two extreme functional states: It must be "fully on" in somatic cells (to preclude all aspects of reproductive development), but it must be "fully off" in gonidia (to permit reproductive development).

2. The transition between these two states involves a cyclic DNA-level rearrangement, such as the kind of cyclic inversion of the promoter region of a flagellin gene that is involved in *Salmonella* phase transition (Zeig, Hillmen & Simon 1978).

3. Shortly after the end of embryogenesis, and just before overt cytodifferentiation begins, the gene undergoes an inactivating rearrangement in presumptive gonidia to convert it to the fully off

state; then, just before the next round of cleavage is to begin, this rearrangement is reversed, so that the gene will be inherited by all embryonic cells in the potentially active configuration.

4. If agents that cause error-prone recombination or interfere with normal recombination are present in the gonidia at the time these rearrangements are taking place, one of two things may occur: Either the rearrangement may be prevented, causing the *regA* gene to be expressed in the presumptive gonidia (which in turn will preclude them from continuing to develop as gonidia), or the rearranging segment may be lost, causing an irreversible *regA* mutation.

All attempts that have been made thus far to test the predictions of this hypothesis indirectly have yielded observations consistent with the hypothesis. But the only direct and definitive test will be to determine whether or not the DNA sequences of the *regA* locus are different in gonidia and somatic cells. Steps toward performing this test will be discussed in the next chapter.

6.5.2 Late-gonidia mutants

A mutation at any one of a set of at least four related but unlinked loci will cause a breakdown of the germ–soma division of labor in presumptive gonidia that is the reciprocal of that which occurs in somatic cells carrying a *regA* mutation. These are called the *lag* (late-gonidia) loci (Kirk 1988, 1990; Kirk, Kirk et al. 1990; Kirk, Kaufman et al. 1991). In Lag mutants, asymmetric division appears to occur quite normally. But whereas the larger cells (presumptive gonidia) of wild-type embryos begin growing rapidly as soon as they start to differentiate, the larger cells of Lag mutants do not (Figure 6.5). Instead, they first develop as larger-than-normal somatic cells, with unusually long flagella and large eyespots. Only after functioning for a day as somatic cells do these larger cells resorb their flagella, dedifferentiate, and redifferentiate as gonidia. Thus, just as a *regA* mutation causes the smaller cells produced by asymmetric division to follow the ancestral pathway of "first vegetative and then reproductive," a *lag* mutation causes the larger cells produced by asymmetric division to do this.

No change in the behavior of somatic-cell initials can be detected in Lag mutants. Therefore, just as we believe that the *regA* gene is not normally expressed in gonidia, we believe that the *lag* genes are not normally expressed in somatic cells. Mutations at all four *lag* loci will cause the same abnormal gonidial phenotype, both individually and in various combinations;

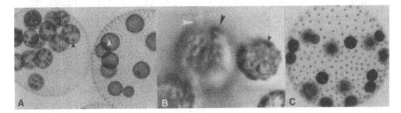

Figure 6.5 A "late-gonidia" (Lag) mutant. **A**: A comparison of a wild-type spheroid (left) with a Lag spheroid (right) that is about 12 hours older. Whereas gonidia are readily apparent in the younger wild-type juveniles (black arrowhead), in the older Lag juveniles they are not. However, on closer examination it is seen that the Lag juveniles do have some slightly larger cells (white arrowhead) in the positions where gonidia are normally located. At higher magnification it can be seen (**B**) that these larger cells (presumptive gonidia) have differentiated as larger-than-normal somatic cells with long flagella (white arrowhead) and large eyespots (large black arrowhead; compare to the eyespot in the adjacent somatic cell, marked with the smaller arrowhead). Within another day, these larger-than-normal somatic cells undergo the same transformation as the somatic cells of Reg spheroids: They resorb their flagella and eyespots and redifferentiate as gonidia that will subsequently cleave (**C**).

thus we conclude that they all act through a common pathway. The function of the *lag* pathway in wild-type spheroids appears to be to act in the large cells produced by asymmetric division to suppress expression of those genes required for differentiation as somatic cells.

6.5.3 Gonidialess mutants

Experimental analysis indicates that any cell in the *V. carteri* embryo that is more than about 8 μm in diameter at the end of cleavage – no matter where or how it has been produced – will pursue the gonidial pathway of development, whereas any cell below this threshold size at the end of cleavage will follow the somatic-cell pathway of development (Kirk, Ransick et al. 1993). Thus, genes regulating asymmetric division, and hence cell size at the end of cleavage, appear to play central roles in the programming of germ–soma differentiation in this species.

As noted earlier (Section 6.3.1), many loci play roles in determining where and when asymmetric divisions will occur in the cleaving embryo. Mutations in such "pattern-switching" genes affect the numbers and locations of the large cells that will differentiate as gonidia, but they have no discernible effect on the phenotype of these large cells and hence no significant effect on the basic germ–soma dichotomy of the species. But one class of mutations

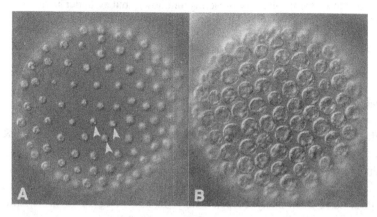

Figure 6.6 A gonidialess/regenerator (Gls/Reg) double mutant viewed by Nomarski differential-interference microscopy. In Gls/Reg embryos there are no asymmetric divisions; all cells are of approximately the same size, and all differentiate initially as somatic cells (**A**); arrowheads identify eyespots (flagella are not readily apparent at this magnification). Within a day or so, all cells of the Gls/Reg mutant then redifferentiate as gonidia (**B**). This resembles the asexual life cycle of a colonial volvocacean such as *Eudorina*.

abolishes the germ–soma dichotomy by abolishing asymmetric division. These are the gonidialess (*gls*) mutations (Huskey & Griffin 1979; Kirk, Kaufman et al. 1991).

Because a Gls embryo fails to set aside the large cells required for formation of gonidia, "tight" Gls mutants can be neither recovered nor maintained on a wild-type genetic background. They can be recovered and maintained only on a Reg background, in which regenerating somatic cells compensate for the absence of gonidia. A Gls/Reg double mutant (Figure 6.6) is a "homocytic" organisms in which all cells follow the ancestral volvocalean pathway of development: They all first differentiate as biflagellate somatic cells and then later redifferentiate as reproductive cells. Thus (except for its greater size and cell number) the Gls/Reg mutant bears a strong resemblance to a colonial volvocacean such as *Eudorina*.

In Gls/Reg mutants there is no discernible defect in cell division; indeed, cell division in these embryos proceeds with unrivaled symmetry and precision. But there are no asymmetric divisions. Thus we conclude that the *gls* loci[12] encode products that are required for shifting the division plane from the center of the cell to one side. The cytological basis for such a programmed

---

[12] We have evidence that there is more than one *gls* locus, but because of difficulties in crossing Gls strains, we do not yet know how many loci may mutate to yield a Gls phenotype.

shift of the division plane is presently unknown, but is under investigation, now that one of the *gls* loci has been cloned (Section 7.5.2).

6.5.4 Modifiers and suppressors

Although there is but a single locus at which a mutation leads to the Reg phenotype, several strains with spontaneous mutations that act as second-site suppressors of the Reg phenotype (called *regA*-modifier, or *ram* mutations) have been recovered from Reg cultures by phototactic selection,[13] and some of these have been analyzed genetically. In an M-Reg strain (a "modified regenerator" with a *regA⁻ram⁻* genotype), the following features are seen that distinguish it from the parental Reg strain (Kirk 1990): Many somatic cells remain small and biflagellate, with no apparent inclination toward re-differentiation or division; some enlarge only slightly and divide while still possessing both flagella and eyespots, forming small clusters containing tiny somatic cells but lacking any gonidia; other somatic cells enlarge somewhat more, lose their flagella but not their eyespots, and divide to form spheroids much smaller than those normally produced by standard Reg mutants, and within which one cell is seen to have inherited the eyespot of the precursor somatic cell.

Crosses of the first M-Reg strain with the wild type yielded standard Reg, M-Reg, wild type, and a fourth category of progeny in about a 1:1:1:1 ratio, indicating that the strain possessed *regA* and *ram* mutations that segregated independently. The fourth category of germlings obtained from those crosses lived and reproduced for no more than one or two asexual generations, because most of the gonidial initials that they produced failed to enlarge or divide normally (Kirk 1990). We inferred that those individuals had the genotype *regA⁺ram⁻*, which suggested that the *ram* mutation is lethal in the absence of a *regA* mutation. Because *regA⁺* is believed to encode a negative regulator of the genes required for reproduction, we postulated that the *ram⁺* locus encodes a positive regulator that is required for normal expression of such genes in gonidia (Kirk 1990). We presume that a hypomorphic variant of such a positive regulator is not fatal in a Reg background, in which all of the cells are capable of trying to express reproductive genes, but is fatal in

---

[13] Reg somatic cells resorb their eyespots and flagella as they begin to regenerate, and hence the spheroids become immotile. Thus phototactic selection of mature spheroids from a Reg culture yields a population enriched for mutants in which some or all somatic cells fail to regenerate. Such mutants include those that appear to have undergone reversion at the *regA* locus, as well as those with unlinked second-site suppressers of the *regA* mutation.

a wild-type background in which only a few cells (the gonidia) are able to do so.

Subsequently another M-Reg strain with a somewhat more complex phenotype was successfully crossed with the wild type. It produced (in addition to many inviable progeny) small numbers of Reg and wild-type progeny plus three different types of offspring with distinguishable differences in the details of the M-Reg phenotype (H. Gruber & D. Kirk, unpublished data). Although the numbers of viable progeny were small, the outcome of the cross was consistent with the hypothesis that this strain possesses two different *ram* mutations (*ramA* and *ramB*) that have slightly different modifying effects on the Reg phenotype, but both of which are lethal in the absence of the *regA* mutation. Although many independent M-Reg strains have now been isolated, how many *ram* loci may exist is presently unknown and will be difficult to ascertain because of the difficulties encountered in mating these M-Reg strains.

Many independent secondary mutations that partially suppress the Gls phenotype (in addition to many complete reversions) have also been recovered (Kirk 1990; Kirk, Kaufman et al. 1991; S. M. Miller personal communication). Some of these may represent intragenic suppressors, but others appear to be second-site suppressors that may define loci encoding products that interact with the *gls* product.

6.6 The genetic program for germ–soma differentiation: a working hypothesis

Study of the three major categories of mutations described in the preceding section has led us to formulate a working hypothesis regarding the way in which germ–soma differentiation is programmed in the genome of modern *V. carteri*, and how this program may have evolved (Kirk & Harper 1986; Kirk 1988, 1990, 1994; Kirk, Kirk et al. 1990; Kirk, Kaufman et al. 1991; Tam & Kirk 1991b). This hypothesis is based on three assumptions: (1) that all members of the family Volvocaceae have inherited the ancestral genetic program for cellular differentiation in which cells differentiate first as biflagellate cells that execute vegetative functions and then redifferentiate as cells that execute reproductive functions, (2) that any deviations from this ancestral program of sequential development that are observed in various volvocaceans are results of novel functions that have been added to the genome in the course of evolution, and hence (3) that these added functions can be identified by defining the loci at which mutations cause cells to revert to the ancestral program.

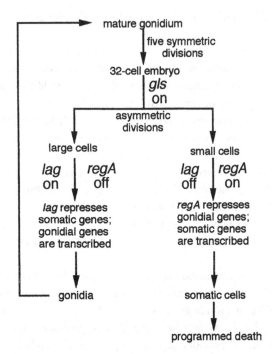

Figure 6.7 A working hypothesis regarding the key elements of the genetic program for germ–soma differentiation in *V. carteri*. Underlying this hypothesis is the assumption that the ancestral program of volvocacean differentiation ("first vegetative and then reproductive") is also the default program in *V. carteri*. However, the three types of genes listed in this diagram appear to act to convert this default program to a program for dichotomous differentiation, as follows: First the *gls* loci act to cause asymmetric division and the production of large and small cells; then the *lag* genes act in the large cells to repress expression of somatic genes, while the *regA* gene acts in the small cells to repress expression of gonidial genes.

In *V. carteri* we postulate that the three major functions that now act to modify the ancestral program are *gls*, *lag*, and *regA*. In skeletal form, the hypothesis is as follows (Figure 6.7): First *gls* acts in the cleaving embryo to cause asymmetric division and the production of large and small cells. Then, shortly after the end of embryogenesis, the *lag* loci are activated in the large cells – but *regA* is not – which causes these cells to bypass the biflagellate phase of the life cycle and differentiate directly as gonidia. Meanwhile, the *regA* locus is expressed in the small cells – but the *lag* loci are not – which causes these cells to develop as biflagellate cells in which all functions that would be required for redifferentiation as reproductive cells are repressed.

This hypothesis is obviously too simplistic to account for every aspect of germ–soma differentiation in *V. carteri*, but it is not intended to do that. Genetic data indicate, for example, that loci such as *mulA, B, C,* and *D* probably are involved in specifying the times and places at which the *gls* function will act to cause asymmetric division, thereby controlling the number of germ cells that will be produced by an embryo. Moreover, the fact that more than one locus can mutate to yield a *gls* phenotype, and the fact that second-site suppressors of such mutations can be identified, as discussed earlier, indicate that several gene products (some of which may also be involved in symmetric cell division) probably must interact to effect a shift in the division plane. Similarly, *regA* modifiers (*ram* loci) have been defined that may encode positive regulators that are required for expression of the genes over which the wild-type *regA* allele exerts negative control, and Mendelian analysis indicates that several *lag* genes must somehow act in concert if somatic development is to be suppressed in the presumptive gonidia. Nevertheless, the basic fact remains that mutations at the *gls, regA,* and *lag* loci are the only ones discovered so far by ourselves or others that have the effect of abrogating the fundamental germ–soma division of labor, thereby causing cells to follow the ancestral pathway of "first vegetative and then reproductive." Thus, we view these three classes of loci as crucial genes involved in the evolution of cellular differentiation in the lineage leading to *V. carteri.* Accordingly, we have made them prime targets of our efforts to understand the molecular-genetic basis for both the development and the evolution of the *V. carteri* germ–soma dichotomy, as will be discussed toward the end of the next chapter.

7

Molecular Analysis of *V. carteri* Genes and Development

I believe that . . . the offspring owes its origin to a peculiar substance of extremely complicated structure, viz., the "germ–plasm". This substance can never be formed anew; it can only grow, multiply and be transmitted from one generation to another.

. . . we may now hope to succeed in recognizing the *probable* explanations among the many *possible* ones. . . . This will assuredly be the work of time, and our approach to the truth will be a very gradual one. . . . What . . . appears to afford additional promise of success is that we can . . . approach from two sides; – namely, by observations, firstly on the *phenomena* of heredity, and secondly, on the hereditary *substance* itself, with which we are now of course acquainted.

Weismann (1892b)

. . . understanding cell differentiation in *Volvox* will require knowledge of its nucleic acid metabolism. . . .

Kochert and Sansing (1971)

When he was writing his magnum opus, it is unlikely that August Weismann had any notion just how gradual the "approach to the truth" regarding the role of the hereditary substance in embryonic development would be in the next eight decades – or how different the outcome would be from the one that he was predicting! On the other hand, it is equally unlikely that when

Kochert published the first studies of *Volvox* nucleic acids (only three years after he had described development in the forma of *Volvox carteri* that carries Weismann's name) he could have anticipated how dramatically the pace – and indeed the very nature – of studies of "nucleic acid metabolism" was destined to change in the next two decades.

Since Kochert initiated molecular-level studies of *V. carteri* (Hutt & Kochert 1971; Kochert 1971; Kochert & Sansing 1971; Yates & Kochert 1972), molecular biology (indeed all of biology) has undergone a profound revolution. However, during this extraordinarily vibrant period the study of *Volvox* molecular genetics has been nearly as insular a pursuit as the study of *Volvox* transmission genetics, and thus it is in a similarly inchoate state relative to developments in other model systems that have attracted many more investigators. Nevertheless, recent developments indicate that the study of *V. carteri* molecular genetics has passed a threshold and is about to yield substantial insights into the molecular basis for the ontogeny and phylogeny of this exquisitely simple form of germ–soma differentiation.

7.1 Some general features of the *V. carteri* genome and its replication

A *V. carteri* somatic cell contains about 0.12 pg of DNA (Kochert 1975; Kirk & Harper 1986); most of it is nuclear DNA, 5–7% is chloroplast DNA, and probably less than 1% is mitochondrial DNA (Kochert & Sansing 1971; Coleman & Maguire 1982a). The sperm of the species contain less chloroplast DNA than somatic cells, but the same amount of nuclear DNA (Coleman & Maguire 1982a), confirming the assumption that somatic cells are arrested in the G1 (or G0) phase and contain the normal haploid complement (1C quantity) of nuclear DNA. Thus, the haploid nuclear genome of *V. carteri* contains about 1.2×10^8 bp of DNA, which is very similar in size to the genome of *Chlamydomonas*, about 50% larger than that of *Caenorhabditis*, and about 70% as large as that of *Drosophila*.

Based on a value of 120 Mb of DNA per genome, the 14 chromosomes of *V. carteri* should average around 8 Mb in length, which is toward the upper end of the size range in which chromosomes have been separated by one or another form of pulsed-field gel electrophoresis (Chu, Vollrath & Davis 1986; Smith, Kloc & Cantor 1988). However, when we attempted to separate *V. carteri* chromosomes by such methods, we observed that under a variety of conditions in which the 7-Mb chromosome of *Schizosaccharomyces pombe* moved a substantial distance into the gel and was well resolved, all of the *V. carteri* DNA behaved as if it were all of substantially larger size, and it could not be resolved into discrete bands (M. M. Kirk and D. L. Kirk, unpublished observations). The reason for this remains to be determined.

The buoyant densities of nuclear and chloroplast DNAs of *V. carteri* have been reported to be 1.715 and 1.705 g/cm³, respectively (Kochert 1975; Margolis-Kazan & Blamire 1976), which yields estimated base compositions of 56% and 46% G+C, respectively. This value of 56% G+C for *V. carteri* nuclear DNA agrees closely with an estimate based on melting temperature (Margolis-Kazan & Blamire 1976); it is significantly lower than the value of 62% G+C for the *C. reinhardtii* nuclear genome (Sager & Ishida 1963), but is much closer to that value than to the value of 44% reported for another species of *Chlamydomonas, C. geitleri* (Tetík & Zadrazil 1982). The difference between *V. carteri* and *C. reinhardtii* in G+C content of the nuclear genome as a whole is reflected in the codon biases of the two species, as will be discussed in more detail later: Although *V. carteri* exhibits a distinct bias toward the use of C or G in the third position of each codon, this bias is not nearly as intense as the bias exhibited by *C. reinhardtii* in homologous genes (Schmitt et al. 1992).

It was argued on the basis of an early study of renaturation kinetics that, apart from the genes encoding rRNA, the *C. reinhardtii* genome consisted almost exclusively of unique sequence elements (Howell & Walker 1976). However, several categories of middle-repetitive DNA elements (some of which are highly transcribed and are postulated to have regulatory implications) have since been identified by more direct approaches (Day & Rochaix 1989; Wakarchuk, Müller & Beck 1992; Hails, Jobling & Day 1993). Quantitative studies of the renaturation kinetics of *V. carteri* DNA have not been reported, but nevertheless it is clear that this genome is well supplied with repetitive elements, some of which are also extensively transcribed (Kirk & Harper 1986; Harper et al. 1987; our unpublished observations) The best studied of these is, of course, rDNA; it is estimated that 10% of the genome is occupied by the approximately 1,200 copies of the 9.6-kb rDNA repeat unit, each of which includes a 6.3-kb transcription unit and a 3.3-kb nontranscribed spacer (Rausch et al. 1989; Schmitt et al. 1992). Using the synthetic polymer poly(dT-dG)•poly(dC-dA), a probe that detects repetitive elements with a propensity to form Z-DNA in a wide range of eukaryotes (Hamada, Petrino & Kakunaga 1982; Morris, Kushner & Ivarie 1986), a family of dispersed, hypervariable elements that was estimated to contain more than 1,000 members was detected in *V. carteri* (Kirk & Harper 1986). Another equally large, equally dispersed, but less variable family of repetitive elements was detected with the synthetic probe poly(dG)•poly(dC); this family was found to be well represented in all volvocaleans examined, including

Chlamydomonas, but to be much less abundant in a wide range of other eukaryotes that were tested (Kirk & Harper 1986). Many other families of middle-repetitive elements have been detected in *V. carteri* with cloned genomic fragments (Kirk & Harper 1986; Harper et al. 1987), and some of these exhibit polymorphisms that have been used for genome-mapping purposes (Harper et al. 1987; Adams et al. 1990), as will be discussed later.

Whereas the chloroplast genome of its simpler cousin *Chlamydomonas* has been subjected to extensive and detailed genetic, physical, and molecular analyses (as reviewed by Harris 1989), the chloroplast genome of *V. carteri* has not. Quantitative microspectrophotometry of DAPI-stained specimens indicated that the two to four nucleoids within the chloroplast of a *V. carteri* somatic cell together contain about 5% as much DNA as is present in the nucleus (Coleman & Maguire 1982a). Therefore, on the assumption that the *V. carteri* chloroplast genome is about the same size as that of *C. reinhardtii* ($\sim 1.2 \times 10^8$ Daltons, or ~ 180 kb), Coleman and Maguire estimated that each somatic cell contains about 50 chloroplast genomes. As we will see later, however, a mature gonidium contains much more chloroplast DNA, perhaps as much as 10^5 copies, which would be adequate to supply all somatic cells of the progeny with all the chloroplast DNA that they would ever have.

The existence of as many as 10^5 copies of the chloroplast genome in each gonidium undoubtedly accounts for the fact no chloroplast mutations have yet been identified in *V. carteri*, despite repeated searches[1] (Huskey, et al. 1979a; Kirk & Harper 1986; our unpublished data). Even in *C. reinhardtii*, which has fewer than 100 chloroplast genomes per cell, chloroplast mutations are not easily recovered unless the redundancy of the chloroplast genome is reduced by treatment with fluorodeoxyuridine and/or the recovery of nuclear mutants of similar phenotype is suppressed by using vegetative-diploid cells (Harris 1989).

The mitochondrial genome of *V. carteri* has been studied even less than the chloroplast genome. However, polymorphic DNA markers have been used to show that during sexual reproduction in *V. carteri* the chloroplast and the mitochondrial genomes are both transmitted maternally (Adams et al. 1990). This is in distinction to *C. reinhardtii*, in which the chloroplast and mitochondrial genomes are both inherited uniparentally, but from opposite parents (Harris 1989).

---

[1] Because *V. carteri* (in distinction to *C. reinhardtii*) is an obligate photoautotroph, photosynthetic mutants cannot be expected to be viable, but resistance to antibiotics that inhibit chloroplast gene expression should not suffer from this constraint. Nevertheless, our repeated attempts to isolate such mutants of *V. carteri* have yielded only nuclear mutations that appear to affect the permeability of gonidia to the selective agent (unpublished data).

7.1.1.   A curious uncertainty: Is nuclear DNA replication
tied to the mitotic cycle?

As noted in Section 5.4.1.6, DNA replication is required during cleavage in *Volvox*, as it is in all mitotic systems. However, a number of studies over many years have raised the intriguing possibility that the relationship between nuclear DNA synthesis and mitosis is not as simple and direct in the volvocaceans as it is in the "typical" eukaryotic cell, and that these algae may enter cell division having already undergone several rounds of endoreduplication of nuclear DNA, so that they do not have to complete a full doubling of DNA in every division cycle.

One of the first indications that the mitotic behavior of the volvocaceans might be rather unusual came from the first karyological analysis of representative members, including 10 species of *Volvox* (Cave & Pocock 1951b). In that report it was stated that "this description of mitosis has been drawn from embryos which contain no more than 16 cells. In older embryos the individual chromosomes are too small to make out . . . [because] with successive mitoses . . . chromosomes become smaller." This observation was later confirmed and quantified with respect to *Pandorina* (Coleman 1959); indeed, the measurements reported in the latter study have been calculated to mean that chromosome volume is rather precisely halved in each of the first three mitotic cycles of *Pandorina* (Tautvydis 1976). The observation that chromosomes get both thinner and shorter in each of the early cleavage divisions has been confirmed with respect to both well-studied formas of *V. carteri* (G. Kochert personal communication; R. C. Starr personal communication).

When Kochert (1975) reported the first data bearing on the nuclear DNA content of *V. carteri* cells, he appeared to provide support for the concept of precleavage endoreduplication of the chromosomes. By chemical assay of total cellular DNA, he found that by six to eight hours before the onset of cleavage the gonidia of *V. carteri* f. *weismannia* already contained 25 times as much DNA as somatic cells. Then, using analytical ultracentrifugation to determine the ratio of nuclear DNA to chloroplast DNA, he estimated that chloroplast DNA accounted for 30% of the gonidial DNA, whereas it constituted 6.5% of somatic-cell DNA. As he emphasized at the time, those data indicated that chloroplast DNA was accumulated during gonidial maturation. But his data also indicated that those gonidia (which were not yet mature) contained about 17 times as much nuclear DNA as somatic cells and (if taken at face value) suggested that by six to eight hours prior to their first mitotic division *V.*

carteri gonidia had already undergone four rounds of nuclear DNA replication.[2]

The following year, Tautvydis (1976) was the first to suggest explicitly that maturing volvocacean reproductive cells undergo chromosomal endoreduplication and enter cleavage in a polytene state. Using colorimetric and fluorometric assays, he concluded that gonidia of *Eudorina* (*Pleodorina*) *californica* undergo an approximately 60-fold increase in DNA content during the time that they enlarge sufficiently to produce a 64-cell embryo. He then used isopycnic centrifugation to separate nuclear DNA from chloroplast DNA and found no evidence for a change in the ($\sim$4:1) ratio of these two components during development. Putting those data together, he concluded that a gonidium of this species synthesizes all of the nuclear DNA that will be required to supply a 1C-equivalent to each of its daughter cells before it ever begins to divide. He further proposed that a parallel process might characterize other volvocaceans.[3]

Tautvydis's hypothesis was, however, distinctly at odds with measurements reported at about the same time by Lee and Kemp, who used a chemical assay of extracted nucleic acids and microspectrophotometry of Feulgen-stained nuclei to estimate the DNA content per cell and per nucleus during the asexual life cycle of *Eudorina elegans*. Those authors concluded that the nuclear DNA content never exceeded the 2C value and hence that duplication of DNA occurred just before each division, as it should in any well-behaved eukaryotic cell (Lee & Kemp 1975, 1976).

A conclusion perfectly paralleling that of Lee and Kemp was drawn by Coleman and Maguire when they used microspectrofluorometric analysis of DAPI-stained specimens to test the validity of Tautvydis's hypothesis with respect to *Volvox*. Coleman and Maguire (1982a) reported that the nuclei of mature gonidia, two-cell embryos, and four-cell embryos of *V. carteri* all exhibited levels of DAPI fluorescence very similar to (and never more than twice as great as) the fluorescence level of somatic-cell nuclei.[4] Thus, they concluded that nuclear DNA duplication occurs in *Volvox* as it does in other eukaryotes: just before each nuclear division. Coleman and Maguire duly noted the striking discrepancy between their findings and Kochert's and stated

---

[2] Using the same methods as Kochert, but studying gonidia that were only about one hour from cleavage, we have obtained an even higher gonidial-to-somatic ratio for nuclear DNAs in *V. carteri* f. *nagariensis*: about 32:1 (M. M. Kirk & D. L. Kirk unpublished data).

[3] Implicit in this hypothesis, of course, is the assumption that the volvocaceans must have some unprecedented mechanism for splitting polyneme chromosomes (i.e., chromosomes containing multiple chromatids) precisely in half at each anaphase.

[4] These findings were consistent with those obtained in a preliminary study that employed Feulgen microspectrophotometry (Coleman 1979a).

that "the basis for this discrepancy is unknown." It remains unknown to this day.[5]

In short, measurements obtained with different methods of analysis engender profoundly different pictures of the temporal relationship between DNA replication and nuclear division in the *Eudorina-Pleodorina* complex and in *V. carteri*. It requires no great insight to see that somewhere in this series of studies there must have been some serious flaws in experimental design or analytical methods, but identifying the nature of those flaws is quite another matter.

Let us consider only the case of *V. carteri*. Both of the studies reviewed earlier indicated that gonidia accumulated substantial quantities of DNA as they enlarged and matured: Coleman and Maguire reported that the total DAPI-DNA fluorescence of a mature gonidium was more than 100 times that of a somatic cell, whereas Kochert obtained a figure of 25 times by bulk chemical assay. For discussion purposes, let us assume that the lower value is correct and also that not more than a doubling of nuclear DNA occurs during this period, as Coleman and Maguire concluded. In that case, ultracentrifugal analysis of total gonidial DNA should have revealed a chloroplast-to-nuclear DNA ratio of about 12:1, not 0.3:1 as observed by Kochert. There have been suggestions that certain extraction methods may not solubilize cytoplasmic DNA with the same efficiency at all stages of the *V. carteri* life cycle (Margolis-Kazan & Blamire 1976, 1979a), but it seems highly unlikely that such differences could be severe enough to result in full recovery of chloroplast DNA from somatic cells, but less than 3% recovery from gonidia (as would be required to account for a 36-fold error in the estimated chloroplast-to-nuclear DNA ratio for the gonidium). On the other hand, it is not

[5] The question whether or not significant quantities of nuclear DNA are synthesized by gonidia in advance of cleavage is not readily addressed by conventional DNA labeling studies, because volvocaleans lack a cytosolic or nuclear thymidine kinase, and thus labeled thymidine or BUDR is incorporated almost exclusively into chloroplast DNA (Swinton & Hanawalt 1972; Tautvydis 1976; Harris 1989). Margolis-Kazan and Blamire (1976) exposed *V. carteri* f. *weismannia* cultures to ³H-adenine and then determined incorporation into nuclear DNA (separated from other nucleic acids by isopycnic centrifugation); they concluded that "nuclear DNA is made to some degree throughout the life cycle." However, the 16–20 hour labeling periods employed in their study made it impossible to resolve the time course of label incorporation with any degree of clarity. Yates and Kochert (1976) studied incorporation of ³²P during short (one-hour) pulses into DNA (separated from RNA electrophoretically). They concluded that, as anticipated, DNA synthesis was maximal during cleavage; however, a substantial increase in DNA synthesis had already occurred several hours before cleavage was initiated. Unfortunately, however, their methods did not permit them to determine how much of that incorporated radioactivity was in nuclear DNA versus chloroplast DNA. In short, the two studies of DNA synthesis in *V. carteri* that were done two decades ago contributed little toward a resolution of the discrepancy between the findings obtained by chemical and microspectrophotometric determinations of the DNA content of the mature gonidial nucleus. This problem clearly is worthy of reexamination.

inconceivable that the fluorescence of DAPI-stained DNA could deviate from linearity under some conditions, particularly when the nuclear DNA is as diffuse as it is in the very large nuclei of precleavage gonidia (Coleman & Maguire 1982a). But, again, it is extremely difficult to believe that such deviations from ideality could be severe enough to result in nearly a 10-fold underestimate of DNA content per gonidial nucleus, which is what would be required to explain the discrepancy on this basis alone. Moreover, the conclusion drawn by Coleman and Maguire (1982a) that gonidia do not accumulate excess nuclear DNA during development was strongly reinforced by their observation that each of the relatively condensed nuclei of two-cell and four-cell embryos exhibited a level of total fluorescence very similar to that of the much larger gonidial nucleus (and also simliar to that of the condensed somatic nucleus).

In short, there appears to be no way to resolve this discrepancy with the evidence presently at hand. Thus, although the possibility that volvocaceans may initiate division with endoreduplicated chromosomes is extremely intriguing, it is probably safer to assume that they are innocent of such an infraction of eukaryotic canon law, unless they are proved guilty beyond reasonable doubt by further evidence.[6]

7.1.2 Is all chloroplast DNA of somatic cells of gonidial origin?

All studies indicate clearly that chloroplast DNA is accumulated to very substantial levels during gonidial maturation in *V. carteri*. However, estimates of the magnitude of this accumulation are inextricably linked to estimates of the degree to which nuclear DNA is accumulated during this same period.

Based on ultracentrifugal data indicating that only 30% of the DNA in a gonidium is chloroplast DNA, Kochert (1975) estimated that six to eight hours before cleavage a gonidium contains only enough chloroplast DNA to supply 100–150 somatic cells with the normal complement. However, Coleman and Maguire (1982a) came to a very different conclusion. Having observed that the DAPI-DNA fluorescence of gonidial nucleoids increases logarithmically as a function of cell diameter, reaching a value 2,300 times

---

[6] As this manuscript was about to go to the printer, I learned from Rüdiger Schmitt that his student Iris Kobl had begun to reexamine this issue, had preliminary data consistent with the hypothesis that *V. carteri* gonidial nuclei contain much more than a 1C complement of DNA, and was attempting to resolve this long-standing controversy by obtaining accurate estimates of the number of copies of "unique-sequence" genes (such as the gene encoding actin) that are present in the nuclei of mature gonidia and in the nuclei of embryos at various stages of cleavage (R. Schmitt personal communication).

that of a somatic cell in the largest gonidia, they concluded that "the mature gonidium is almost fully stocked with the ctDNA necessary for all the daughter cells it will form."[7] The concept that somatic cells contain only the chloroplast DNA they have happened to inherit – and do not need to replicate it in order to develop normally – was reinforced by the observation that about 5% of all mature somatic cells (which otherwise appeared perfectly normal) lacked any detectable chloroplast nucleoids (Coleman & Maguire 1982a).

The potential significance of these conclusions will become apparent at the end of this chapter, when we discuss recent insights into the role that may be played by chloroplast biogenesis in control of germ–soma differentiation in *V. carteri*.

7.2 Progress in establishing a DNA-based map of the *V. carteri* nuclear genome

As noted in Section 6.1, the first provisional map of the *V. carteri* genome (Huskey et al. 1979a) was based largely on morphological mutations that interfered to varying extents with sexual reproduction, and most of which have subsequently been lost. More recently, we have been attempting to develop a revised map that will include a few morphological and biochemical markers that were included in the original map, will add a number of drug-resistance markers, but will be based primarily on DNA polymorphisms of two types: restriction-fragment length polymorphisms (RFLPs) and random amplified polymorphic DNAs (RAPDs).

7.2.1 Strains used for mapping DNA polymorphisms

Strain HK 10, which Richard Starr isolated in 1967 from a pond near Kobe, Japan (Starr 1969), remains the "standard female" used in most investigations of *V. carteri* biology and genetics. In 1969, however, the male partner of HK 10, strain HK 9, developed a mutant phenotype and was therefore replaced as the "standard male" by strain 69-1b, an F_1 male derived from an HK 9 × HK 10 cross (Starr 1970a; Meredith & Starr 1975). Strain 69-

---

[7] The only published attempt to count *V. carteri* cells accurately indicated that in cultures maintained under conditions that permit synchronous growth and development, a majority of adult spheroids contain 1,983 ± 17 somatic cells (very nearly the number predicted if somatic cell initials divide a total of 11 times), and the rest contain twice that many (Green & Kirk 1981).

1b served adequately as the male parent in studies examining the transmission genetics of conventional Mendelian markers (Starr 1970a; Huskey, Griffin et al. 1979a), but it turned out to be less than ideal when attempts were initiated to expand the *V. carteri* genetic map by addition of DNA polymorphisms.

When subclones of 69-1b and HK 10 (called ADM and EVE, respectively) were used as sources for DNAs to be analyzed on Southern blots, no polymorphisms could be detected with any of the probes then available that hybridized to unique sequences or small families of related sequences (Harper et al. 1987). That was not particularly surprising, because HK 9 and HK 10 (having come from the same small pond) presumably were so closely related that they did not possess many DNA-level differences, and when HK 9 was replaced by 69-1b, about half of that variation undoubtedly was lost.[8] Therefore, the only DNA polymorphisms of *V. carteri* that we were able to detect and study initially were those present in repetitive-sequence elements. When Southern blots of ADM and EVE DNA digests were hybridized to six probes detecting different repetitive-element families, 24% of all fragments detected (nearly 300 restriction fragments altogether) were present in only one of the two strains and thus appeared to represent markers useful for mapping purposes (Harper et al. 1987). Later work was to cast a serious shadow over such markers, however, as discussed in the next section.

Subsequently we obtained from Starr another pair of *V. carteri* f. *nagariensis* strains that had been isolated more recently from a pond near Poona, India, and subclones of the male and female members of that pair (named PM1 and PF1, respectively) were introduced into our analysis. The extent of polymorphism within the pair isolated from India (called the "I" strains) was not significantly different from that observed within the pair isolated from Japan (the "J" strains). In sharp contrast, when comparisons were made between the I and J strains, it was found that more than 90% of all DNA fragments detected on Southern blots differed between the I and J strains, irrespective of the type of restriction enzyme used to digest the samples, whether the probe used was one that detected unique sequences, members of a small gene family, or middle-repetitive elements, or whether it detected nuclear, chloroplast, or mitochondrial genes (Adams et al. 1990). Not surprisingly, however, that extreme degree of DNA polymorphism was accompanied by genetic incompatibility between the I and J strains such that

---

[8] We have recently obtained from Starr a sample of the original HK 9 strain, as well as a sample of HK 12 (another male isolated at the same time and place) (R. C. Starr personal communication), but unfortunately we have been unable to induce sexuality in either strain. Apparently, while being maintained for nearly three decades in asexual cultures, these strains have lost the ability to respond to the sex-inducing pheromone and are sterile.

post-zygotic lethality in the range of 95% was observed, and thus considerable effort has been required to build up a progeny pool adequate to initiate meaningful mapping studies.

In early studies, it appeared that genetic incompatibilities between the I and J strains caused significant distortions of segregation ratios and that linkage values obtained with these strains would have to be interpreted with extreme caution (Adams et al. 1990). More recently, however, it has become clear that those apparent distortions were connected principally to one class of DNA polymorphisms, a class that is subject to problems of a very different kind, as will be discussed next.

7.2.2 DNA modifications complicate analysis of some V. carteri polymorphisms

With the limited numbers of strains and probes that were available when RFLP mapping of *V. carteri* was first attempted, the only RFLPs that could be detected were in middle-repetitive DNA, and of the first 300 such RFLPs identified, the vast majority were detected in DNA samples that had been digested with restriction enzymes with tetranucleotide recognition sequences (hereafter called "four-base cutters" for simplicity) (Harper et al. 1987). It was assumed that the reason that four-base cutters were much more useful than six-base cutters for revealing repetitive-element polymorphisms was that they digested most of the DNA into small fragments that ran off the gel, leaving behind only those exceptional fragments that had a dearth of cognate restriction sites. Indeed, whereas four-base cutters should produce genomic DNA fragments averaging 256 bp in length, the fragments selected for segregation analysis in that first study ranged up to 70 times that length. It was stated that "examination of this minor fraction of exceptionally long members of each restriction digest was based on the empirical observation that it worked" and thus the use of a four-base cutter (primarily *Sau*3AI) and repetitive-element probes was continued when the I strains were added to the analysis (Adams et al. 1990). In the first study, in which the progeny cohort analyzed was small, it appeared that *Sau*3AI/repetitive-sequence RFLPs segregated in a regular Mendelian manner (Harper et al. 1987). In a second study with more progeny and different markers of the same general type, however, transmission anomalies were observed (Adams et al. 1990). As both the number of such markers and the size of the progeny cohort were increased, enough additional transmission anomalies were discovered to call into question the validity of such RFLPs as linkage markers (D. L. Kirk and others unpublished data).

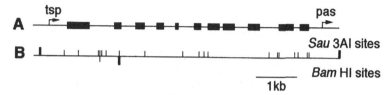

Figure 7.1 *Sau*3AI and *Bam*HI sites within the *nit*A gene. **A**: Diagram of the *nit*A gene, showing the locations of the transcription start point (tsp), exons (boxes), and the polyadenylation signal (pas). **B**: Locations of 21 *Sau* AI sites (above the line) and three *Bam*HI sites (below the line) that are predicted from the nucleotide sequence (Gruber et al. 1992). The heavy lines identify the only *Sau*3AI and *Bam*HI sites in DNA from *V. carteri* J strains that are recognized and cut by the cognate restriction enzymes. In contrast, all these sites appear to be cut normally in DNA from I strains. (H. Gruber & D. L. Kirk unpublished data.)

The source of these anomalies became clear much more recently when it was discovered that *Sau*3AI and certain other restriction endonucleases do not always cut *V. carteri* genes of known sequence as predicted from the sequence and that the deviations from expectations are heritable, but not as alleles of the loci being analyzed![9]

This phenomenon was first discovered by analysis of the *Sau*3AI restriction pattern of the *V. carteri nit*A gene (which encodes the enzyme nitrate reductase) in the J and I strains. On the basis of the published sequence, which reveals 19 *Sau*3AI sites within (and 2 closely flanking) the approximately 7-kb *nit*A coding region (Gruber et al. 1992), one would predict that *Sau*3AI digestion of genomic DNA should reduce the *nit*A gene to a collection of 20 small fragments ranging in size from a few base pairs to 1.5 kb in length (Figure 7.1). Genomic DNA from the I strains exhibited such a pattern on Southern blots. In marked contrast, in *Sau*3AI digests of genomic DNA from the J strains (from which the gene was cloned and sequenced), the *nit*A gene was reproducibly left as a single 7.4-kb fragment, indicating that *Sau*3AI is able to recognize and cut at only the outermost 2 of its 21 cognate sites! When approximately 200 progeny from several J × I crosses were examined in a parallel manner, it was found that that difference segregated as a Mendelian character: Each of the J × I progeny exhibited either the J phenotype (a single 7.4-kb *Sau*3AI fragment carrying the entire *nit*A gene) or the I phenotype (a smear of small *Sau*3AI fragments of the *nit*A gene). However, that "*Sau*3AI-digestibility" trait of the *nit*A gene did not map to the *nit*A locus. The J females used in these crosses carried a stable *nit*A mutation that conferred resistance to chlorate and inability to grow on

---

[9] All of the findings discussed in this section will be presented in more detail in a paper by D. L. Kirk, H. Gruber and others that is now in preparation.

Table 7.1. *Methylation sensitivity of isoschizomers that fail to cut internal cognate sites in the* nitA *gene of J strains of* V. *carteri*

Restriction enzyme	Recognition sequence	Methylated sites cut	Methylated sites not cut
*Sau*3AI	^GATC[a]	G[m6]ATC[b]	GAT[m4]C GAT[m5]C GAT[hm5]C
*Mbo*I	^GATC	GAT[m4]C GAT[m5]C	G[m6]ATC GAT[hm5]C
*Dpn*II	^GATC	GAT[m4]C GAT[m5]C	G[m6]ATC

[a]The carat mark indicates the site where the sequence is cut.
[b]Methylated bases are represented in the conventional manner: [m6]A = 6-methyladenine, [hm5]C = 5-hydroxymethylcytosine, and so forth.

nitrate (see Section 6.2), and the J and I strains exhibited a *nitA* RFLP after digestion with *Hin*dIII (which cuts the *nitA* gene as predicted from the DNA sequence). In J × I progeny the *Hin*dIII RFLP co-segregated with the chlorate-resistance/nitrate-auxotrophy trait, indicating that it revealed a DNA difference at the *nitA* locus per se. In contrast, the *Sau*3AI-digestibility trait segregated completely independently of those other three traits, indicating that it was not associated with the gene itself, but was mediated by a gene in a separate linkage group,[10] presumably a gene whose product is involved in modification of the *nitA* DNA. That was the first clear evidence that certain *V. carteri* RFLPs segregated independently of the DNA used to detect them.

Two methylation isoschizomers of *Sau*3AI that were tested (*Mbo*I and *Dpo*II) yielded the same pattern: They left the *nitA* gene of J strains intact, but reduced the gene of the I strains to the predicted set of small fragments. Based on the differences in methylation sensitivity of these enzymes (Table 7.1) (Nelson & McClelland 1991; Roberts & Macelis 1991) it was possible to rule out all three types of DNA methylation that had been detected in eukaryotic DNA (Ehrlich & Zhang 1990) as the cause for the heritable difference in *Sau*3AI digestibility of the *nitA* gene. Those findings did not rule out 5-hydroxymethylation of cytosine, but that type of modification (although present in certain prokaryotic DNAs) has never been detected in eukaryotic DNA (Ehrlich & Zhang 1990).

To our even greater astonishment we then found that several other four-base cutters with very different recognition sequences also cut the gene from

[10] As shown in Figure 7.3, the *nitA*-*Sau*3AI-modification (*nsm*) gene resides on the chromosome bearing the mating-type locus. However, because these two loci are more than 100 cM apart, *nitA* modification and mating type assort completely independently in crosses.

Table 7.2. Other restriction enzymes tested for ability to digest cognate sites in the nitA gene

A. Enzymes with demonstrably impaired cutting[a]

Restriction enzyme	Recognition sequence	Methylated sites cut	Methylated sites not cut
Hha I	GCG^C	?	$G^{m5}CGC$ $GCG^{m5}C$ $G^{hm5}CG^{hm5}C$
Hinfl	G^ANTC	$GANT^{m5}C$	$G^{m6}ANTC$ $GANT^{hm5}C$
Msp I	C^CGG	$^{m4}CCGG$ $C^{m4}CGG$ $C^{m5}CGG$	$^{m5}CCGG$ $^{hm5}C^{hm5}CGG$
Rsa I	GT^AC	$GTA^{m5}C$	$GT^{m6}A^{m5}C$
Taq I	T^CGA	$T^{m5}CGA$ $T^{hm5}CGA$	$TCG^{m6}A$
Scr Fl	CC^NGG	$^{m5}CCNGG$	$C^{m5}CNGG$ $C^{m4}CNGG$
Dra I	TTT^AAA	$TTTA^{m6}AA$	?
Nsi I	ATGCA^T	?	$ATGC^{m6}AT$ $ATG^{m5}CAT$

[a]In general, all of these enzymes failed to cut any internal sites in the *nitA* gene and left it in one large fragment.

B. Enzyme with impaired cutting of one internal site[b]

Restriction enzyme	Recognition sequence	Methylated sites cut	Methylated sites not cut
Bam HI	G^GATCC	$GG^{m6}ATCC$ $GGATC^{m4}C$ $GGATC^{m5}C$ $GG^{m6}ATC^{m5}C$	$GGAT^{m4}CC$ $GGAT^{m5}CC$ $GAT^{hm5}C^{hm5}C$

[b]Cut only one of two internal sites; no examples of impaired cutting yet detected with other genes of known sequence.

C. Enzymes with no detected cutting impairment[c]

Restriction enzyme	Recognition sequence	Methylated sites cut	Methylated sites not cut
Alu I	AG^CT	?	$^{m6}AGCT$ $AG^{m4}CT$ $AG^{m5}CT$ $AG^{hm5}CT$
Hae III	GG^CC	$GGC^{m5}C$	$GG^{m5}CC$ $GG^{hm5}C^{hm5}C$
Eco RI	G^AATTC	$GAATT^{hm5}C$	$G^{m6}AATTC$ $GA^{m6}ATTC$ $GAATT^{m5}C$
Hin dIII	A^AGCTT	?	$^{m6}AAGCTT$ $AAG^{m5}CTT$ $AAG^{hm5}CTT$
Pst I	CTGCA^G	?	$CTGC^{m6}AG$ $^{m5}CTGCAG$
Sal I	G^TCGAC	$GTCGA^{m5}C$	$GT^{m5}CGAC$ $GTCG^{m6}AC$
Ssp I	AAT^ATT	$^{m6}AATATT$	?

[c]No impairment yet detected with *nitA* or other genes of known sequence. Most other commonly used restriction enzymes with 6-base recognition sequences appear to fall in this category, but no others have been as thoroughly tested as those listed in the table.

an I strain into the expected small fragments, but left the gene of the J strain in one large piece, having failed to cut internal sites (Table 7.2A). In contrast, several six-base cutters cut the *nitA* gene of the J strains as predicted from the sequence (Table 7.2C). Examination of sequences surrounding restriction sites in the vicinity of the *nitA* gene that could and could not be cut by various four-base cutters failed to uncover any contextual motifs distinguishing sites that are always modified in the J strains from those that are not.

At that point it appeared that the modification system of *V. carteri* caused a generalized impairment of cutting by enzymes with four-base recognition sequences. However, that proved not to be a valid generalization when we subsequently found that two four-base cutters (*Alu*I and *Hae*III) cut the J gene as predicted from the sequence, but two six-base cutters (*Dra*I and *Nsi*I) did not (Table 7.2A,C).

*Bam*HI fell in a category by itself: It cut one of the internal cognate sites in *nitA*, but not the other (Table. 7.2B). Because *Bam*HI sites always span *Sau*3AI sites, this observation indicates that within the *nitA* gene there are at least two kinds of modified *Sau*3AI sites: one that can be recognized and cut by *Bam*HI, and one that cannot.

The *Sau*3AI restriction patterns of two additional genes, those encoding the α-tubulins, were examined next. That study revealed that the DNA modification problem was even more complex, and had a much more complex genetic basis, than analysis of the *nitA* gene had indicated.

The two α-tubulin genes in the *V. carteri* genome encode identical polypeptides, and analysis of segregation patterns of RFLPs detected following digestion with either *Bam*HI or *Hind*III (both of which digest these genes as predicted from the sequence) has been used to map these unlinked genes reliably (Harper & Mages 1988; Mages et al. 1988, 1995). Both the J and I strains reproducibly exhibit impaired cutting of some, but not all, *Sau*3AI sites within the coding regions of these two genes. Between the two strains, there are 14 α-tubulin *Sau*3AI fragments that are significantly larger than any predicted from the sequence data: 7 in the J strains, and 8 (including 1 shared fragment) in the I strains. All J × I progeny that have been examined (>50) have exhibited from 4 to 11 of these anomalously large *Sau*3AI fragments (Figure 7.2), and no others. These RFLPs, which we postulate are identifying tubulin-Sau-modification' (*tsm*) loci, segregate as 13 distinct loci in four different linkage groups. Although five *tsm* loci are linked to the α1-tubulin gene (as defined by the *Hind*III or *Bam*HI RFLPs), they map to positions as far as 88 cM from it. The remaining eight *tsm* loci map to three other linkage groups, none of which includes any α- or β-tubulin genes (Figure 7.3). Nor are any of these 13 loci linked to the locus that controls the *Sau*3AI digestibility of the *nitA* locus!

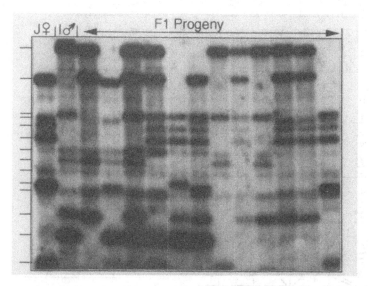

Figure 7.2 Restriction-fragment length polymorphisms (RFLPs) observed when a Southern blot of *Sau*3AI-digested DNAs from a J female, an I male, and 12 of their F₁ progeny was hybridized to a radioactive probe for α-tubulin. The marks along the left edge call attention to 14 fragments, all of which are observed reproducibly in such digests, but all of which are larger than any *Sau*3AI fragments that are predicted by the known nucleotide sequences of the α1- and α2-tubulin genes of *Volvox*. All but one of these fragments (the sixth from the bottom) is either J- or I-specific, and all of them segregate independently of one another (and independently of the α1- and α2-tubulin loci themselves!), among more than 50 J × I progeny that have been examined. (Gruber & Kirk in preparation.)

Taken together, then, those studies of three sequenced genes indicate that at a minimum there are 14 distinct genes involved in some novel kind of modification of *Sau*3AI sites at specific locations within the *V. carteri* genome. Preliminary analysis of the segregation patterns for middle-repetitive RFLPs detected with *Sau*3AI indicates that there may be considerably more and that some of them may interact in complex ways to generate *Sau*3AI RFLPs in progeny that are not seen in either parent.

Many other details of these DNA modification systems (such as whether the same or different loci are responsible for the modifications detected with other enzymes listed in Table 7.2A, and what the chemical nature of the modifications may be) remain to be established – as does their significance. However, one thing is already abundantly clear: RFLPs observed in DNA samples digested with *Sau*3AI (and several other enzymes listed in Tables 7.1 and 7.2A) cannot be assumed to reflect genetic differences residing within

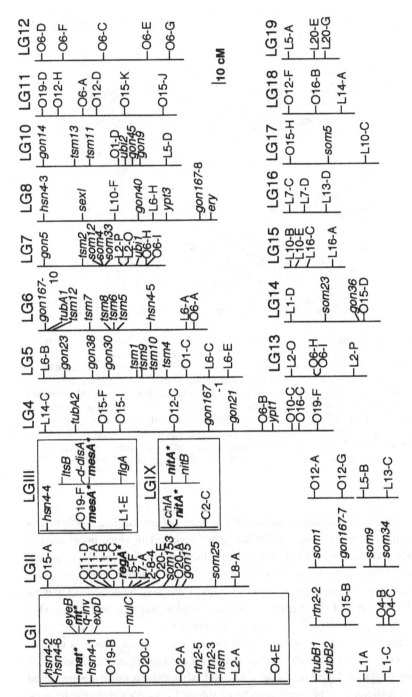

Figure 7.3 A provisional linkage map of the *V. carteri* genome, with the scale in centi-Morgans (cM) given at the bottom. The linkage groups (LGs) designated with roman numerals can be equated to linkage groups defined earlier by Huskey et al. (1979a) by virtue of markers (identified by boldface type and asterisks) that are still

the DNA sequences that are used to detect them, because in most cases they segregate completely independently of those sequences.

The preceding findings have particularly serious implications for attempts to perform positional cloning of *V. carteri* genes. Earlier we described two different *Sau*3AI RFLPs that mapped very close to the *regA* gene (Harper et al. 1987; Adams et al. 1990). We cloned first one and then the other of the DNA fragments that exhibited a *regA*-linked *Sau*3AI polymorphism and attempted to use those cloned fragments as starting places for a chromosome walk to the *regA* locus (J. F. Harper, J. K. Miller & D. L. Kirk unpublished data). However, in each case we found that when genomic clones overlapping the starting fragments were isolated (from either of two different libraries), they invariably contained sequences that could be shown to be derived from some other region of the genome. At the time, that distressing outcome was attributed to cloning artifacts. But now an alternative explanation seems much

Caption to Figure 7.3 *(cont.)*
available for inclusion in the analysis. In three of these cases the linkage group defined by Huskey and co-workers is shown to the right of its current equivalent; in the fourth case (LGII) it is not, because their "linkage group" contained only one marker (*regA*). Other LGs containing three or more markers have been arbitrarily assigned arabic numerals, and several other LGs that contain only two markers so far are shown at the lower left, without numerals. Genetic markers given in roman type (e.g., L2-A) represent RAPDs. The convention used in naming RAPDs is as follows: The letter and number before the hyphen indicate which standard RAPD primer (purchased in kits of 20 from Operon Technologies, Inc., Alameda, CA) was used in the polymerase chain reaction (thus, "L2" identifies oligomer 2 from kit L), and the letter after the hyphen indicates which of the polymorphic fragments obtained using that primer was analyzed (thus, "A" indicates the largest polymorphic product obtained with that primer; "B" would indicate the second largest, etc.). Markers given in italics fall into the following four categories: (1) "Conventional" Mendelian markers. These include: *chl*, chlorate-resistant; *ery*, erythromycin-resistant; *exp*, nonexpander; *d-dis*, delayed dissolver; *eye*, eyespots abnormally oriented; *flg*, flagella absent; *lts*, temperature-sensitive lethality; *mat* (formerly *mt*), mating type (male vs. female); *mes*, methionine sulfoximine-resistant; *mul*, multiple gonidia; *nit*, nitrate auxotroph; *q-inv*, quasi-inverter; *reg*, somatic regenerator. Capital-letter suffixes are those assigned by Huskey et al. (1979a). (2) RFLPs detected with probes for genes of known function. These include genes encoding histone H4 (*hsn4*), nitrate reductase (*nitA*), retrotransposon 2 (*rtn2*); the sex-inducing pheromone (*sexI*), α- and β-tubulins (*tubA* and *tubB*); ubiquitins (*ubi*); and "yeast protein two" homologues (*ypt*). (3) RFLPs detected with cDNAs that define genes expressed preferentially in gonidia (*gon*) or somatic cells (*som*); the number immediately following one of these abbreviations is the number given to the corresponding cDNA by Tam and Kirk (1991a). (4) Loci involved in modifying *Sau*3AI recognition sites within genes encoding α-tubulins (*tsm*) or nitrate reductase (*nsm*). In categories 2–4, numerical suffixes identify distinct polymorphic fragments (in order of decreasing molecular weight) in cases in which more than one is identified by a single probe.

more likely: One of the genes involved in modification of *Sau*3AI sites in middle-repetitive DNA is closely linked to the *regA* locus, but its activity is detected as strain-specific differences in the digestibility of DNA elements that are located elsewhere in the genome. Thus, it is unlikely that positional cloning will ever succeed in *V. carteri* when fragments produced by digestion with the enzymes listed in Tables 7.1 and 7.2A are used to define the starting place.

The *Sau*3AI RFLPs detected with the *nitA* and α-tubulin-encoding genes appear to define "modification" loci that segregate in a reliable Mendelian manner (albeit not with the DNA sequences used to detect them), and thus they are tentatively included in our current genetic map. However, the approximately 150 *Sau*3AI/middle-repetitive-DNA RFLPs that were described in the early stages of our analysis of the genome (Harper et al. 1987; Adams et al. 1990) exhibit so many transmission anomalies (which we now attribute to complex interactions among various *Sau*3AI-modification alleles) that we have withdrawn from our linkage analysis all *Sau*3AI/middle-repetitive-DNA RFLPs. We now focus our RFLP mapping efforts on polymorphisms that are detected with cloned unique or small-gene-family sequences in DNA samples that have been cut with restriction enzymes that reliably digest all, or virtually all, cognate sites that have been examined (Table 7.2B,C). Segregation patterns for more than 50 such markers have now been analyzed in a standard progeny cohort of approximately 130 strains, and about 30 others have been identified but have yet to be analyzed.

7.2.3 RAPDs provide many additional markers

Mapping of many eukaryotic genomes has been greatly accelerated in recent years by the use of random amplified polymorphic DNAs, or RAPDs (Williams et al. 1990, 1993; Reiter et al. 1992; Postlethwait et al. 1994).[11] Using two sets of 20 commercially available primers (Operon Technologies, Inc.) that were arbitrarily selected for testing, we have been able to define more than 100 markers that segregate in a Mendelian manner. In parallel with the experience of others, we find that genetic data can be accumulated more

---

[11] The hundreds of commercially available nucleotide decamers that are designed for use as primers in RAPD analysis (Operon, Inc.) are not random in sequence, but are systematic variants on certain empirically established themes, such as high G+C content (Williams et al. 1993). Thus "arbitrary amplified polymorphic DNAs" (AAPDs) would be a more appropriate name for the markers detected with this technique. However, the more euphonious acronym RAPDs is unlikely to be displaced from the literature.

quickly (by one or two orders of magnitude) with the RAPD technique than by analysis of RFLPs. Not only does the RAPD analysis itself proceed much more rapidly, but also it requires much smaller quantities of progeny DNA, which greatly reduces the time and effort that must be expended in DNA isolation and purification.

Our current genetic map of the *V. carteri* genome, which incorporates certain conventional Mendelian markers with many more RFLP and RAPD markers, is shown in Figure 7.3. The number of morphological markers that we have attempted to include in the analysis so far is small, because we believe that mapping of such genes will proceed much more efficiently after a map rich in other types of markers has been generated and used to develop tester strains that will permit placing a new morphological marker on the map with a single cross.

7.3   General features of *V. carteri* protein-coding genes

Despite the fact that only four laboratories have participated in the effort so far, more than 30 protein-coding genes of *V. carteri* have been sequenced to date (Table 7.3). Some of these encode proteins known only in *V. carteri* (such as the sexual pheromone and various components of the ECM), but the majority encode proteins of widespread distribution and known functions (such as the histones, tubulins, and actin). Like the organism itself, the genes of *Volvox* possess certain animal-like and plant-like features, but they combine such features with other features that are distinctively volvocalean, thereby speaking to broader issues regarding the evolutionary history of eukaryotic protein-coding genes.

7.3.1   Exon–intron structure and the question of intron origins

V. carteri genes generally are richly endowed with introns. Indeed, in most cases in which genes of homologous sequences and equivalent functions have been sequenced in *V. carteri*, vascular plants, animals, and/or fungi, *Volvox* presently holds the record for the largest number of introns.[12] For example,

---

[12] This relationship appears to distinguish the Volvocales as a group, because in cases in which *Chlamydomonas* genes are available for comparison, they usually have about the same number of introns as *V. carteri* (Schmitt et al. 1992).

Table 7.3. *Protein-coding genes of* V. carteri *sequenced as of early 1996*

Gene family	Gene product	Copy # (# seq'd)	Coding region (bp)	Introns	Polypeptide (aa)	Ref.[a]
actA	actin	1	1131	9	377	1
ars	arylsulfatase	1	1950	16	649	2
cam	algal CAM	1	1323	8	217-440[b]	3
fnr	ferredoxin-NADP⁺ reductase	?(1)[c]	1044	?	347	4
hstH1	histone H1	≥ 4 (2)	783; 723	3	260; 240	5
hstH2A	histone H2A	~15 (2)	387	0	128	6
hstH2B	histone H2B	~15 (2)	471; 465	0	156; 154	6
hstH3	histone H3	~15 (2)	405	1	134	7
hstH4	histone H4	~15 (2)	312	0	102	7
isg	inversion-specific glycoprotein	1	1398	2	465	8
lhcA	light-harvesting complex A	?(1)	798	?	265	4
nitA	nitrate reductase	1	2592	10	864	9
oee1	oxygen-evolving enhancer 1	?(1)	609	?	202	4
phe	pherophorins I, II & II	≥12 (6)	1455-1485	7	484-494	10
sexI	sexual pheromone	~6 (1)	624	4	208	11
ssgA	sulfated glycoprotein 185	1	1458	7	485	12
tubA	α-tubulin	2 (2)	1353	3	451	13
tubB	β-tubulin	2 (2)	1329	3	443	14
ubqA	polyubiquitin	?(1)	1143	5	381[d]	15
ypt	small G proteins (YPTV 1-5)	5 (5)	609-648	5-8	203-217	16

[a]References: 1. Cresnar et al. (1990). 2. Hallmann & Sumper (1994a). 3. Huber & Sumper (1994). 4. Choi, Przybylska & Straus (1996). 5. Lindauer, Müller & Schmitt (1993b). 6. Müller, Lindauer et al. (1990). 7. Müller & Schmitt (1988). 8. Ertl et al. (1992). 9. Gruber et al. (1992). 10. Sumper et al. (1993); Godl et al. (1995). 11. Mages, Tschochner & Sumper (1988). 12. Ertl et al. (1989). 13. Mages et al. (1988, 1995). 14. Harper & Mages (1988); Mages et al. (1995). 15. Schiedlmeier & Schmitt (1994). 16. Fabry, Nass et al. (1992); Dietmaier et al. (1995).
[b]Alternative splicing yields three different polypeptides.
[c]Question mark indicates that copy number has not yet been determined for that gene family.
[d]Processed to produce four peptides 77 aa long and one 76 aa long.

whereas the vast majority of histone H1 genes that have been sequenced in all other taxa are free of introns (and only three had previously been identified that have one intron), each of the two *V. carteri* histone H1 genes that have been sequenced has three introns, of which only one is in the same position in both genes (Lindauer, Müller & Schmitt 1993b). Similarly, whereas the genes encoding nitrate reductase in plants and fungi contain three and six introns, respectively, in *V. carteri* the homologous gene (which is otherwise well conserved) has 10 introns (Gruber et al. 1992) (the gene encoding nitrate reductase is the only one known in which *C. reinhardtii* has more introns than *V. carteri*: The *C. reinhardtii* version contains 15 introns) (Zhang 1996). Whereas actin genes are intron-free in several other eukaryotic microorganisms, and the largest number of actin introns known previously was seven (in a human gene), the *V. carteri* actin gene was found to contain nine introns: eight in the coding region and one in the 5'-untranslated region (5'-UTR) (Cresnar et al. 1990). Seven is the greatest number of introns found in genes encoding arylsulfatase in other taxa, but the *Volvox* arylsulfatase gene has 16 introns (again, including one in the 5'-UTR), resulting in a gene in which nearly three-fourths of the transcription unit consists of intervening sequences (Hallmann & Sumper 1994a). A third *V. carteri* gene known to have an intron in the 5'-UTR is the one encoding SSG 185 (Section 5.4.4.3; Ertl et al. 1989).

The exceptional abundance of introns in *Volvox* genes has stimulated investigators to use them to evaluate two widely discussed theories regarding the origins and significance of introns. In simplified form, one of these, the "introns-early" theory, postulates that the primordial gene in any given category had introns in all the positions in which introns are present in genes of that category in modern eukaryotes and that differences among modern taxa with respect to numbers and locations of introns reflect lineage-specific losses during evolution. A corollary of this hypothesis is that because introns were preferentially located between regions encoding discrete functional domains of polypeptides, they facilitated a process of "exon shuffling" that generated proteins with novel properties (e.g., Gilbert, Marchionni & McKnight 1986). In equally simplified form, the "introns-late" theory postulates that the "spliceosomal" introns of protein-coding genes originated relatively late in evolution, only after the appearance of the eukaryotic nucleus, and that such introns have been gained and lost repeatedly during eukaryotic diversification as a result of movement of mobile elements into and out of coding sequences, with little regard for domain boundaries (e.g., Cavalier-Smith 1991).

Most *V. carteri* introns are more readily accommodated within the introns-late hypothesis. For one thing, many of the numerous *V. carteri* introns are in novel locations – locations not occupied by introns in homologous genes of any other organisms that have been studied – and thus increase sharply the number of primordial introns that would have to be postulated under a strict application of the introns-early model. For example, if the primordial actin gene had introns at the six novel locations defined by the *V. carteri* gene, plus the other locations at which introns are now found in just nine other organisms, the gene would have consisted of 31 exons, with an average length of only 12 codons (Schmitt et al. 1992). Similarly, had introns been present in a primordial histone H3 gene at the two different positions at which they are now found in the *V. carteri* H3 genes, that gene would have consisted of three exons, one of which would have been only one nucleotide long (Müller & Schmitt 1988; Schmitt et al. 1992; Fabry, Müller et al. 1995a). A second point is that in cases in which functional protein domains can be clearly identified, there is no obvious tendency for *V. carteri* introns to fall at or near the borders of such domains. For example, although strong sequence similarities suggest that the gene encoding nitrate reductase might have originated by assembly of exons encoding functional domains that are found separately in sulfite oxidase, cytochrome b_5 and cytochrome b_5-reductase (plus two "hinge" domains) (Crawford et al. 1988), the 10 introns in the *V. carteri* version of this gene mostly fall within (rather than near the boundaries of) the regions encoding those five domains (Gruber et al. 1992).

Nevertheless, detailed analysis of introns in genes encoding small G proteins (SGPs) in *Volvox*, *Chlamydomonas*, and 10 other species has supported the possibility that although most modern introns appear to be of the recent/insertional type, a few may represent leftovers from a much earlier, exon-switching era (Dietmaier & Fabry 1994). The Ras superfamily of SGPs comprises several families of related guanosine-nucleotide-binding proteins that all consist of analogous arrays of structural and functional domains, and that all function as molecular switches in various eukaryotic cellular processes (Bokoch & Der 1993). The regularity of SGP domain structure, combined with the observation that the nine sequenced genes of *V. carteri* and *C. reinhardtii* that encode members of the Ypt ("yeast protein 2") family of SGPs all have more introns than their closest homologues in other species (Dietmaier et al. 1995),[13] provided the stimulus for a detailed analysis of the locations of introns with respect to domain boundaries (Dietmaier & Fabry 1994). That analysis showed that the introns present in the 28 SGP genes for which appropriate sequences were available defined 60 different intron locations, and if introns had been present at all those positions in the primordial SGP-encoding gene, its exons would have averaged only three codons in length, and several exons would

---

[13] That analysis of volvocalean *ypt* genes led to an interesting secondary conclusion. Although genes encoding the same Ypt isotype in *V. carteri* and *C. reinhardtii* had very high sequence homologies in coding regions, had introns at identical locations in every case, and often had introns of similar lengths at equivalent positions, they had no detectable sequence similarities within any of those introns. That not only confirmed the long-established opinion that such intron sequences evolve much more rapidly than coding sequences but also suggested the possibility that such introns might be useful for developing a fine-grained molecular phylogeny of volvocalean isolates, formas, and species that are very closely related. That has turned out to be the case: Sequences of these introns have, for example, provided the first statistically robust phylogentic tree relating the various natural isolates of *V. carteri* f. *nagariensis* to one another and to the other formas of *V. carteri* discussed earlier (Liss et al. 1997).

have been a codon or less in length. Most of these intron positions (80%) are occupied in only a single sequenced gene, or in two or three genes of extremely similar sequences from quite closely related organisms. Moreover, most such introns fall in an apparently random pattern within regions encoding highly conserved domains. These two considerations make it seem unlikely that such introns are ancient ones that participated in exon shuffling, and much more likely that they are the result of quite recent insertions. However, only 3 of the 60 intron locations identified so far are shared by genes encoding members of different SGP families. Provocatively, these three intron locations not only lie between regions encoding recognizable protein-folding domains (albeit within guanosine-nucleotide-binding functional domains in two cases) but also are shared by SGP genes of organisms as distantly related as *Volvox* and fungi, or *Volvox* and humans. Thus, these three are candidates for positions at which introns may have been present very early in evolution and where introns may have facilitated exon shuffling that produced the genetic and functional diversity that now exists within the Ras superfamily of SGPs.

Be that as it may, the overview that emerges from such studies is that although a minority of the many introns now present in *V. carteri* genes may represent residua of an ancient exon-shuffling epoch, most such introns probably represent much more recent modifications of the genome. More to the point, perhaps, no amount of analysis has yet provided any significant insight into the biological significance of the fact that introns are so abundant in the genes of volvocaleans in general and *V. carteri* in particular. Data discussed in the following section indicate that (as in other eukaryotes) introns play a role in gene expression, but give no insight into why volvocalean genes have so many introns.

7.3.2 *The role of introns in regulating gene expression*

The genomic copy of the *V. carteri nitA* gene that has been used as a homologous selectable marker for transformation (Schiedlmeier et al. 1994) (see Section 7.4.1) is 7.6 kb in length and contains 10 introns (Gruber et al. 1992). In an effort to enhance transformation efficiency, Gruber et al. (1996) constructed a vector in which the *nitA* cDNA was ligated in frame to the upstream and downstream elements of the genomic copy of the gene, which reduced the length of the cloned element by nearly half. They then compared the transformation efficiency and expression level of that construct with the values for the construct carrying the native gene with its 10 introns and with the values for constructs carrying the cDNA version to which either one or two introns had been added back. The construct carrying the intronless version of the *nitA* gene was as effective a transformation agent as the one bearing the genomic version, and following nitrate induction it was transcribed with about the same efficiency as all other constructs that were tested. However, the intronless gene apparently was expressed with only marginal efficiency at the protein level, because strains carrying that version of the gene grew only about 1% as fast in nitrate medium as strains carrying a wild-type *nitA* gene (Table 7.4). In marked contrast, constructs containing the cDNA to which either intron 1 or introns 9 and 10 had been added back were both much more efficient as transformation vectors than were the constructs containing either the genomic version of the gene or the intronless cDNA, and they both supported vigorous growth on nitrate (Gruber et al. 1996). These data appear to indicate that efficient posttranscriptional processing (possibly including transport from the nucleus) and subsequent translation of *V. carteri* transcripts may depend on the presence of at least one intron. However, these data yield no insight into the reason that volvocalean genes contain so many introns.

7.3.3 Codon bias and its evolutionary implications

The news that bias is now politically incorrect apparently has not reached the Volvocales. Nearly all volvocalean nuclear genes examined thus far exhibit a strong bias in codon usage, with the order of the third-base preference always being C>G>U>A.[14] However, in most cases the magnitude of this bias is not as great as was once believed, based on the first few genes that happened to be sequenced.

In the first four volvocalean nuclear genes that were sequenced, namely, those encoding the α- and β-tubulins of *C. reinhardtii* (Brunke et al. 1984; Youngblom, Schloss & Silflow 1984; Silflow et al. 1985), 93% of all codons end in C or G (with C being favored by more than 2:1 over G) (Figure 7.4); 20 sense codons, including 13 of 15 ending in A, are not used at all, and the two remaining codons ending in A are used only once each. In the next *C. reinhardtii* nuclear genes to be sequenced, namely, two encoding variant small subunits of rubisco (Goldschmidt-Clermont & Rahire 1986), codon bias

---

[14] The chloroplast and mitochondrial genes of *Chlamydomonas* also exhibit strongly biased codon usage, but the preferred codons are entirely different from those preferred in nuclear genes. In the chloroplast, third-base preference is described by the sequence U>A>G>C, whereas in the mitochondrion it follows the sequence U>C>G>A (Harris 1989).

Table 7.4. *Transformation and growth rates observed with* nitA *transgenes containing different numbers of introns*[a]

Volvox Strain	nitA genotype	Transgene	Transgene introns	Transforma- tion rate	Relative growth rate on nitrate
Gls22	nitA+	none	NA[b]	NA	100
153-81	nitA-	none	NA	NA	0.01
153-81	nitA-	pVcNR4	all ten	~4 X 10-5	ND[c]
153-81	nitA-	pVcNR13	none	~5 X 10-5	1.2
153-81	nitA-	pVcNR15	#1	~50 X 10-5	122
153-81	nitA-	pVcNR16	#9 & 10	~50 X 10-5	76

[a]After Gruber et al. (1996).
[b]NA: Not applicable
[c]ND: Not determined

turned out to be even more pronounced: 95% of all codons in these two genes end in C or G, and fully 30 codons (including all of those ending in A, and 9 of those ending in U) are never encountered. The *C. reinhardtii* histone genes exhibit only slightly less bias, with individual histone genes failing to use 20–29 codons (Schmitt et al. 1992; Fabry, Müller et al. 1995a). However, as sequences have been accumulated for a number of *C. reinhardtii* genes that encode proteins that are less abundant in the cell than those in the preceding three categories, significant relaxation of this codon bias has been observed (Figure 7.4A). In such genes, the same pattern of preference with respect to third bases is observed, but fewer codons go unused in most cases, and all possible codons are found to be used at least occasionally (De Hostis, Schilling & Grossman 1989; Schmitt et al. 1992). At the extreme, it has recently been discovered that two genes encoding important mating-type proteins of *C. reinhardtii* exhibit no discernible codon bias whatsoever (Ferris, Woesnner & Goodenough 1996; P. J. Ferris and U. W. Goodenough personal communication).

V. carteri genes exhibit codon bias that is qualitatively similar to, but quantitatively less stringent than, that observed for the homologous *C. reinhardtii* genes (Figure 7.4B). In the four *V. carteri* genes encoding the α- and β-tubulins, for example, only 78% of all codons end in C or G, and only eight codons go unused (Harper & Mages 1988; Mages et al. 1988, 1995). Similarly, on average the histone genes of *V. carteri* exclude about five fewer codons than do the homologous *C. reinhardtii* genes (Müller & Schmitt 1988; Müller, Lindauer et al. 1990; Fabry, Müller et al. 1995a). As with *Chlamydomonas*, the fact that all possible codons are used at least occasionally clearly indicates that *V. carteri* must have the ability to translate all codons

All amino acids   Leucine   Other 'ACGT' aa's

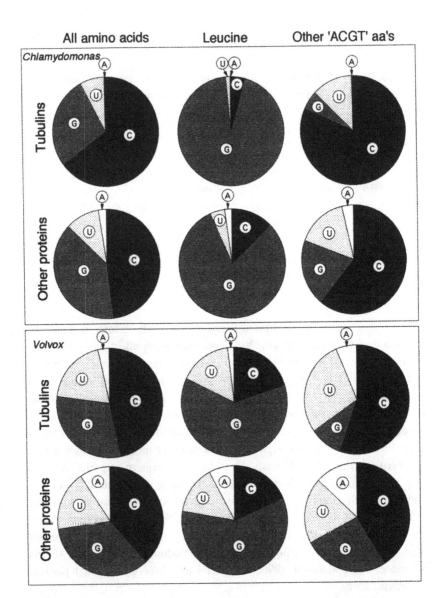

Figure 7.4 Relative frequencies of codons ending in the indicated nucleotides in *C. reinhardtii* and *V. carteri* nuclear genes encoding the α- and β-tubulins and a variety of less abundant proteins. In both species the genes encoding the "Other proteins" included all of the nuclear protein-coding genes for which sequences were available in international data bases in mid-1995, except for those encoding the tubulins, the histones, and the small subunit of rubisco (which are all quite similar to the tubulin genes in their codon biases). In both species there is a bias in favor of codons ending in C>G>U>A. However, this bias is significantly more intense in *Chlamydomonas* than it is in *Volvox*, and in both species it is more intense in genes encoding abundant proteins than it is in those encoding less abundant proteins. The concept that these patterns are the sum of specific biases

in the standard code book (Mages et al. 1988; Gruber et al. 1992; Hallmann & Sumper 1994a).

In other species, codon usage patterns have been found to be strongly correlated with abundance patterns for the corresponding iso-accepting tRNAs (Ikemura 1985). No studies of tRNA abundance patterns in volvocaleans have yet been reported; however, the available data are most easily rationalized within such a framework. Perhaps this is best exemplified by the regular differences in the third-base preference patterns that are observed within codon sets representing different amino acids. For example, if one first considers all amino acids other than valine and leucine for which codons ending in all four bases are available, it is seen that the bias in favor of C relative to G is usually stronger than it is for all amino acids considered together (Figure 7.4, left versus right columns). In marked contrast, however, in both *Chlamydomonas* and *Volvox* genes the valine and leucine codons ending in G are generally used much more frequently than those ending in C (Figure 7.4, center columns). The fact that this is not just some sort of generalized third-base preference, but is codon-specific, is well illustrated by the leucine codons: Of the two codons ending in G that signify leucine, CUG is greatly preferred to UUG in nearly all volvocalean genes that have been analyzed. For example, in the six *C. reinhardtii* genes encoding the tubulins and the rubisco subunits, CUG is used 180 times, and UUG is never used! Even in the *V. carteri* genes encoding low-abundance proteins (where codon bias is relatively relaxed), CUG is used about 3.5 times as often as UUG (and about 2.5 times as often as CUC) to encode leucine (data not shown). The simplest hypothesis to explain these data is that they reflect the relative abundance of different leucine-accepting tRNAs in the volvocaleans.

Additional insight into the skewed codon usage patterns of these two species comes from careful consideration of four cases in which two genes of equivalent function have been sequenced in both *C. reinhardtii* and *V. carteri* (two pairs of tubulin genes and two pairs of histone genes in each species) (Schmitt et al. 1992). In each of these cases the vast majority of the differences in nucleotide sequences that are observed between cognate genes, both

Caption to Figure 7.4 *(cont.)*
favoring the use of particular codons (rather than a reflection of some more generalized property, such as base composition of the genome) is supported by a comparison of the codons used for leucine and valine to those used for all other amino acids that can be represented by codons ending in all four nucleotides ("Other 'ACGT' aa's"). Although codons ending in C are used much more frequently than codons ending in G to encode the "other" amino acids, the situation is reversed with respect to codons for leucine (shown) or valine (not shown).

within and between species, represent silent, third-codon-position nucleotide exchanges. In each of these gene pairs the two *Chlamydomonas* genes are much more similar to one another than the genes of *Volvox* are, and (quite surprisingly) in each case the two *Volvox* genes are nearly as different from one another in nucleotide sequence as each is from either of its *Chlamydomonas* homologues (Figure 7.5). Moreover, the nucleotide differences that exist between cognate *Volvox* and *Chlamydomonas* genes are distinctly non-random: Fully two-thirds of the differences represent cases in which a codon ending in C in *C. reinhardtii* ends in some other nucleotide in *V. carteri*, and 80% of the remaining differences represent cases in which a codon ending in G in *C. reinhardtii* ends in some other nucleotide in *V. carteri*. Thus, the vast majority of the nucleotide sequence differences that exist between cognate genes in these two volvocaleans can be accounted for by the fact that the preference for codons ending in C and G is substantially stronger in *C. reinhardtii* than it is in *V. carteri*.

All such details of the codon usage patterns of *Volvox* and *Chlamydomonas* are consistent with the hypothesis that selection acts to balance codon usage patterns with iso-acceptor tRNA abundance patterns in each species and that the intensity of such selection increases with the amount of the encoded protein that must be synthesized to fulfill normal cellular functions (Ikemura 1985; Schmitt et al. 1992). In this framework, the fact that codon bias is significantly stronger in *Chlamydomonas* than in *Volvox* clearly implies either that the iso-acceptor tRNA pools are more strongly skewed, or that the selection for high translational efficiency is stronger in *Chlamydomonas* than in *Volvox*.

Does the stringent codon bias of modern *C. reinhardtii* imply the action of stabilizing selection to maintain a stringent ancestral pattern of codon usage? Or does it imply the action of directional selection to convert a less stringent ancestral bias (perhaps similar to that seen today in *V. carteri*) to a more stringent bias? This interesting question cannot be addressed until suitable data become available regarding codon usage patterns in other volvocine algae.

7.3.4 Putative regulatory signals

As noted, the first volvocalean genes that were sequenced were the *Chlamydomonas* tubulin genes. Those genes had already been shown to be co-ordinately expressed and regulated during the cell cycle and during flagellar regeneration (Sarma & Shyam 1974; Brunke et al. 1982), and even before

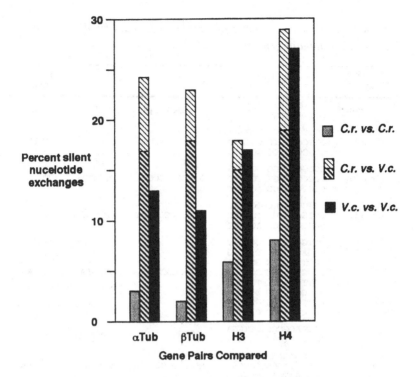

Figure 7.5 Abundance of silent substitutions within and between cognate gene pairs of *C. reinhardtii (C.r.)* and *V. carteri (V.c.)*. Within each species, the two α-tubulin genes and the two β-tubulin genes encode identical polypeptides. Between the two species the amino acid sequences of the cognate tubulins are more than 99% identical; but the nucleotide sequences of the cognate genes differ by as much as 24%, as a result of silent substitutions. Such silent substitutions are more than five times as abundant in *V. carteri* as in *C. reinhardtii*, however. A-similar pattern is seen when the two histone H3 and two histone H4 genes that have been sequenced in each of these species are compared. In the histone case, the cognate genes of *V. carteri* differ as much or more from one another as they differ from the corresponding *C. reinhardtii* genes. In the interspecific comparisons (*C.r.* vs. *V.c.*), the minimum and maximum frequencies of silent substitutions calculated for four pair-wise comparisons are indicated by bars with heavy and light cross-hatching, respectively.

the genes had been fully sequenced Brunke et al. (1984) made the striking observation that each of them possessed multiple (frequently overlapping) copies of a 16-bp motif upstream of the putative TATA box (Table 7.5). They immediately postulated that these elements might be involved in co-ordinating expression of the tubulin genes. Consistent with this idea, rather

Table 7.5. *Sequence elements of putative regulatory significance in volvocalean tubulin genes*[a]

Element / Gene	Tub box[b]		GC-rich	TATA box	Poly(A) signal
Consensus [c]	G T GCTCCAAGGC	T T GCGNNG	A(G/C)$_{9-11}$A	?	TGTAA
C.r. α-1T	GCTCGAAGGC GCTCGAAGTC GCCCCATTCC	TCGAAG GGCATC GGGGCG	ACGGCGCCCGA	TTATAA	TGTAA
C.r. α-2T	GCTGCATGGG GCGCGATGGT GCCCGATGC	GCCGCG CCTTAG ACTACT	AGGCCCCGCCCA	TACTTAA	TGTAA
C.r. β-1T	CTTCGAATGC GCTGCAATGC GCTCGAAGGC	TCGAAG TCGAAG AGTCGT	AGCCCCCCCGCA	TTCAAATT	TGTAA
C.r. β-2T	GCTGCGAGAC GCGAGACGGC GCTTCCCGGC CCCCGAAGCC CCTTCGGGGC GCTGCATGGG GCTCCGTGCC GCTCCAGGGC	GGCTTC TTCCCG GCTGCA CCTTCG TGCATG GCGCTC GCTCCA CAGCGC	ACGCCGGCCCCGA	TTTAAAT	TGTAA
V.c. α-1T	GCTGAGGCCC GCCCCCTGGA	CCTGGA GCGCCA	AGCGCCCTTCA	TTCAAAT	TGTAA
V.c. α-2T	GTTCCAAAGA TATGCTAGGC GTACGATGAT	GCGTAA GCGGAG TCAAAA	ATGTTGCCGCGCA	TACAAAA	TGTAA
V.c. β-1T	CTCAAAAGGC	ACTCGC	ACTCGCTGGCA	TCAAAA	TGTAA
V.c. β-2T	TCCAGCCTGC CCATGATGGC	TGGACG GCCATG	AGCTGCGGGGCA	TACTTAA	TGTAA

[a]Compiled from Brunke et al. (1984), Youngblom, Schloss & Silflow (1984), Silflow et al. (1985), Harper & Mages (1988), and Mages et al. (1988, 1995).

[b]Brunke et al. (1984) noted that of the 16 bp in the repeating motif they found upstream of the TATA box in each tubulin gene, the first 10 bp were most strongly conserved. It is this 10-bp motif (here set off by a space) that is now usually referred to in discussions of tub boxes.

[c]Common alternative bases are given above the consensus sequence for the tub box, except where N is used to indicate that A, C, G, or T occur with nearly equal frequencies.

similar motifs (≥60% identity) were subsequently found upstream of the TATA boxes of the *V. carteri* tubulin genes (Harper & Mages 1988; Mages et al. 1988). A second recurring feature detected in the tubulin genes of both algae was a GC-rich sequence, 11–13 bp long, in the short stretch between the presumptive TATA box and the transcription start site (Table 7.5) (Brunke et al. 1984; Harper & Mages 1988; Mages et al. 1988).

Neither of those elements turned out to be uniquely present in tubulin genes, however. So-called tub boxes (the more strongly conserved first 10 bp of the 16-bp motif) and/or GC-rich elements like those of the tubulin genes have now been identified in a range of other volvocalean genes (Goldschmidt-Clermont & Rahire 1986; Imbault et al. 1988; Williams et al. 1989; Schloss 1990; Müller, Igloi & Beck 1992; Mages et al. 1995).

The hazards of attempting to deduce the possible regulatory functions of recurring sequence motifs from the results of certain kinds of functional tests – let alone in the absence of such tests – are well illustrated by consideration of the GC-rich elements. The sheer ubiquity of these particular elements in volvocalean genes was widely interpreted to mean that they must be involved in transcriptional regulation, and the first functional assay of putative regulatory signals in *C. reinhardtii* appeared to substantiate that conclusion (Bandziulis & Rosenbaum 1988). It was observed that when a plasmid bearing the native α1-tubulin gene was injected into *Xenopus* oocytes it was transcribed efficiently and faithfully, whereas plasmids in which the GC-rich element had been either deleted or interrupted by a 12-bp insertion were not transcribed at all. Thus, it was concluded that the GC-rich element "is absolutely required for transcription" (Bandziulis & Rosenbaum 1988). Subsequently, however, a diametrically opposed conclusion was drawn when the putative regulatory sequences of the *C. reinhardtii* β2-tubulin gene were analyzed in *C. reinhardtii* strains that had been transformed with hybrid constructs containing native or mutated versions of the β2-tubulin promoter fused to an arylsulfatase reporter gene (Davies, Weeks & Grossman 1992; Davies & Grossman 1994). In the latter study it was observed that when the GC-rich element was converted to an AT-rich element, there was no significant effect on the basal transcription rate nor on the changes in transcript abundance that occur throughout the life cycle nor even on the magnitude of the induction that followed deflagellation (Davies & Grossman 1994). Although it cannot be ruled out that the diametrically opposed findings obtained in those two studies may have been due to differences in the promoters that were analyzed or the systems in which the constructs were tested, it is at least as likely that the difference was due to the types of mutations that were examined and that a modification that changed the distance between the TATA box and the transcription start site had an entirely different (and less meaningful!) effect than a change in the nucleotide sequence of that element. In any case, it now appears clear that this relatively ubiquitous GC-rich element has no substantial sequence-specific role to play in regulating transcription of the *Chlamydomonas* β2-tubulin gene, and unless compelling evidence to the contrary should be reported, it is probably safest to assume that the same is probably true for the other volvocalean genes in which it is present.

The studies of Davies and Grossman (1994) did produce evidence of a regulatory role for the tub boxes. There are more tub boxes in the *Chlamydomonas* β2-tubulin gene than in any other yet analyzed: Seven or eight (depending on the criteria used) occur in three clusters within 140 bp of the transcription start site. Studies with truncated genes have indicated that none of these tub boxes are required for basal transcription, but at least four of them are required for normal transcriptional modulation during the cell cycle and for the transcriptional response to deflagellation (Davies & Grossman 1994). Both of those responses were observed when the middle tub-box cluster and either of the end ones were present in the promoter, but neither was observed with a promoter containing just the middle cluster or one containing both terminal clusters. That set of observations led to the conclusion that both cell-cycle regulation and the deflagellation response require at least two clusters of tub boxes that are appropriately spaced and/or oriented with respect to one another (Davies & Grossman 1994).

The generality of the foregoing conclusion is left in doubt, however, by observations such as the following: The gene encoding radial-spoke protein 3 is also induced by deflagellation, despite the fact that it has only one tub box (Curry, Williams & Rosenbaum 1992). In contrast, a number of other genes that have multiple tub boxes – such as those encoding a rubisco small subunit (Goldschmidt-Clermont & Rahire 1986), HSP70 (Müller, Igloi & Beck 1992), and a G protein (Schloss 1990) – have never been observed to be induced by deflagellation. Therefore, it seems unlikely that the tub-box elements function in the same way in all of the volvocalean genes where they are present.

A study by Kropat et al. (1995) of *cis*-regulatory elements of the *C. reinhardtii* gene encoding HSP70 strongly underscored the potential pitfalls of taking shortcuts in an attempt to define the *cis*-regulatory elements responsible for transcriptional regulation of a gene of interest. Those authors first concluded, from differences in induction kinetics and inhibitor sensitivity, that heat shock and light induced *hsp70* transcription via different pathways. In an attempt to locate the *cis*-regulatory elements involved in each of those responses, they transformed *C. reinhardtii* with a series of *hsp70* constructs that were truncated at different upstream sites. When they observed that a construct truncated at −138 gave a normal heat-shock response, whereas light induction required that sequences between −138 and −209 also be present, it may have been tempting to terminate the analysis and conclude that these two responses were mediated by completely different *cis*-regulatory elements. However, Kropat et al. next explored the region between −196 and −60 in greater detail, using more closely spaced truncations, and found that the situation was more complex than their original set of truncations had indicated. Surprisingly, they found that both heat and light induced full and vigorous expression of a gene truncated at −108, but not one that was truncated at −87. That implicated an element in the −108 to −87 region in mediating both types of induction. However, very different responses to heat and light were then observed with genes truncated at position −138 or −141; those two constructs were fully inducible by heat, but not (in any of eight independent transformants) by light. Full induction by light was restored, however, in constructs truncated at −146 or anywhere farther upstream. The simplest interpretation of those observations is that the region between −87 and −108 is adequate to confer normal heat-shock induction on the gene, but at least three regions are involved in mediating its light-induced expression: two positively acting elements located in the regions −146 to −138 and −108 to −87, plus a negative control element lying between them, in the region −138 to −108 (Kropat et al. 1995).

Had Kropat and co-workers determined only the effects of truncating *hsp70* at −138 and −208, they might have concluded that its responses to heat and light were mediated by completely different *cis*-regulatory elements. On the other hand, if they had analyzed only the effects of truncation at −108 and −87, they might have concluded just the opposite: that the two responses were mediated by the same element. The lesson to be learned from this is one that has been repeatedly emphasized in studies of other organisms: Because *cis*-regulatory regions often contain binding sites for multiple kinds of *trans*-acting factors that interact in complicated ways, analysis of the details of transcriptional regulation is seldom simple; it frequently requires extensive and systematic mutational studies, and even then one can never be sure that all of the regulatory complexities have been uncovered. Functional analyses of various *V. carteri* promoters are now in progress (M. Sumper, R. Schmitt & D. Straus independent personal communications), but none has yet been published. Thus, the recurring motifs that will be discussed next should be regarded as prime candidates for careful functional analysis, rather than as elements of known regulatory significance.

An interesting set of coordinately regulated volvocalean genes that have been analyzed in considerable detail in search of recurring sequences of potential regulatory significance comprises the histone genes (Müller & Schmitt 1988; Müller, Lindauer et al. 1990; Lindauer et al. 1993b; Fabry, Müller et al. 1995a). Such an analysis is facilitated by the fact that genes encoding the nucleosomal histones are arranged in H3/H4 and H2A/H2B pairs, in which members of each pair are divergently transcribed from a compact, shared upstream region. Of some 60 histone genes believed to be present in each of these two species, 10 from *V. carteri* and 13 from *C. reinhardtii* have now been sequenced, and they have been found to have several interesting arrays of recurring sequence elements of possible regulatory significance (Table 7.6). Figure 7.6 represents diagramatically the genomic arrangements of these elements, as well as the genes that contain them.

Table 7.6. *Sequence elements of putative regulatory significance in volvocalean histone genes*[a]

Element Genes[b]	"E"[c]	"A"	"B"	TATA	3' Palindrome[d]
C.r. H2B/2A (3)	(6) GCGAGGM-GAGCGGTT GCCARGSCGAGSCGCC	(6) TGGCCAGGSCGAG TGGCCARGSCGAG	(5) AGCGTTGACC AGCGTTGACY	(6) TWWWTKA TTATAT	(6) AAAACTCGGTGTTTC |||||| CCACCACAACT
C.r. H3/H4 (3)	(3) RCCAGGCTGAGGCGTT	(3) TGRCCAGGCTGAG	(12) MGTTGACC CGTTGACC	(6) TTAARW TTAWSA	(6) AAAACTCGGTGTTTC |||||| CCACCACAACT
C.r. H1 (1)	(0)	(0)	(3) CGTTGACC	(1) TTATAT	(1) AAAACTCGGTGTTTC |||||| CCACCACAACT
V.c. H2A/2B (2)	(4) CCCGGTGAGCCMWACC CCCRGTGAKCYGMYCS	(0)	(0)	(4) KTTAMY WAWYAK	(4) WMMWYYCGGTGTTTT |||||| HCACCACAACT
V.c. H3/H4 (2)	(2) CCCGSTGAGCCGAWCY	(0)	(0)	(4) WATAAY TATAAW	(4) AMWAHYCGGTGTTTY |||||| RCACCACAACT
V.c. H1 (2)	(2) GSGGRTCASCGGGCCA	(0)	(4) CGTTGACC	(1) TAWAWM	(2) WWAAYCCGGTGTTTT |||||| CCACCACAACT

[a]Compiled from Müller & Schmitt (1988), Müller, Lindauer et al. (1990), Lindauer et al. (1993b) and Fabry, Müller et al. (1995a). In cases where two or more similar elements are present per shared intergenic region of an H2B/H2A or H3/H4 gene pair (Figure 7.6), the sequences of the copies found in the left half of such regions are given above those found in the right half. Letters other than those identifying specific nucleotides follow the standard convention: H=A/C/T, K=G/T, M=A/C, R=A/G, S=C/G, W=A/T, Y=C/T; a dash represents a gap. Not included in this table are a "C" element (CGGTTG) that occurs twice in each of the H2B/H2A pairs of *C. reinhardtii* (but not in any other genes listed here) and a "GC" element [(G/C)₁₀] that is found only in the H1 genes.

[b]The numbers in parentheses in this column indicate the number of gene pairs (or genes in the case of H1) that have been sequenced to date from *C. reinhardtii* (*C.r.*) and *V. carteri* (*V.c.*). Numbers in parentheses in the other columns indicate the total numbers of various elements found in the genes that have been sequenced to date.

[c]Putative enhancer elements.

[d]The vertical lines at the right are to indicate that these palindromic elements are represented as if folded back on themselves at that point; vertical lines connecting base pairs represent potential hydrogen bonds. The consensus sequences for these elements are identical in all *Volvox* and *Chlamydomonas* histone genes, but any deviations from the consensus that may have been observed in various *Chlamydomonas* histone genes were not reported.

The intergenic region shared by each H3/H4 or H2B/H2A gene pair contains one or two copies of a highly conserved element ("E") that is presumed to be an enhancer element. A similar E element is also found upstream of the *Volvox* histone H1 genes, but not in the one *Chlamydomonas* histone H1 gene that has been sequenced. All of the *Chlamydomonas* nucleosomal-histone genes also contain an array of other shorter recurring motifs of possible regulatory significance ("A," "B," and/or "C") (Figure 7.6). Homologous A, B, and C motifs have not been identified in *Volvox* nucleosomal-histone genes, but the B motif is found twice in each *Volvox* H1 gene, and each of the two *Volvox* H3/H4 gene pairs that have been sequenced contains a conserved 14-bp element ("U," Figure 7.6) similar in sequence

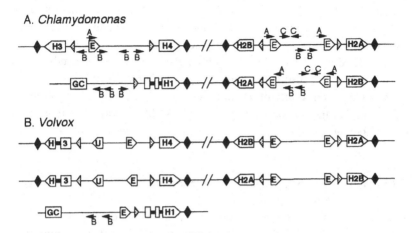

Figure 7.6 Organization of histone loci that have been sequenced in *C. reinhardtii* and *V. carteri*, drawn approximately to scale. Pointed boxes labeled H3, H4, etc., indicate the locations and transcription directions of various histone-coding regions; heavy lines interrupting such boxes represent introns. Shaded triangles indicate locations of TATA boxes. Pointed boxes labeled E represent locations and orientations of putative enhancer elements. Black diamonds indicate locations of palindromic sequences presumed to act as transcription-termination signals. Arrows labeled A, B, etc., indicate locations of other recurring-sequence motifs suspected to be of regulatory significance. For further details, see text and Table 7.6. (Adapted from Fabry, Müller et al. 1995a, with permission.)

to a yeast "upstream activating sequence" (Müller & Schmitt 1988; Müller, Lindauer et al. 1990).

As can be seen from Tables 7.5 and 7.6, most putative TATA boxes of volvocalean genes diverge to varying degrees (and in some cases not shown, more widely) from canonical TATA box sequences. Putative polyadenylation signals appear to be much more regular, however: Most of the *V. carteri* mRNAs that have been analyzed to date have the sequence UGUAA at approximately 15 bp upstream of the polyadenylation site, although variation in one or both of the last two bases is not unknown (Schmitt et al. 1992; Fabry, Jacobsen et al. 1993; Dietmaier et al. 1995). Like the mRNAs encoding the major nucleosomal histones of animals, volvocalean histone mRNAs are not polyadenylated; instead of a poly(A) signal they contain a 3' transcription-terminating palindrome very similar in sequence to those present in the equivalent vertebrate transcripts (Fabry, Müller et al. 1995a).

Selection of 3' splice sites in *V. carteri* introns is presumed to resemble that of animal genes, because, like the latter, most *Volvox* genes have a polypyrimidine tract (rather than the AU-rich motif found in plants) just upstream of each 3' splice site. It is of some interest, however, that introns in two unrelated *V. carteri* genes were found to possess the previously unknown border sequences 5'GC . . . AG3' (in place of the canonical 5'GT . . . AG3') (Harper & Mages 1988; Fabry, Jacobsen et al. 1993), and recently such a splice junction has also been found in a *C. reinhardtii* HSP70 gene (Kropat et al. 1995). Transcripts of all three of these genes are spliced properly, indicating that the volvocalean spliceosome is able to cope with this noncanonical border sequence.

7.3.5 Volvox *genes: plant-like or animal-like?*

Whereas several nineteenth-century biologists imagined *Volvox* to be the "missing link" between plants and animals, twentieth-century botanists have generally claimed *Volvox* and its relatives as their own (as "simple, motile plants," if you will), while zoologists have been equally persistent in treating the volvocaleans as a collection of protozoa (or "little green animals"). As noted in Chapter 2, modern evidence yields little support for any of those points of view: Many lines of evidence now suggest that the green-flagellate lineage leading to modern volvocaleans diverged from that leading to the higher plants well before the beginning of the Cambrian period, and not long after green flagellates last shared a common ancestor with the flagellates that would eventually give rise to fungi and animals (Figures 2.3 and 2.13). If this phylogenetic hypothesis is even approximately correct, one might anticipate that volvocalean genes encoding ubiquitous eukaryotic proteins might share certain distinguishing features with each of these other major taxonomic groups and that their similarities to plant genes might be little greater than their similarities to the genes of animals and fungi. That expectation has been confirmed.

Such a relationship is perhaps best illustrated by the genes encoding the nucleosomal histones. Whereas the volvocalean H3, H4, and H2B histones are all slightly more similar in amino acid sequences to the corresponding genes of plants than they are to those of animals, in the case of histone H2A the situation is reversed (Fabry, Müller et al. 1995a). Moreover, the regular clustering of H3/H4 and H2B/H2A gene pairs in quartets (Figure 7.6) is a feature that the volvocaleans share with many animals, but not with plants, fungi, or other protists. The H3 histones exhibit a particularly interesting combination of traits (Table 7.7). The H3 histones of *V. carteri* and *C. reinhardtii* share one more amino acid residue with higher plants than they do with animals. However, volvocalean H3 genes differ from those of plants in

Table 7.7. *Properties of histone H3 genes of* Volvox *and other taxa*

Taxon	Gene type	H3/4:H2B/A quartets?	S-phase synthesis?	Introns?	Poly(A) mRNA?	3' palin-drome?
Volvox carteri	Major	Yes	Yes	Yes	No	Yes
Ciliates	Major	No	Yes	Yes	Yes	No
Animals	Major (H3.1)	Yes[a]	Yes	No	No	Yes
Animals	Replacement (H3.3)	No	No	Yes	Yes	No
Fungi	Major	No	Yes	Yes	Yes	No
Plants	Major	No	Yes	No	Yes	No

[a]True of invertebrates and certain lower vertebrates only.

two other important regards: (1) Like all other volvocalean histone genes, their transcripts are terminated just downstream of a 3' palindrome and lack a poly(A) tail. (2) All *Volvox* H3 genes (and about half of all *Chlamydomonas* H3 genes) contain introns (Müller & Schmitt 1988; Müller, Lindauer et al. 1990; Schmitt et al. 1992). In possessing introns, these volvocalean genes resemble the H3 genes of ciliates and fungi, as well as the genes encoding the H3.3, or "replacement," type of H3 histones of vertebrates. But in possessing a 3' palindrome instead of a polyadenylation signal, and in being transcribed during DNA replication, they resemble the genes encoding the H3.1, or "replication-dependent," type of animal H3 histones. In short, whereas they share selected features with the H3 genes of a variety of other major taxa, the *V. carteri* histone H3 genes combine these features in an overall pattern that is shared only with other volvocaleans (Fabry, Müller et al. 1995a).

Similar evidence for mixed relationships is found in other protein-coding genes of *V. carteri*. For example, although the tubulin genes of *V. carteri* encode polypeptides that are slightly more similar to those of higher plants than to those of animals and fungi (Harper & Mages 1988; Mages et al. 1988, 1995), the actin encoded by the *V. carteri* gene is significantly more similar in sequence to animal actins than it is to plant actins (Cresnar et al. 1990). Of the five small G-protein genes of *V. carteri* that have been sequenced so far, one is most similar to a plant gene, one to a slime-mold gene, one to a yeast gene, and one to a mammalian gene, and the fifth is about equally similar to a plant gene and a mammalian gene – and more similar to both of those genes than the plant and animal genes are to each other (Fabry, Jacobsen et al. 1993).

Thus, the comparative anatomy of the limited number of *V. carteri* protein-coding genes sequenced to date lends no support to the idea that the volvocaleans are either simple plants or little green animals. Instead, such data are consistent with the view expressed in Chapter 2, namely, that the green flagellates represent one of the major lineages that diverged more or less simultaneously during the explosive Precambrian radiation of eukaryotes and that they have had an independent history nearly as long as that of plants, animals, or fungi. At the same time, however, the extreme similarities of most protein-coding genes of *V. carteri* and *C. reinhardtii* reinforce the concept drawn from comparative analysis of rRNA sequences (Figure 2.12) that the emergence of the volvocaceans within this ancient group was a quite recent event.

7.3.6   Gene duplication and/or homogenization

Several of the *V. carteri* genes that have been characterized show evidence of having undergone tandem duplications quite recently and/or having been subjected to a "homogenization" process that keeps them extremely similar. This was first pointed out in the case of the polyubiquitin gene (Schmitt et al. 1992; Schiedlmeier & Schmitt 1994). The polyubiquitin locus contains five tandem repeats of a coding unit, each of which encodes the 76-residue peptide that constitutes the active form of ubiquitin produced by proteolysis of the initial translation product. One clue that this gene may have arisen by tandem replication is that whereas the polyubiquitin genes of most species are intronless, the *V. carteri* gene is split into six exons by five introns, each of which is located within the codon for Gly_{35} in one of the coding units (Figure 7.7), which is the same site where an intron interrupts the monoubiquitin genes of plants and animals. With the exception of the first, the successive coding units of this gene are surprisingly similar in both exon and intron sequences. Whereas the first coding unit differs from the other four by 13 silent substitutions (12 of which are in exon 1), the remaining four units differ only by 2 silent substitutions in the last exon, just upstream of an additional Leu codon that is present at the C-terminus. Even more surprising is the extreme similarity of the introns: Except for a 19-bp insertion and two nt exchanges in intron V, introns II–V are all perfectly identical in sequence. In contrast, intron I is entirely different in both length and sequence from all the others.

The scenario that has been proposed to account for this unusual gene structure (Schmitt et al. 1992; Schiedlmeier & Schmitt 1994) is as follows:

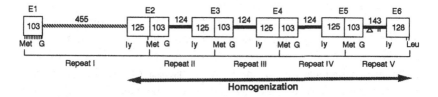

Figure 7.7 Organization of the polyubiquitin locus of *V. carteri*. Boxes indicate locations of exons (E1, E2, etc.), and lines between them represent introns; numbers indicate the lengths (in base pairs) of various sequence regions. The regions encoding the five ubiquitin polypeptides (which are produced by proteolysis after translation has been completed) are indicated below (I, II, etc.). Each of these repeat units is interrupted by an intron at the same location within a Gly codon (G. . . . ly). The intron between E1 and E2 (shown hatched) is entirely different in both length and sequence from the other introns. Otherwise, the repeat units are identical except at the 15 locations (indicated by vertical lines) where silent nucleotide substitutions occur, the point where a 19-bp insertion is present in intron five (triangle), and the very end of the coding region, where an additional amino acid (Leu) is present. The double-headed arrow at the bottom indicates the region within which it is speculated that some homogenization process may have been acting to maintain the high level of identity among both exons and introns. (From Schmitt et al. 1992, with permission.)

A monoubiquitin gene underwent a single tandem duplication and persisted in that form long enough for both the exons and introns to diverge in sequence. Then the second coding unit was duplicated three more times. To account for the extreme similarity of the exons and introns of units 2–5, it is postulated that either the second round of duplication occurred very recently or that some form of gene homogenization (such as gene conversion) is acting to keep both the exons and introns of these units extremely similar in sequence (Figure 7.7).

A second locus at which it appears either that a series of tandem duplications occurred very recently or that an efficient homogenization process is active is *sexI*, the locus encoding the sexual pheromone (Tschochner et al. 1987; Mages et al. 1988). The basic structure of a pheromone-encoding gene is a 3.8-kb transcribed region containing five exons and four introns that vary from 99 to 1,526 bp in length. However, six tandem repeats of this basic unit, separated by approximately 3-kb untranscribed spacer regions, appear to be present in the genome of standard laboratory strains (M. Sumper & D. L. Kirk unpublished data, summarized by Schmitt et al. 1992). The only feature distinguishing the coding regions of these six genes is that certain of them possess an $(ATT)_n$ insertion in intron IV. Similarly, the 5' leaders of these genes are distinguished only by the presence of a 290-bp insertion element (with features like a transposon) in the second and sixth copies within the array. The concept that tandem replication of the *sexI* gene may have occurred quite recently is supported by the observation that strains from a different geographical region (the I strains discussed in Section 7.2.1) have only a single copy of the *sexI* gene. However, the existence of a homogenization system acting to maintain the similarity of the tandemly repeated *sexI* genes in the J strains cannot be ruled out.

7.4 Development of molecular-genetic tools

A decade ago it had become abundantly clear to those in the field that detailed analysis of the molecular-genetic program underlying *V. carteri* development would never be feasible without the use of tools that had become commonplace in other developmental-genetic systems – methods such as transformation, transposon tagging, and gene replacement. As one attempt after another to transform *V. carteri* with bacterial or plant genes met with failure (Mages 1990), it became obvious that what was needed was a homologous selectable marker: a cloned *Volvox* gene for which one could select both loss-of-function mutants and regain-of-function mutants. In principle, such a gene could be used not only as a transformation vector but also to trap transposons that might be suitable for tagging genes of developmental interest. The only *V. carteri* gene for which suitable selection methods existed was *nitA*, the gene encoding nitrate reductase (NR): Mutants lacking NR activity can be selected by growth in the presence of chlorate, whereas mutants with restored NR activity can be selected by growth on nitrate as a sole nitrogen source (Huskey et al. 1979b) (see Section 6.2). Therefore, cloning and sequencing of the *nitA* gene became high priorities. When success was finally achieved as a result of a collaboration between our research group and that of Rüdiger Schmitt in Regensburg, it was a more important advance in *Volvox* molecular genetics than may have been apparent to readers of the published report (Gruber et al. 1992). Although the cloning and sequencing of *nitA* had turned out to be plagued by many unforeseen technical difficulties, and thus had required far more time and effort than had ever been anticipated, that effort was soon repaid when the gene proved to be suitable both for transformation and for trapping a transposon that is now being used as a gene-tagging tool.

7.4.1   Transformation

Transformation studies can be greatly simplified if a recipient strain with a nonrevertible mutation of the selectable marker can be used, to preclude appearance of false positives. As luck would have it, the *nitA⁻* allele that had already been used extensively for Mendelian genetic studies (Adams, Stamer et al. 1990), and that therefore was already present in many strains carrying other markers of interest, such as *regA⁻* and *gls⁻*, turned out to be exceptionally stable, with a reversion rate of less than 10^{-8}. Subsequently it has been established that this allele carries a point mutation (a G-to-A transition) in a splice junction that results in production of aberrantly spliced, nonfunctional products (Section 7.3.2) (Gruber et al. 1996). In initial studies, a Gls/Reg strain carrying this allele was used as the recipient, because of the ease with which large, homogeneous populations of reproductive cells could be obtained from this strain (Schiedlmeier et al. 1994). However, the transformation methods developed with Gls/Reg are now used routinely with Reg and wild-type strains carrying the same allele. Of the many methods of DNA delivery that were tested, bombardment with DNA-coated gold particles propelled by flowing helium (Takeuchi, Dotson & Keen 1992) has proved most effective. With this method, it is possible to produce 100–200 stable Nit⁺ transformants in a series of bombardments that take about one hour. Molecular evidence indicates that such transgenes are generally integrated by heterologous recombination at random locations in the genome, rather than by homologous recombination with the corresponding host gene (Schiedlmeier et al. 1994).

In our initial studies we observed that when lambda DNA carrying a nonselectable marker was introduced along with the plasmid carrying the *nitA* gene, co-transformation (as measured by integration of the lambda markers into the host genome) occurred with a frequency of 40–80% (Schiedlmeier et al. 1994). More importantly, we have recently observed that when lambda clones containing wild-type *Volvox* genes as inserts are introduced along with the plasmid carrying the *nitA* gene, co-transformation that results in phenotypic rescue of mutant recipients can occur with a frequency as high as 50% (S. M. Miller, M. M. Kirk & D. L. Kirk unpublished data). This is parallel to the experience of at least some investigators who have employed the *Chlamydomonas* transformation system: The first report of successful co-transformation in *C. reinhardtii* indicated that phenotypic rescue by the unselected marker occurred with a frequency of about 40% (Diener et al. 1990).[15] The fact that expression of co-transformed genes occurs at a simi-

---

[15] Many investigators have reported considerably lower levels of expression of co-transformed

larly high frequency in *Volvox* has opened up a whole new range of analytical possibilities, as will be discussed in several of the following sections.

7.4.2 Reporter genes, inducible promoters, and anti-sense constructs

Bacterial genes that are widely used as reporter genes in other organisms (such as those encoding β-galactosidase, β-glucuronidase, or chloramphenicol acetyltransferase) have not proved useful for either *Chlamydomonas* or *Volvox*, because such genes generally are not expressed in either alga, even if they are coupled to host promoters (Mages 1990; Blankenship & Kindle 1992). The reason for this lack of expression is not yet certain; it may be the result of a specific silencing mechanism that selectively methylates genes of foreign origin (Blankenship & Kindle 1992). But in any case it means that reliable reporter constructs require the use of suitable homologous genes.

The feasibility of using the gene encoding arylsulfatase (Ars) for such purposes was first demonstrated with *C. reinhardtii* (De Hostis et al. 1989; Davies et al. 1992; Davies & Grossman 1994). Following induction of the endogenous *ars* gene by sulfur deprivation, Ars is synthesized and secreted into the extracellular space, where it is capable of hydrolyzing a variety of arylsulfates, including X-SO₄ (5-bromo-4-chloro-3-indoyl sulfate), the hydrolysis product of which is a readily detected blue pigment. As discussed earlier, the *C. reinhardtii ars* gene was used as the reporter in studies of the roles of putative upstream regulatory regions in regulating transcription of the *Chlamydomonas* β2-tubulin gene (Section 7.34) (Davies & Grossman 1994).

When the *V. carteri* gene encoding Ars was cloned and characterized (Hallmann & Sumper 1994a), it turned out to be useful not only for producing

markers, however. In one study, unselected markers carried on a plasmid were expressed in only about 5% of the cases in which they were found to have been incorporated into the genome by co-transformation (Davies & Grossman 1994), and in at least some cases the frequency with which *Chlamydomonas* genes are expressed appears to be even lower when they are introduced as lambda clones, rather than on plasmid vectors (L.-W. Tam personal communication). In the one case in which nonexpressing clones were analyzed in detail, the genes that failed to be expressed were all found to have undergone rearrangements and/or partial deletions during their integration into the host genome (Davies & Grossman 1994). Presumably such disabling integration events also occur with a similar frequency in the case of the selectable marker used in the transformation system, but strains containing such rearranged genes would be recovered only if one or more additional copies of the gene had integrated in an active configuration. Indeed, in *V. carteri* it has been found that in cases in which multiple copies of the *nitA* gene have been integrated, some of them appear to have been inactivated by recombining with the host genome within the coding region, or by undergoing more complex types of rearrangements (Schiedlmeier et al. 1994).

a reporter construct but also for producing an inducible-promoter cassette (Hallmann & Sumper 1994b).[16] A chimeric gene containing the Ars-coding region fused to the promoter region of the *isg* gene (which encodes ISG, an ECM glycoprotein selectively expressed in inverting embryos) was constructed and transformed into *V. carteri* together with a *nitA* plasmid. Among the co-transformants recovered as strains capable of growing on nitrate and producing Ars in sulfur-sufficient medium, one was analyzed further. It was found to express the *ars* gene selectively (at both the RNA and enzyme-activity levels) at the time of inversion, when the *isg* gene is normally expressed. The corollary to that experiment was one in which the sequence encoding the sexual pheromone was fused to the promoter region of the *ars* gene, and the resulting chimera was transformed into *V. carteri* together with a *nitA* plasmid. Among the Nit$^+$ transformants that were recovered, five were found that developed sexually in sulfate-free medium. One of those was studied further and was shown to develop asexually in sulfur-sufficient medium, but to secrete pheromone and undergo autoinduction when deprived of sulfur (Hallmann & Sumper 1994b).

The *ars* gene is also being used in attempts to produce anti-sense vectors for *V. carteri*. When a construct was produced in which the Ars-coding region had simply been inverted and reattached to its own promoter, co-transformants were recovered in which the expression of the endogenous *ars* gene in response to sulfur starvation was suppressed by nearly 80% (I. Kobl, B. Schiedlmeier & R. Schmitt personal communication). Attempts are now being made to apply anti-sense technology to other *V. carteri* genes of interest (I. Kobl & R. Schmitt personal communication).

7.4.3 Introduction and expression of foreign transgenes

As previously noted, early attempts to transform *Volvox* with heterologous selectable markers employed genes derived from bacteria, vascular plants, or fungi, and were uniformly unsuccessful. More recently, however, co-transformation was used to demonstrate that a selectable marker derived from another green alga could be incorporated and expressed in *Volvox* with high efficiency (Hallmann & Sumper 1996). The foreign gene employed was the HUP-1 gene of *Chlorella* (a green alga in the order Chlorococcales) (Section

---

[16] An interesting sidelight to the analysis of the *V. carteri* Ars protein is that, in parallel with two human sulfatases that had previously been studied, it undergoes a posttranslational modification that converts one particular, highly conserved cysteine residue to a serinesemialdehyde residue (Selmer et al. 1996). It is postulated that this modification may be essential for establishing sulfatase catalytic activity.

2.2), which encodes a glucose/H^+ symporter. When HUP-1 was coupled to a *Volvox* β-tubulin promoter and introduced together with the *Volvox nitA* gene, an astonishing 86% (12 of 14) of the Nit^+ transformants that were recovered also had incorporated and expressed the transgene! Whereas wild-type *V. carteri* are unable to utilize exogenous glucose or other sugars, the transgenic strains took up glucose and glucosamine with high efficiency and incorporated both into glycoproteins, thereby providing access to many kinds of metabolic studies that previously had been inaccessible, because *V. carteri* lacks a glucose uptake system. The fact that 75% of all HUP-1-expressing cells survived a prolonged dark incubation that was uniformly fatal for wild-type spheroids not only indicated that the transgene might also confer the ability to use exogenous glucose for energy metabolism to some limited extent but also indicated that HUP-1 might provide a useful second selectable marker for use in transformation.

Of potentially greater importance, however, is the fact that HUP-1 provides a potent negative-selection system: Transformants are killed by concentrations of 2-deoxyglucose that are more than 300-fold lower than those in which wild-type spheroids are able to flourish. This is significant, because "gene-knockout studies" of the sort that have been used so successfully to study gene functions in mice and other organisms are facilitated by using constructs that contain a copy of the target gene that is interrupted by a positive selectable marker and flanked by a negative selectable marker. With such a construct, one first uses positive selection to select target cells that have incorporated the transgene and then uses negative selection to kill off transformants in which the negative marker has not been lost by homologous recombination with the target locus of the host (Mansour, Thomas & Capecchi 1988).[17] The possibility of using the HUP-1 gene as the negative selectable marker in such a protocol was explicitly discussed by Hallmann and Sumper (1996).

7.4.4 Gene modification by homologous recombination

In light of the point that was made at the very end of the preceding paragraph, it came as a bit of a surprise that when those investigators subsequently reported the first example of gene modification by homologous recombination

---

[17] A successful experiment in gene targeting and disruption in *Chlamydomonas* has recently been described (Nelson & Lefebvre 1995), but because the gene that was used for negative selection in that case was the one whose disruption was being targeted, the protocol used is not of general applicability.

in *Volvox*, their successful protocol did not involve the use of the HUP-1 gene as a negative selectable marker.

Hallmann, Rappel, and Sumper (in press) first used two plasmids carrying different portions of the *ars* gene to determine whether or not *V. carteri* has a system capable of mediating homologous recombination of transgenes that have been introduced on plasmid vectors. The two vectors employed carried different, overlapping portions of a chimeric gene in which the *ars* coding sequence had been fused to a β-tubulin promoter; in one construct, part of the 5' end of the coding sequence and the promoter had been removed, whereas in the other construct the 3' end of the coding sequence had been removed. Following bombardment of Nit⁻ recipient cells with a mixture of these two plasmids and a *nitA* plasmid, 6 of the 20 Nit⁺ transformants that were recovered produced ARS constitutively (in sulfur-sufficient medium). The investigators then used reverse transcription, polymerase chain reaction (PCR) amplification, and DNA sequencing to demonstrate that in each of those six co-transformants the *ars* message had been transcribed from a chimeric gene in which a complete *ars* coding region was attached to the β-tubulin promoter. Because details of the sequence information obtained indicated that such a gene could have arisen only as a result of homologous recombination between the two truncated genes that had been introduced as plasmid inserts, the experiment provided clear evidence that *V. carteri* has a homologous-recombination system capable of operating on transgenes.

Having established the existence of such a recombination system, Hallmann, Rappel, and Sumper (in press) next explored whether or not that system could be exploited to make a targeted modification of a host gene. For that experiment they designed a vector potentially capable of repairing the known defect in the *nitA⁻* allele discussed in Section 7.4.1, which is a G-to-A transition in an intron splice site that prevents correct processing of *nitA* transcripts (Gruber et al. 1996) The plasmid that they constructed contained a *nitA* fragment that would be inadequate to encode a functional enzyme if it were incorporated into the genome outside the *nitA* locus, and that differed in sequence from the corresponding region of the host *nitA* locus at five nucleotide positions. One of these differences involved the G that they hoped to introduce into the intron splice site to repair the mutant defect; the other four were "marker" nucleotide replacements: ones that would cause synonymous changes in four nearby codons of the adjacent exon if they were incorporated into the host gene by homologous recombination. Following bombardment of Nit⁻ cells with gold particles coated with this construct, six Nit⁺ transformants were recovered. Sequencing of the *nitA* loci of those transformants revealed that in all six of them the G-to-A mutation of the recipient had been repaired. Two of the transformants had also incorporated all four of the marker nucleotides present in the transforming vector, a third had incorporated only the marker nucleotide closest to the repaired mutation site, but the remaining three had undergone a change only at the mutation site itself. The incorporation of all four targeted nucleotide replacements in two of the transformants demonstrated unambiguously that gene modification by homologous recombination is feasible in *Volvox carteri*. The importance of this conclusion for the future of *Volvox* molecular genetics can hardly be exaggerated. It means that a whole new era of functional analysis, via targeted gene modification, can now begin.

The fact that three of the transformants recovered by Hallmann, Rappel, and Sumper did not incorporate any of the flanking markers of the transforming plasmid adds only a minor note of caution to the preceding assessment: It implies that the extent of the sequence modifications that it will be possible to introduce into *Volvox* via transformation and homologous recombination probably will be quite small in most cases, and therefore gene-disruption vectors will have to be designed with considerable precision to be effective.

7.4.5 Transposon trapping

As soon as the *nitA* locus had been characterized, efforts to use it as a transposon trap were initiated by analysis of "spontaneous" chlorate-resistant mutants that appeared under standard culture conditions or after cultures had been heat-shocked in midcleavage (which increases the mutation rate). Fully 10% of such mutants were found to have insertions of a similar 1.6-kb element in the same restriction fragment of the *nitA* gene, and one of those strains that was studied more extensively was found to be capable of reverting to the wild type with high frequency, which is a hallmark of certain transposon-induced mutations. Moreover, all revertants of that strain that were analyzed were found to have lost the insertion element. That insertion element, which was subsequently cloned, sequenced, and found to resemble certain higher-plant transposons, was named *Jordan*, in recognition of its superb jumping ability (Miller et al. 1993).

In the initial studies, heat shock increased the rate of *Jordan* transposition by about threefold. Subsequent studies have shown that the rate can be elevated even more (from a basal level of about 3×10^{-5} to an induced level of about 10^{-4}) by cultivation at low temperature (24°C, which is 8°C below our standard incubation temperature and nearly the minimum temperature at which our laboratory strains will grow) or by exposing the culture to very low levels of UV irradiation (60 J/m²) (L. Fortner, S. M. Miller & D. Kirk unpublished data). The ability of *Jordan* to cause revertible mutations in some cases is attributed to the fact that when it excises, it leaves behind a target-site duplication of three (or occasionally a multiple of three) base pairs, which adds one (or more) amino acid to the coding sequence, but does not change the reading frame (Miller et al. 1993). *Jordan* has been used to tag and recover several genes of interest, as will be discussed in the next section.

Three other transposable elements of *V. carteri*, called *Osser* (Lindauer et al. 1993a), *Lückenbüsser* (G. Kohl & R. Schmitt personal communication), and *Jackie* (M. B. Karr & S. M. Miller personal communication), have been identified and characterized. However, in contrast to *Jordan*, which moves by a classic DNA-level excision and reinsertion, the other three transposons all appear to be of the retrotransposon type, and methods for using them to tag genes of interest have not yet been discovered.

7.5 Transposon tagging of developmentally important genes

Two properties, in particular, recommended *Jordan* as a prospective gene-tagging agent: (1) Although it is moderately stable under standard culture conditions, it can be induced to transpose at a significantly higher frequency by exposing a culture to mild stress. (2) At least some of the mutations that it causes are revertible. The major weakness of *Jordan* as a gene-tagging agent, on the other hand, is that there are about 50 copies of the transposon in the wild-type genome (Miller et al. 1993), which complicates identification and cloning of novel copies that have been generated by transposition (Figure 7.8).

Several approaches have been tried in the attempt to circumvent that potential weakness: First, a search was made for a strain with significantly fewer copies of *Jordan*. None was found. Second, variants of *Jordan* carrying unique-sequence elements were constructed and introduced into the genome by transformation. But none of those engineered versions transposed with sufficient frequency to be useful. Third, to increase the probability that mutants generated under conditions that stimulate *Jordan* transposition would possess *Jordan*-induced mutations, a strain with a mobile copy of *Jordan* in the *nitA* locus was subjected to stress, Nit⁺ revertants were selected for, and then they were screened for morphological mutations that had appeared when *Jordan* left the *nitA* locus. Although the more than 1,000 Nit⁺ mutants recovered with that strategy did include some morphological mutants that may eventually be of interest, none were of the mutant classes we were most interested in, presumably because the genes of interest constitute such a small

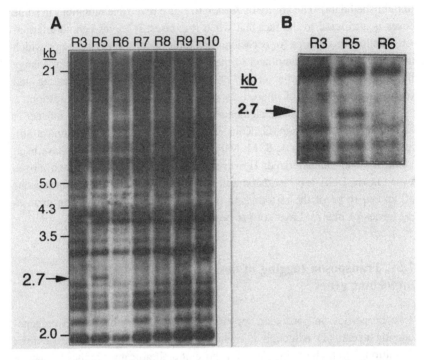

Figure 7.8 Restriction fragments cross-hybridizing to the transposon *Jordan*. A Southern blot of *Sal*I-digested DNAs from seven Reg mutants (R3–R10) that had been isolated under conditions that stimulated *Jordan* transposition was hybridized with a radioactive *Jordan* probe. Because of the great number of genomic fragments present in the genome (**A**), not every *Jordan* transposition event would be detected by this procedure using any single type of restriction digest. However, in this case a particularly clear and well-isolated novel *Jordan*-containing *Sal*I fragment was detected in strain R5; this is shown at higher magnification in **B**. The genomic DNA flanking *Jordan* in this fragment was cloned, yielding a plasmid, pVR2, that was used for additional experiments described in the text and in Figure 7.9. (M. M. Kirk & D. L. Kirk unpublished data.)

part of the total target area available for *Jordan* insertions (i.e., the entire genome).

Thus, the strategy that was used in the successful gene-tagging experiments to be described next was simply to incubate a culture at 24°C for a prolonged period, screen it at intervals for morphological mutants of interest, and then analyze all such mutants for (1) reversion ability and (2) the presence of novel restriction fragments containing *Jordan*.

7.5.1 Tagging and cloning of the regA locus[18]

One Reg mutant produced by the strategy just described exhibited a particularly clear novel restriction fragment on a Southern blot that had been probed with *Jordan* (Figure 7.8). A fragment of genomic DNA flanking the *Jordan* insert in that novel fragment was cloned and used to probe Southern blots and genomic libraries. That initial probe and/or various genomic clones cross-hybridizing with it detected RFLPs in 28 of the first 36 independent Reg mutants that were tested (Figure 7.9A). Genetic analysis of one such RFLP (a small deletion) showed that it co-segregated perfectly with the Reg phenotype in about 50 progeny (Figure 7.9B). Together, these data provided strong evidence that the *regA* locus is located in this region of the genome, with the region within which 29 Reg-associated RFLPs map providing a preliminary estimate of the boundaries of the locus. Then a 15-kb genomic clone spanning this region (Figure 7.10) was used to rescue a nonrevertible *regA⁻* mutant by transformation, completing the proof that the entire *regA* gene had been cloned.

Genomic fragments from the *regA* region, as well as a set of cross-hybridizing cDNA clones, all detect the same single, stage-specific 7.6-kb transcript when used to probe developmental northern blots. This transcript appears in young somatic cells at the time when, as shown by previous studies, the *regA* gene exerts its effects on somatic cell development (Huskey & Griffin 1979; Tam & Kirk 1991b); it remains abundant for about 12 hours and then disappears. The 3' end of the ~12-kb *regA* transcription unit was defined by cDNA sequencing and its 5' end was located by northern-blot hybridization with small genomic subclones, and by primer extension (Figure 7.10).

Sequencing of genomic and cDNA clones revealed that the *regA* transcription unit is about 12-kb long (Figure 7.10), comprises eight exons and seven introns (four of which are in the 5'-UTR), and potentially encodes a 113 kDa polypeptide (RegA) that contains two putative helix-turn-helix (potential DNA-binding) motifs. Although the deduced RegA polypeptide sequence is not highly homologous to the sequence of any particular protein in various international data bases when it is considered as a whole, it contains many sequence motifs that are extremely similar to motifs present in a wide range of different transcription factors. Among the most interesting of such motifs are several that are particularly rich in the amino acids glutamine,

---

[18] A manuscript providing details of the work summarized in this section is in preparation by M. M. Kirk, W. Müller, B. Taillon, S. M. Miller, H. Gruber, K. Stark, R. Schmitt & D. L. Kirk.

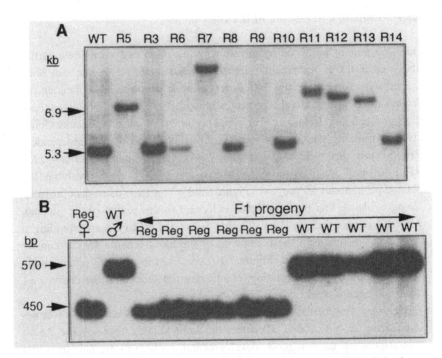

Figure 7.9 Southern blots probed with pVR2, the cloned genomic DNA derived from the novel *Jordan* fragment identified in Figure 7.8. **A**: DNAs from a wild-type strain (WT) and 11 Reg mutants (R3–R14) that were independently derived from this WT strain under conditions that stimulated *Jordan* transposition were digested with *Sca*I and used to prepare a Southern blot that was then probed with pVR2. Five of these Reg strains exhibited RFLPs that are visible here, and a sixth (R9) had an RFLP in a region of the blot that is not shown. When genomic clones isolated on the basis of cross-hybridization to pVR2 were used to probe such Southern blots, 28 of 36 such Reg strains that were analyzed were found to have insertions in this region of the genome (see Figure 7.10). **B**: DNAs from a Reg female, a wild-type male, and 11 of their F₁ progeny were digested with *Sal*I and used to prepare a Southern blot that was probed with pVR2. This probe detected a bleomycin-induced deletion in the Reg mutant female that co-segregated perfectly with the Reg phenotype among more than 50 progeny, including the 11 shown here. (M. M. Kirk & D. L. Kirk unpublished data.)

proline, and/or alanine. Indeed, glutamine, proline, and alanine account for more than a third of the amino acid residues in the deduced protein. This is considered significant because an abundance of those three amino acids is the only structural feature that appears to be shared by many "active" or "direct" transcriptional repressors (i.e., repressors that inhibit transcription in a locus-specific way by interacting with the basal transcription complex, rather than by interacting with and blocking the function of transcriptional activators) (Cowell 1994). In short, the deduced amino acid sequence is con-

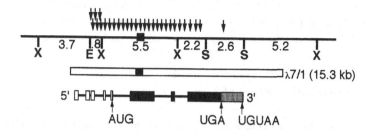

Figure 7.10 The *regA* locus. The upper line represents a partial restriction map of the *regA* genomic region; for simplicity, only the *Eco*RI (E), *Sac*I (S), and *Xho*I (X) sites are shown, and numbers below the line indicate the sizes (in kilobases) of the fragments defined by these restriction sites. The black box indicates the location of the sequence homologous to the pVR2 probe that was initially cloned by transposon tagging. The arrows above the line indicate the approximate locations where insertions have been found in 28 different Reg mutants that were isolated under conditions that promote *Jordan* transposition. The bar immediately below the restriction map indicates the portion of the genome that is represented by the 15.3-kb insert present in a lambda clone (λ7/1) that is capable of rescuing a nonrevertible Reg mutant. Below that bar is a diagram representing the structure of the 12-kb *regA* primary transcript and the eight exons that are spliced together to produce the 7.6-kb mature transcript. The mature *regA* mRNA consists of a 5' UTR derived from the first four exons plus most of the fifth (white boxes), a coding region (black boxes) capable of encoding a 113-kDa polypeptide, and an unusually long 3' UTR (stippled box). The arrows indicate the locations of the initiation codon (AUG), the stop codon (UGA), and the polyadenylation signal (UGUAA). (M. M. Kirk, W. Müller, B. E. Taillon, S. M. Miller, H. Gruber, K. Stark, R. Schmitt & D. L. Kirk unpublished data.)

sistent with our long-held hypothesis that RegA acts as a transcription factor (and possibly an active repressor) that regulates expression of genes required for germ-cell development. However, confirmation of that hypothesis will require detailed functional analysis.

With substantial information about the structure of the *regA* locus and its products in hand, we are currently attempting to determine how *regA* expression is regulated. The first hypothesis being tested is, of course, the iconoclastic idea that *regA* is inactivated in presumptive gonidia, and later reactivated in mature gonidia, by a cyclical DNA rearrangement (Section 6.5.1; Kirk, Baran et al. 1987). To that end, we are using PCR to amplify sequential fragments (~1 kb) of the locus, searching for regions that have different structures in gonidia and somatic cells. Shortly we expect to begin testing a new, equally provocative and testable hypothesis that has recently been formulated regarding the molecular mechanism by which the RegA protein represses reproduction (Section 7.8).

Ever since we learned about the Reg mutant that Starr (1970a) first described, we have considered the locus defined by it – the *regA* gene – to be among the most interesting "master regulatory genes" in developmental biology. Now, at last, the tools to unwrap its mysteries appear to be in hand.

7.5.2 Tagging and cloning of a gls locus[19]

In one of our studies, more than 100 Gls/Reg double mutants were isolated from Reg strains cultured at 24°C. In one subset of those mutants, nearly a third had readily detected *Jordan* RFLPs (relative to the starting Reg strain), and about one-tenth were capable of reverting to *gls*$^+$ when cultivated at 24°C.[20] From one strain having both of these attributes, the novel restriction fragment containing *Jordan* was cloned, and genomic DNA fragments flanking the transposon were then isolated and used to probe Southern blots of DNA from the starting Reg strain, the Gls/Reg mutant, and several revertants. The results were clear-cut: The genomic fragment that had increased in size as a result of a *Jordan* insertion when the Reg strain mutated to Gls/Reg had been restored to very near the wild-type length in each of the revertants (Figure 7.11A).

The data in Figure 7.11 provide compelling evidence that this *Jordan* insertion was the cause of the Gls mutation and also that the flanking DNA fragments that had been isolated represented parts of the *gls* locus. The latter conclusion was reinforced when several overlapping genomic clones that hybridized to those flanking fragments were isolated and used successfully for transformation rescue of the mutant. Southern-blot analysis of such cotransformants has demonstrated that (in addition to possessing a new copy of the putative *gls* gene) they retain the *Jordan* insertion at the host *gls* locus (representative data, Figure 7.11B). Those findings ruled out the possibility that such transformants had reverted to *gls*$^+$ as a result of excision of *Jordan* from the *gls* locus of the host genome, and they supported the conclusion that the mutant phenotype had truly been rescued by the transgenes (S. M. Miller & D. L. Kirk unpublished data).

Because most Gls strains that were isolated under conditions that enhanced *Jordan* transposition did not exhibit RFLPs on Southern blots that were

[19] A manuscript providing details of the work summarized in this section is in preparation by S. M. Miller and D. L. Kirk.

[20] The proportions of all mutants isolated under those conditions that possessed *Jordan* insertions were found to be extremely different with different starting strains, suggesting that there may be heritable factors regulating *Jordan* transposition (S. M. Miller personal communication).

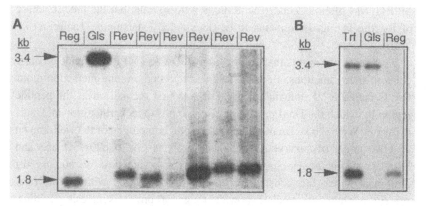

Figure 7.11 Southern blots of genomic DNAs digested with *Bam*HI and probed with a putative *glsA* fragment that had been isolated by a *Jordan*-tagging strategy similar to that used for *regA*. **A:** In the Gls/Reg double mutant (Gls), this probe detects the fragment from which it was isolated, which is 1.6 kb larger than the corresponding fragment in the parental Reg strain (Reg) because of a 1.6-kb *Jordan* insertion. In six independent phenotypic revertants of the Gls mutant (Rev) the hybridizing fragment has reverted to essentially its initial length, indicating that phenotypic reversion was accompanied by relatively precise excision of the *Jordan* element. **B:** The same type of Southern blot was used to compare the starting strain (Reg), the Gls/Reg double mutant (Gls), and a transformed strain (Trf) in which the Gls phenotype of the mutant had been cured following bombardment with a genomic clone believed to contain the *glsA* locus. The transformant retains the 3.4-kb *Bam*HI fragment of the Gls mutant from which it was derived (indicating that the host *glsA* locus retains the *Jordan* insert that inactivated it), but it also has a new band of the same length (1.8 kb) as that in the starting Reg strain, which can be attributed to the transforming DNA. The difference in relative intensity between the two bands in the Trf strain suggests that (as is common in *Volvox* transformants) multiple copies of the transforming DNA may have been incorporated into the genome. (S. M. Miller & D. L. Kirk unpublished data.)

probed with the cloned gene described in the paragraph above, we conclude that (as had previously been suspected) there is more than one *gls* locus in the *V. carteri* genome. Thus, the cloned gene has been named *glsA*, and attempts to tag and recover other *gls* loci are in progress.

Sequencing of *glsA* cDNA revealed that it encodes a polypeptide (GlsA) that is extremely similar in amino acid sequence to a human protein, "M phase phosphoprotein 11" (MPP11) (S. M. Miller & D. L. Kirk unpublished data). MPP11 is one of several mammalian proteins that are selectively phosphorylated by a cyclin-dependent kinase early in the mitotic phase of the cell cycle (and thereby activated in some cases, or inactivated in others) as interphase structures and activities are disrupted and assembly of the mitotic apparatus begins. Immunocytological studies indicated that in mitotic cells

MPP11 was often associated with the mitotic spindle, but the precise nature of its role in mitosis remains to be determined (Matsumoto-Taniura et al. 1996).

Two sequence motifs that are shared by GlsA and MPP11 provide possible clues to their functions. Both possess a ubiquitous protein-interaction motif that is called a "J domain" because it was first recognized as the peptide region by which the DnaJ protein of *E. coli* binds to its partner protein, DnaK (Silver & Way 1993). DnaK is the prokaryotic homologue of the well-known 70-kDa family of eukaryotic heat-shock proteins (the Hsp70s). DnaKs and Hsp70s are "molecular chaperones" that are now known to function not just after heat shock but constitutively and, in conjunction with particular J-protein partners, to mediate a wide range of protein-synthesis, -folding, -targeting, -translocation, and -assembly reactions. In such partnerships the function of the J proteins (which characteristically possess at least one type of protein-interaction domain in addition to the J domain) is to target the Hsp70-type chaperones (which are themselves relatively non-specific in their actions) to specific membrane sites, protein substrates, and so forth (Rassow, Voos & Pfanner 1995). Both GlsA and MPP11 possess a second protein-interaction motif, an "Id-binding" (Idb) domain, a protein-binding domain that was first characterized in the mouse "MIDA1" protein, a J protein that is as similar in amino acid sequence to GlsA and MPP11 as they are to each other (Shoji et al. 1995). Thus, GlsA and MPP11 both appear to have the potential to link an Hsp70 partner to some particular intracellular target.

Given that it possesses two protein-binding domains, what function might GlsA have in *Volvox* asymmetric cell division? An interesting possibility arises from the recent demonstration that an Hsp70 molecule plays a critical role in the structure, assembly, stability, and microtubule organizing functions of mammalian centrosomes (Brown et al. 1996a,b). We can imagine, for example, that GlsA might function in the midcleavage embryo by binding to an Hsp70 that is associated with the *Volvox* centrosome equivalent (= the basal apparatus, or BA; see Section 4.4) thereby tethering the BA to a novel, off-center plasma membrane site in preparation for asymmetric division. We anticipate that immunocytological studies with antibodies directed against the J protein encoded by *glsA* should permit us to determine if this protein does indeed interact with the *Volvox* BA and/or should provide other clues about how it may contribute to the asymmetric division process. Such a study will be initiated soon (S. M. Miller personal communication). We also anticipate that identification and characterization of additional *gls* genes and their products may identify the proteins with which GlsA interacts to promote asymmetric division.

Although the molecular studies summarized here have not progressed far enough to provide a full understanding of the mechanism by which either the RegA or the GlsA protein plays its role in *Volvox* development, they appear to have brought us much closer to that goal. At this point, therefore, it is appropriate to consider what is known about the nature of the downstream genes that are expressed differentially in the two cell types produced by the combined actions of the products of the *gls, lag,* and *regA* genes.

7.6 Patterns of differential gene expression accompanying germ–soma differentiation

If there is a central dogma of developmental biology, it is that differentiated cells differ not with respect to the genes that they possess but with respect to the genes that they express. A first attempt to define some of the differences in gene expression that distinguish germ and soma of *V. carteri* used differential hybridization to identify clones within a cDNA library that hybridized selectively to transcripts present in one of the two cell types. By that process we readily identified 31 distinct genes (or small gene families) whose transcripts were accumulated in a cell-type-specific and stage-specific manner (Tam & Kirk 1991a; Tam et al. 1991); 19 of those were preferentially expressed in developing gonidia and/or dividing embryos, whereas 12 were somatic-cell-specific.

7.6.1 Early and late gene-expression patterns in somatic cells

The somatic-cell-specific genes identified in our studies fell into two distinct categories. As mentioned in Section 5.4.3.1, transcripts of certain genes begin to be accumulated preferentially in presumptive somatic cells of embryos very shortly after they have been set aside from presumptive gonidia by asymmetric division and while the two cell types are still interconnected by an extensive network of cytoplasmic bridges. Such early-somatic-gene transcripts then rise to peak levels during early somatic-cell differentiation, are maintained at more or less constant levels for about two more days, and then slowly decline as the cells approach programmed death (Figure 7.12A) (Tam & Kirk 1991a; Tam et al. 1991). Genes exhibiting this expression pattern included, in addition to five early-somatic genes identified by cDNAs isolated in this study, one previously characterized gene: the gene encoding SSG 185, the 185-kDa sulfated glycoprotein of the ECM that forms the walls of the compartments that surround all somatic cells in the adult (see Section 5.4.4).

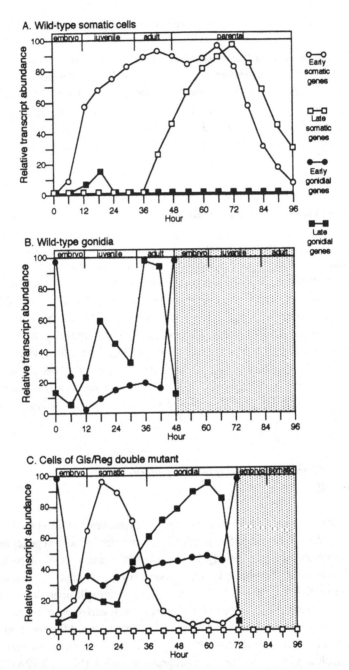

Figure 7.12 Expression patterns of cell-type-specific genes in three types of *V. carteri* cells. **A:** In wild-type somatic cells (which have a four-day life cycle, see Figure 5.6), two patterns of expression of somatic genes are seen. Transcripts of early-somatic genes appear during the embryonic period, accumulate as the cells differentiate during day 1, remain at high levels during days 2 and 3, and then decline in abundance during day 4. Transcripts of late-somatic genes, in contrast,

Other early-somatic genes that have been identified via cDNAs (as well as ones that have not yet been identified) may, like the *ssg185* gene, encode products required for assembly of the ECM surrounding the somatic cells. But undoubtedly some of them also will be found to encode products required for elaboration of cell-type-specific features of the somatic cells themselves, such as flagella and eyespots. Indeed, the genes encoding α1-tubulin and a second component of the flagellum were shown to be expressed in somatic cells in patterns very similar to the early-somatic genes (although they differed from the early-somatic genes by also being expressed, but at lower levels, in gonidia and embryos) (Tam & Kirk 1991a). Sequencing of certain of the early-somatic genes defined by Tam's study is in progress (D. Straus & G. Choi personal communication); so information about the nature of their products may be available soon.

As the term implies, the seven late-somatic genes defined by cDNAs in Tam's studies are not expressed until quite late in the asexual life cycle: after the spheroids have hatched and the somatic cells are already fully differentiated and are executing their motile and phototactic responsibilities. During the third day of the life cycle, somatic cells progress from the "adult" to the "parental" state as the gonidia within the spheroid cleave to produce a new cohort of juveniles. During this transitional period, transcripts of the late-somatic genes accumulate progressively, reaching peak abundance by the time the somatic cells are about 72 hours old, a day before they will begin to die (Figure 7.12A) (Tam et al. 1991). Thus, these late-somatic genes appear to encode products required by somatic cells not for their visible cytodiffer-

Caption to Figure 7.12 *(cont.)*
appear only after the cells are fully differentiated, reaching maximum abundance only during day 4, as the cells approach the period of programmed cell death. Transcripts of late-gonidial genes begin to accumulate in the somatic cells during the early differentiation period, but then rapidly disappear before the end of day 1, and early-gonidial transcripts are never detected in somatic cells. **B**: In wild-type gonidia (which live only two days before they divide to produce a new cohort of embryos), transcripts of somatic genes are never detected. Early-gonidial transcripts are maximally abundant at the beginning of cleavage, then decline rapidly, remain at low levels during gonidial development, and return to maximal levels again as the next round of embryogenesis begins. Transcripts of the late-gonidial genes are present in gonidia at all times, but exhibit one abundance peak as the gonidia are differentiating (day 1), and a second as they are maturing (day 2). **C**: The cells of the Gls/Reg mutant (which have a three-day life cycle) simultaneously accumulate transcripts of the early-somatic genes and both categories of gonidial genes during day 1, while they are differentiating as somatic cells. In the second day, however, transcripts of the early-somatic genes decline rapidly, while transcripts of late-gonidial genes increase in abundance, and the cells begin to redifferentiate. Late-somatic transcripts never appear in these cells. Data from Tam & Kirk (1991a,b) and Tam et al. (1991). (Adapted from Schmitt et al. 1992, with permission.)

entiation but for functions specific to the late stages of the life cycle. In the published account, we stated that "it is unlikely that any of these genes play a role in senescence, because the period within which these transcripts were abundant well [preceded] the senescent phase" (Tam et al. 1991). That statement was ill-advised. Although it is true that *V. carteri* f. *nagariensis* somatic cells do not begin to lose viability (as measured by dye exclusion) until they are about 96 hours old (Pommerville & Kochert 1982), degenerative cytological changes were detected in somatic cells of *V. carteri* f. *weismannia* more than a day before they began to lose viability (Pommerville & Kochert 1981). Moreover, it was shown that inhibition of protein synthesis in the period from 72 hours to 96 hours delayed somatic-cell death in forma *weismannia* by a day or more, suggesting that proteins required for cell death are normally synthesized during that period (Pommerville & Kochert 1982). Thus it is by no means ruled out that some late somatic genes may be involved in the cell-suicide program; this possibility deserves investigation.

The only late-somatic gene that we have begun to characterize is the most highly expressed representative of the group: the one defined by cDNA "S12" (Tam & Kirk 1991a). The S12 cDNA turned out to be incomplete, but when a homology search was performed with the first 600 bp of the coding-region sequence that became available (K. Otipoby & D. L. Kirk unpublished data), we were astonished to find that this sequence was more than 99% identical, at the nucleotide level, with that of a *V. carteri* gene that had already been sequenced: the gene encoding pherophorin II (Sumper et al. 1993; also see Section 5.5.4). Of course, the reason we found that surprising was because pherophorin II had been identified as a gene product that was undetectable in asexual spheroids, but was rapidly induced in response to the sexual pheromone, whereas S12 had been identified as one of the most abundant transcripts of uninduced, asexual spheroids. Although its sequence clearly places S12 in the pherophorin II subfamily,[21] which is now known to contain about 10 genes (Godl et al. 1995), it resembles pherophorins I and III in the respect that it is very strongly expressed (at the RNA level, at least) in uninduced asexual spheroids. Further study of this interesting family of genes and other late-somatic genes is warranted.

---

[21] The peptide encoded by the 600-bp segment of S12 that has been sequenced to date is 99.5% identical with pherophorin II, 83% identical with pherophorin III, and 39% identical with pherophorin I.

7.6.2   Two patterns of gene expression in gonidia

The 19 gonidial genes that were defined by cDNAs also fell into two categories: 18 were "late," or "maturation-abundant" gonidial genes that were expressed maximally during late stages of gonidial development. The remaining gene, defined by cDNA G167, was an "early," or "embryogenesis-abundant" gene that was expressed maximally during early cleavage (Figure 7.12B). Most of the late-gonidial genes are expressed for a few hours in very young somatic cells at levels not substantially lower than their levels in very young gonidia (Figure 7.12A). Soon, however, transcripts of all of these genes disappear completely from somatic cells, at the same time that they are being accumulated in a progressive fashion in the developing gonidia (Tam & Kirk 1991a). This pattern is consistent with the idea that the cell-type-specific expression of gonidial genes occurs not because these genes are selectively activated in young gonidia but because they are selectively repressed in somatic cells. We will return to this important theme later.

Three of the maturation-abundant cDNAs, G5, G8, and G40 (Tam & Kirk 1991a), have recently been sequenced, and all three have turned out to encode functionally related products of a wholly unanticipated nature: chloroplast proteins that play key roles in three important steps of photosynthesis (Choi et al. 1996)! G8 encodes a chlorophyll a/b-binding protein that is part of the light-harvesting complex of photosystem II; G5 encodes the oxygen-evolving enhancer protein 1, which is involved in photolysis of water; and G40 encodes a ferredoxin/NADP$^+$ reductase, the enzyme that uses the electrons produced by photosystem I to produce NADPH for use in carbon dioxide fixation and other anabolic processes. Although this set of findings was initially extremely surprising, it has led to a whole new way of looking at the possible basis for germ–soma differentiation in *Volvox*, as will be discussed at some length in Section 7.8.

In distinction to the maturation-abundant gonidial genes, transcripts of G167 (the only embryogenesis-abundant gonidial gene initially identified) can never be detected in somatic cells. Although G167 transcripts are already moderately abundant in developing gonidia, they increase in abundance dramatically just before cleavage begins, composing one of the most abundant classes of transcripts in the early embryo. By about the time asymmetric division is occurring, however, levels of G167 transcript begin to drop as precipitously as they had risen a few hours earlier, so that these transcripts are no longer detectable at the end of embryogenesis (Figure 7.12) (Tam & Kirk 1991a). This unique pattern of expression led us to hope that G167 might encode a protein involved in asymmetric division. However, when the full-length G167 cDNA was sequenced, no open reading frame of significant length could be identified within it (B. E. Taillon & D. L. Kirk unpublished data). On the assumption that the cDNA that had originally been sequenced might represent a transcript of an inactive, mutant member of the G167 gene family (which comprises 14 unlinked copies) (C. R. Velloff & D. L. Kirk unpublished data), or that it might have undergone modification during cloning, 15 additional G167 clones from two independent cDNA libraries were analyzed, but none of them exhibited any significantly greater polypeptide-coding potential than the original clone. Furthermore, PCR analysis indicated that none of the genomic copies of G167 contained any introns, which is a feature that had never been encountered previously in any volvocalean protein-coding nuclear gene. Because G167 is expressed in such a vigorous but tightly controlled temporal and cell-type-specific pattern, we suspect that the G167 gene product may execute some important function in early cleavage as a noncoding, polyadenylated RNA; but comparisons with other gene products known to act at the RNA level give us no clue to what that function may be. Thus the function of the G167 gene family remains mysterious.

7.6.3 Additional embryo-specific genes are readily identified

In the initial cDNA screen, our objective was to identify genes that were expressed selectively in one of the two differentiating cell types. Therefore, the RNA used to construct the library was derived from cultures in which embryonic stages were poorly represented, and the library was then screened with cDNAs derived from relatively mature gonidia and somatic cells. Given that approach, it was not surprising that only one embryogenesis-abundant cDNA, G167, was identified. Indeed, that cDNA probably was detected only because its transcript became moderately abundant again in maturing gonidia (Figure 7.12B).

In a search for additional genes with embryo-specific expression patterns, a second cDNA library was prepared from RNA isolated from early-cleavage embryos. When this library was screened differentially with probes representing the transcript populations of early embryos and postembryonic juveniles, an embarrassment of riches was uncovered (B. E. Taillon & D. L. Kirk unpublished data). Of the first 60 cloned inserts analyzed, very few cross-hybridized. This abundant collection of cDNAs identifying genes preferentially expressed in embryos will be analyzed via northern blots and partial sequencing, in an effort to select the ones with the most interesting expression patterns and/or sequences for more detailed study.

7.7 Patterns of gene expression in a mutant lacking germ–soma differentiation

The cDNAs identifying cell-type-specific genes in wild-type spheroids provided an opportunity to test our ideas about the regulatory network controlling germ–soma differentiation. Our working hypothesis (Section 6.6) is that there are three key steps in this process: First the *gls* genes act in the embryo to cause asymmetric division and production of large and small cells; then the *lag* genes act in the large cells to repress somatic genes, and *regA* acts in the small cells to repress gonidial genes. It follows from this hypothesis that in a Gls/Reg double mutant neither somatic nor gonidial genes should be repressed, because the *regA* gene has been disabled by mutation, and in the absence of the *gls* function there are no asymmetric divisions, and hence no cells are produced that are large enough to activate the *lag* genes. Thus, we anticipated that the Gls/Reg mutant should co-express gonidial and somatic genes.

That was what we observed, at least in the initial phases of cytodifferentiation (Tam & Kirk 1991b). Like wild-type somatic cells, the cells of young Gls/Reg juveniles express both early-somatic and gonidial genes at low levels. But whereas gonidial transcripts suddenly disappeared from wild-type somatic cells at about the time that the *regA* gene was believed to be expressed, no such down-regulation was observed in Gls/Reg cells; instead, both gonidial and somatic gene transcripts continued to accumulate progressively over the first 24 hours, but with somatic transcripts accumulating to somewhat higher levels (Figure 7.12C). During that first 24 hour period, the Gls/Reg cells differentiated as flagellated cells with eyespots – cells that were indistinguishable morphologically from wild-type somatic cells (Figure 6.6). But then, in the second day of life, a dramatic change occurred in both the molecular and the visible phenotypes of these cells. In contrast to wild-type somatic cells (in which early-somatic transcripts remain at peak levels during days 2 and 3, and late-somatic transcripts begin to accumulate by the end of day 2), in Gls/Reg cells early-somatic transcripts fell to very low levels by the end of day 2, and no late-somatic transcripts ever appeared at detectable levels. Meanwhile, gonidial transcripts had continued to accumulate during that period (Figure 7.12C) and at that time the cells underwent morphological redifferentiation: By the end of day 2 they had already lost their flagella and eyespots and had grown 30-fold; then by the end of day 3 they had become mature gonidia and had begun to cleave (Figure 6.6).

The initial aspects of this gene expression pattern were anticipated, but the aspects that had not been specifically predicted in advance were that the early-somatic genes would be so thoroughly down-regulated during the second day and that no trace of late-somatic transcripts would ever be detected in Gls/Reg mutant cells. As noted earlier (Sections 6.5 and 6.6), the Gls/Reg double mutant resembles a colonial volvocacean such as *Eudorina* in its asexual life cycle: It has cells of only a single type, all of which are initially biflagellate cells that execute vegetative functions, and all of which subsequently enlarge, redifferentiate as reproductive cells, and then divide. Thus, "first somatic and then gonidial" appears to be both the ancestral pattern of cellular differentiation that *V. carteri* inherited from its colonial relatives and the default pattern that it still expresses in the absence of *gls, lag,* and *regA* functions. By corollary, we predict that if it becomes possible to study the expression of genes homologous to the somatic and gonidial genes of *V. carteri* in *Eudorina*, they will be expressed in a pattern resembling the pattern seen in the Gls/Reg mutant. Implicit in this prediction is the assumption that as the products of the genes required for reproduction are accumulated in the latter part of the asexual life cycle, one or more of these gene products will act to down regulate expression of somatic genes – an action similar in type, perhaps, but different in timing from the regulatory effects of the *lag* gene products in wild-type *V. carteri*.

7.8 Control of the germ–soma dichotomy: a new molecular-genetic hypothesis

The most provocative fact to emerge from the studies of cell-type-specific genes summarized earlier is that the first three gonidial genes to be sequenced all encode products destined for the chloroplast, where they will play key roles in photosynthesis (Choi et al. 1996). This unanticipated finding was initially interpreted as a rather trivial consequence of the fact that gonidia grow more than somatic cells during the life cycle, and thus, being photoautotrophic, they must make more chloroplast polypeptides, which leads to the corresponding genes being identified as ones that are preferentially expressed in gonidia (Choi et al. 1996). But, one might ask, what if the differential expression of these genes is not a trivial *result*, but rather the *cause* of the differential growth rates of gonidia and somatic cells? What if it is because somatic cells cannot produce new chloroplast components that they cannot grow enough to divide and reproduce?

Following that shift in mind-set, one realizes that several past observations have suggested that somatic cells may be unable to synthesize new chloroplast components and that chloroplast biogenesis may be a fundamental aspect of germ–soma differentiation in *V. carteri*. Among such observations are the following: (1) As *V. carteri* juveniles grow in the postembryonic period, somatic cells become progressively paler while gonidia become darker green. (2) As discussed in Section 7.12, the chloroplast DNA (ctDNA) content of a developing gonidium increases logarithmically with gonidial diameter, so that by the time it is large enough to enter cleavage it appears to contain enough ctDNA to provide all of its descendant somatic cells with all the ctDNA they will ever possess (Coleman & Maguire 1982b). In contrast, the ctDNA content of somatic cells does not increase during development, and a surprising 5% of all somatic cells (which apparently are otherwise normal) have no detectable ctDNA at all (Coleman & Maguire 1982b)! (3) When two-dimensional gels were used to compare the extant polypeptides of somatic cells and gonidia with the polypeptide labeling patterns of those same cells (Baran 1984), it was noted that "many of the proteins synthesized only by the gonidia were present as major components in the patterns of extant proteins in both somatic cells and gonidia." It was speculated at that time that those were "housekeeping proteins" that were synthesized by gonidia and inherited by somatic cells during cleavage (Baran 1984). We now speculate that these housekeeping proteins are actually chloroplast proteins. (4) The first visible change as *regA*⁻ somatic cells begin to diverge morphologically from *regA*⁺ somatic cells and begin their transformation into gonidia is that they become darker green, and then, shortly thereafter, they begin to grow much more rapidly (Starr 1970a; Huskey & Griffin 1979).

On the basis of such information, *our current hypothesis is that the mechanism by which regA prevents somatic cells from entering the reproductive pathway is by repressing transcription of nuclear genes required for chloroplast biogenesis.*

A particularly attractive feature of this hypothesis is that it permits us to simplify our ideas about how the germ–soma dichotomy might have evolved. One of the most problematic aspects of our working hypothesis regarding *V. carteri* evolution (Section 6.6) is its implicit assumption that at some point in history all of the genes that are required for reproduction came under the repressive action of a single novel regulatory gene: *regA*. In candid moments we have been forced to admit that scenarios to explain how that might have come about – such as by introduction of a novel *cis*-regulatory site upstream of each of a sizeable number of otherwise unrelated genes simultaneously – seemed to stretch credulity beyond reasonable limits. However, the concept that the *regA* targets may be nuclear genes required for chloroplast biogenesis simplifies the problem.

In both green algae and higher plants, such genes (of which the best-studied are those encoding the light-harvesting chlorophyll *a/b*-binding proteins, or "Lhcps," and the small subunit of rubisco, or "RbcS") are subject to coordinate transcriptional regulation in response to a variety of different intrinsic and extrinsic signals (Kuhlemeir, Green & Chua 1987).[22] In several cases this coordinate regulation has been attributed to a similar constellation of upstream *cis*-regulatory elements (Gilmartin et al. 1990; Li, Washburn & Chory 1993; Batschauer et al. 1994; Terzaghi & Cashmore 1995; Puente, Wei & Deng 1996). First and foremost, coordinate accumulation of their transcripts can be triggered by light, and in the cases of both a higher-plant gene encoding RbcS (Lam & Chua 1990) and a *Chlamydomonas* gene encoding an Lhcp (Blankenship & Kindle 1992) it has been shown that part of the upstream region is adequate to confer photoregulated transcription on a reporter gene to which it is fused. However, although transcription of such genes can be induced coordinately by light under certain circumstances, the photoresponse also can be modulated or fully repressed in bright light by any one of a number of other signals. In various plants and algae the photoresponse can be modulated or overridden by feedback from photosynthetic products such as glucose or acetate (Sheen 1994), by non-nutritive feedback signals emanating from the chloroplast (Susek & Chory 1992), by cell-cycle events (Shepherd, Ledoight & Howell 1983), by diurnal signals (Kay & Miller 1992; Jacobshagen & Johnson 1994), and so forth. Thus, photoregulated genes of plants and algae appear to be particularly receptive to the imposition of new levels of negative transcriptional control that have adaptive value.

Without any question, of greatest relevance to the present discussion is the observation that in other multicellular photoautotrophs with a division of labor (i.e., vascular plants), regulatory elements exist for cell-type-specific silencing of photogenes (Kuhlemeir et al. 1987). Not only does cell-type-specific repression of photogenes occur in the roots of most plants (Simpson et al. 1986; Puente et al. 1996), it also exists in the leaves of maize, in which neighboring cells with two different types of chloroplasts divide the labor of CO_2 capture and CO_2 fixation (Sheen & Bogorad 1986). It has now been shown that a combination of different types of upstream promoter elements that are found in many such genes is required to establish both light regulation and developmental regulation (i.e., organ- and cell-type-specific regulation) of reporter genes (Puente et al. 1996).

In light of these well-established facts, the concept that the product of a new or modified gene, such as *regA*, that arose in the course of evolution might be able to repress a whole battery of genes that have related functions does not seem nearly as farfetched when those genes are thought of as being

---

[22] There are quantitative differences in the degrees to which transcripts of various nuclear genes encoding chloroplast proteins – even various members of a single gene family – accumulate under any given set of conditions, but space considerations prevent full consideration of all of these subtleties. For a more complete discussion, see Kuhlemeir et al. (1987).

nuclear genes encoding chloroplast proteins as it did when they were thought of under the much more nebulous heading of "genes required for asexual and sexual reproduction." The nuclear genes encoding chloroplast proteins appear to be particularly susceptible to the imposition of new levels of negative regulation, and in an obligate photoautotroph like *V. carteri*, what more straightforward way could there be to control reproduction than by controlling chloroplast biogenesis? During its formation each *V. carteri* somatic cell inherits a fragment (less than 0.05%) of the chloroplast that was present in the gonidium from which it was derived. If unable to augment this modest chloroplast endowment, somatic cells would be unable to grow significantly. And if unable to grow, they clearly would be unable to reproduce.

Another highly attractive feature of this hypothesis is that it is testable. Now that the characterization of the *regA* locus and its product is nearly complete (Section 7.5.1), it should soon be possible to test, rather directly, the hypothesis that the gene encodes a transcriptional regulator that binds to *cis*-regulatory elements of selected nuclear photogenes and represses their transcription.

In any case, the message of this chapter appears clear: Technical developments during the past few years have brought investigators to the point where a much more detailed molecular-genetic analysis of development in *V. carteri* now appears feasible within the next few years. Thus, at long last – three centuries after van Leeuwenhoek (1700) discovered *Volvox*, a century after August Weismann first pointed to *Volvox* as a model system for analyzing the ontogeny and phylogeny of germ–soma differentiation (Weismann 1892b), and a quarter-century after Richard Starr (1970a) pointed to *Volvox carteri* forma *nagariensis* as the isolate of *Volvox* most suitable for such an analysis – we appear to be on the verge of realizing some of the potential of the organism that has been so frequently recognized.

Epilogue

The present contains nothing more than the past, and what is found in the effect was already in the cause.

Bergson (1911)

The ultimate challenge for future *Volvox* research is to visit the past and retrace the pathway that led to the present.

Those of us currently engaged in *Volvox* research are excited by the challenge of capitalizing on the technical advances described in Chapter 7 to elucidate the detailed mechanisms by which cellular differentiation is programmed and executed in modern *V. carteri*. But when this goal has been achieved, the most distinctive attributes of *Volvox* as a developmental-genetic model system will still remain to be exploited. Although *Volvox* does offer a superb opportunity for defining the genetic and molecular basis of a highly interesting form of dichotomous cellular differentiation, it is not this feature that most clearly distinguishes *Volvox* from slime molds, fruit flies, plants, nematodes, zebrafish, mice, and other model organisms currently under intensive developmental-genetic investigation. Rather, it is the potential that it offers for tracing out in some detail the pathway by which such a program for cellular differentiation evolved.

As discussed in Chapter 1, multicellular organisms with a capacity for cellular differentiation clearly have evolved from simple unicellular ancestors numerous times; hence molecular mechanisms leading to dichotomous differentiation must have been independently invented many times in the past.

But in every other group that has been examined thus far, such inventions are buried so deep in antiquity that details of the pathway leading from unicellularity to multicellularity almost certainly have long since been obliterated by tectonic forces and the shifting sands of genetic drift.

To recapitulate the arguments presented in Chapter 2, however, *Volvox* and its relatives are distinguished by three important features in this regard: (1) In distinction to any other known group, the volvocine algae comprise extant organisms at all levels of morphological and developmental complexity from unicellular to multicellular with a complete germ–soma division of labor. (2) Molecular phylogenetic studies indicate that these organisms are all very closely related and also that the family Volvocaceae (the volvocine lineage minus its unicellular relatives) is surprisingly youthful, probably having been in existence only some 5–10% as long as the major multicellular lineages, such as plants, animals, and fungi. (3) Those same studies indicate that although this family as a whole is monophyletic, most of the currently recognized genera and species within it are not. This leads to the conclusion that a pathway leading to multicellularity and germ–soma differentiation probably has been traversed by members of this family not just once, but several times within the quite recent past. Moreover, morphological and developmental criteria indicate clearly that the endpoints reached by the different species of *Volvox* now resident at the tips of various branches of this family tree are quite different from one another (see, e.g., Figures 2.10, 2.11, and 3.7). This clearly implies that different species of *Volvox* must have followed rather different genetic pathways to achieve a similar endpoint: the capacity for germ–soma differentiation. Therefore, the volvocaceans may ultimately provide investigators not just one, but several opportunities to explore the origins of this fundamentally important type of evolutionary novelty. It seems quite likely that if we can only understand the detailed similarities and differences among these various convergent phylogenetic solutions to the common ecological challenges faced by various volvocaceans, our insight into the evolutionary process will be substantially deepened.

What will be required to mine this lode of evolutionary riches offered by the volvocaceans? A sine qua non will surely be a much more extensive and detailed molecular phylogeny of the family than has been established to date. The pioneering studies of Annette Coleman (1959, 1977, 1980; Coleman & Zollner 1977), who studied the mating affinities of various isolates sharing the name *Pandorina morum,* and Melvin Goldstein (1964), who did the same for many isolates that had been pigeonholed as various species of *Eudorina* and *Pleodorina*, clearly indicated that conventional taxonomic characters are sufficiently weak in the family Volvocaceae that in many cases generic and specific epithets fail entirely to reveal the actual phylogenetic relationships among various members of this family. This view has subsequently been reinforced by molecular studies of various kinds (Figures 2.11 and 2.14) (A. W. Coleman personal communication). Thus, it seems evident that before we can hope to trace out the genetic threads that link organisms at different levels of organizational and developmental complexity within this family, we will need to know, for example, which isolates of *Eudorina*, *Pleodorina*, and *Volvox* lie on the same branch of the family tree and which lie on separate branches. A reliable molecular phylogeny will undoubtedly require comparative analyses of more than one type of marker. Fortunately, several candidate molecules are now being explored by colleagues interested in volvocalean evolution and are being found to have the potential, when considered in combinations, of resolving many of the phylogenetic issues of greatest interest. These include the internal transcribed spacers of rRNA genes (Coleman et al. 1994; A. W. Coleman personal communication), the chloroplast genes encoding the large subunit of rubisco (Nozaki et al. 1995), the nuclear and chloroplast genes encoding rRNAs (Larson et al. 1992; Buchheim et al. 1996), and the introns of genes encoding small G proteins (Liss et al. 1997). All in all, it seems reasonable to hope that within the foreseeable future the branching patterns within the volvocacean family tree will become substantially clearer.

When such information becomes available, the volvocacean lineage most likely to be the first subject of a detailed molecular genetic exploration is, of course, the one that includes *V. carteri* forma *nagariensis*, because it is the only species of *Volvox* that has yet been studied at all genetically. Can we make *testable* predictions about what one might expect to find in such an exploration? For the purposes of this discussion, we will assume that the hypotheses presented earlier in this book about the nature of the genetic program regulating germ–soma differentiation in *V. carteri* are at least approximately correct and that our current hopes of understanding this program at a much deeper molecular level will be realized in the not-too-distant future.

First of all, we can predict (as in Section 7.7) that the closest "homocytic" relative of *V. carteri,* such as some isolate of *Eudorina*, will be found to possess genes homologous to many of the early-somatic and gonidial genes of *V. carteri* and will express these genes simultaneously, in a pattern similar to that seen in the Gls/Reg mutant of *V. carteri* (Figure 7.13). Second, we can predict that the isolate of *Pleodorina* that appears to be the closest living relative of *V. carteri* will possess a homologue of the *V. carteri regA* gene and will express this gene selectively in those cells that are destined to remain small, flagellated, and incapable of reproduction (Figure 2.7). But we predict that *Pleodorina* will not have functional equivalents of the *gls* or *lag* genes, because *Pleodorina* species neither divide asymmetrically nor have cells that fail to differentiate as biflagellate motile cells before redifferentiating as gonidia. The most distant relative in which we would expect to find a functional equivalent of the *lag* genes of *V. carteri* is some species of *Volvox*, such as *V. tertius*, in which there are no asymmetric cleavage divisions, but in which nevertheless a discrete set of cells begin to enlarge and differentiate as gonidia immediately after the end of embryogenesis, without first passing through the biflagellate, quasi-somatic condition. Correspondingly, the most distant relative in which we would expect to see functional equivalents of the *gls* loci of *V. carteri* would be *V. obversus*, the only other *Volvox* species that exhibits asymmetric division (albeit in a very different spatial pattern). In short, a specific set of predictions to be tested in such an analysis would be that the three types of genetic functions that are now believed to play central roles in germ–soma specification in *V. carteri* were added to the volvocacean genetic repertoire in the sequence first *regA*, then *lag*, and then *gls*.

However, even if all of the foregoing predictions should be verified, two of the most intriguing evolutionary questions of all would remain to be addressed.

The first of these remaining questions to be considered is the following: How do new regulatory genes (such as *regA*, *lag*, and *gls*) that are the sources of important evolutionary novelties arise? Were they derived from ancient regulatory genes that became modified so that they are now expressed under different conditions and thereby play entirely different roles in the organism? For example, might the *regA* gene be derived from a gene that in other green flagellates participates in repression of photoregulated nuclear genes in the dark, but that in *V. carteri* has become modified so that it represses the same genes in a cell-type-specific, light-independent fashion? Alternatively, are such novel regulatory genes produced by recombining motifs from other genes to generate products with entirely new functions? Conceivably, some important clues to the solution of this riddle may be uncovered when the *V. carteri* versions of these genes are fully sequenced and homologies to previously characterized gene families are discovered. On the other hand, it may well be that such sequence information will give us no significant insight into the possible origins of the regulatory genes of interest, and gaining such insight will take extensive additional detective work. But what group of organisms is more likely to reward such an investigation than the volvocaceans?

The other, and in some ways much more intriguing, question that it should eventually become possible to address when the molecular basis of germ–

soma differentiation in *V. carteri* has been elucidated is the following: Do organisms that evolve similar morphological features by convergent pathways reach their common ends by convergent molecular-genetic routes? Or do they follow entirely different molecular pathways to reach similar morphological ends? In the case of *Volvox*, it should ultimately be possible to rephrase this question in the following very specific way: Does a species such as *V. rousseletii*, which differs from *V. carteri* in many other details of its developmental history (see Sections 2.4.1 and 3.5), use a gene that is homologous to *regA* in structure and function to suppress reproductive development in its somatic cells? Or does it use some entirely different molecular-genetic mechanism to achieve this end?

Superficially, the developmental patterns of these two species of *Volvox* – particularly the ways in which their gonidia are specified – appear extremely different. *V. rousseletii* exhibits no asymmetric division; hence one would not expect it to exhibit any genetic functions homologous to the *gls* functions of *V. carteri*. Moreover, *V. rousseletii* gonidia go through a biflagellate, quasi-somatic phase before revealing their reproductive potentials; hence it is also unlikely that they express functions homologous to the *lag* functions of *V. carteri*. Thus, one might be tempted to predict that terminal differentiation of the somatic cells might have entirely different bases in the two species as well. And so it may. However, molecular studies during the past decade have shown that flies and mice not only use homologous homeobox genes to pattern their very different kinds of bodies along the anterior–posterior axis (Krumlauf 1994) but also use similar *eyeless* genes to trigger development of their very different types of eyes (Halder, Callaerts & Gehring 1995), as well as similar genes in the *otd* (orthodenticle) and *ems* (empty spiracle) classes to pattern their very different looking anterior heads and brains (Finkelstein & Boncinelli 1994). Such unanticipated findings alert us to the fact that lack of homology between final structures (in the classic embryological sense) does not necessarily imply lack of homology in the genetic pathways leading to them. Nature clearly has had a limited number of genetic tools to work with in fashioning evolutionary novelties and has used many of the same tools over and over in slightly different ways. On the other hand, often very different routes have been taken to achieve extremely similar ends. The best evidence for this comes from one of the examples cited earlier: There can now be little doubt that entirely different genetic mechanisms are used to accomplish very similar ends in flies and mice, namely, the patterned expression of related homeobox genes along the anterior–posterior axis of the embryo (Krumlauf 1994). Thus, at this point, one guess appears to be just as good as another with respect to the question whether or not the different species of *Volvox* now resident at the tips of different evolutionary branches used similar genes to establish terminal differentiation of somatic cells. Whichever way it ultimately turns out, the answer is bound to be interesting and undoubtedly will enrich our understanding of the evolutionary process.

Nearly 70 years ago, the incomparable embryologist Hans Spemann assessed the status of research regarding the nature and mechanism of action of his famous "Organizer," using terms that (in translation) seem fully applicable to the state of *Volvox* research in 1997: "What has been achieved is but the first step; we still stand in the presence of riddles, but not without the hope of solving them. And riddles with the hope of solution – what more can a scientist desire?" (Spemann 1927).

References

Aberg, B. (1947). On the mechanism of the toxic action of chlorate and some related substances upon young wheat plants. *Ann. Kungl. Lantbrukshösgkolans*, **15,** 37–107.

Adair, W. S. & Appel, H. (1989). Identification of a highly conserved hydroxyproline-rich glycoprotein in the cell walls of *Chlamydomonas reinhardtii* and two other Volvocales. *Planta*, **179,** 381–6.

Adair, W. S. & Snell, W. J. (1990). The *Chlamydomonas reinhardtii* cell wall: structure, biochemistry, and molecular biology. In *Organization and Assembly of Plant and Animal Extracellular Matrix*. Ed., W. S. Adair & R. P. Mecham. Academic Press, San Diego, pp. 15–84.

Adair, W. S., Steinmetz, S. A., Mattson, D. M., Goodenough, U. W. & Heuser, J. E. (1987). Nucleated assembly of *Chlamydomonas* and *Volvox* cell walls. *J. Cell Biol.*, **105,** 2373–82.

Adams, C. R., Stamer, K. A., Miller, J. K., McNally, J. G., Kirk, M. M. & Kirk, D. L. (1990). Patterns of organellar and nuclear inheritance among progeny of two geographically isolated strains of *Volvox carteri. Curr. Genet.*, **18,** 141–53.

Adams, G. M. W., Wright, R. L. & Jarvik, J. W. (1985). Defective temporal and spatial control of flagellar assembly in a mutant of *Chlamydomonas* with variable flagellar number. *J. Cell Biol.*, **100,** 955–64.

Aitchison, P. A. & Butt, V. S. (1973). The relationship between the synthesis of inorganic polyphosphate and phosphate uptake by *Chlorella vulgaris. J. Exp. Bot.*, **24,** 497–510.

Aitchison, W. A. & Brown, D. L. (1986). Duplication of the flagellar apparatus and cytoskeletal system in the alga *Polytomella. Cell Motil. Cytoskel.*, **6,** 122–7.

Alam, M. J., Habib, A. B. & Begum, M. (1989). Effect of water properties and dominant genera of phyto-plankton on the available genera of zooplankton. *Pak. J. Sci. Ind. Res.*, **32,** 194–200.

Alberch, P. (1982). Developmental constraints in evolutionary processes. In *Evolution and Development*. Ed., J. T. Bonner. Springer Verlag, Berlin, pp. 313–32.

Alberts, B., Bray, D., Lewis, J., Raff, M., Roberts, K. & Watson, J. D. (1989). *Molecular Biology of the Cell*, 4th ed. Garland Press, New York.

Alexopoulos, C. J. (1963). The Myxomycetes II. *Bot. Rev.*, **29**, 1–78.

Alexopoulos, C. J. & Mims, C. W. (1979). *Introductory Mycology*, 3rd ed. Wiley, New York.

Al-Hasani, H. & Jaenicke, L. (1992). Characterization of the sex-inducer glycoprotein of *Volvox carteri* f. *weismannia*. *Sex. Plant Reprod.*, **5**, 8–12.

Andersen, R. A., Barr, D. J. S., Lynn, D. H., Melkonian, M., Moestrop, Ø. & Sleigh, M. A. (1991). Terminology and nomenclature of the cytoskeletal elements associated with the flagellar/ciliary apparatus in protists. *Protoplasma*, **164**, 1–8.

Apte, V. V. (1936). Observations on some species of *Volvox* from Poona with detailed descriptions of *Volvox poonaensis*. *J. Univ. Bombay*, **4**, 1–15.

Baker, G. (1753). *Employment for the Microscope*. Dodsley, London.

Balshüsemann, D. & Jaenicke, L. (1990a). The oligosaccharides of the glycoprotein pheromone of *Volvox carteri* f. *nagariensis* Iyengar (Chlorophyceae). *Eur. J. Biochem.*, **192**, 231–7.

Balshüsemann, D. & Jaenicke, L. (1990b). Time and mode of synthesis of the sexual inducer glycoprotein of *Volvox carteri*. *FEBS Lett.*, **264**, 56–8.

Bancroft, F. W. (1907). The mechanism of galvanic orientation in Volvox. *J. Exp. Zool.*, **4**, 157–63.

Bandziulis, R. J. & Rosenbaum, J. L. (1988). Novel control elements in the alpha-1 tubulin gene promoter from *Chlamydomonas reinhardtii*. *Mol. Gen. Genet.*, **214**, 204–12.

Baran, G. (1984). Analysis of somatic cell differentiation in *Volvox carteri* f. *nagariensis*. Ph.D. thesis, University of Virginia.

Baran, G. J. & Huskey, R. H. (1986). Developmental changes in the sensitivity of *Volvox* to ultraviolet light. *Dev. Genet.*, **6**, 269–80.

Batko, A. & Jakubiec, H. (1989). *Gonium dispersum* sp. nov., a new species of *Gonium* from Poland. *Arch. Hyrobiol. Suppl.*, **82**, 39–47.

Batschauer, A., Gilmartin, P. M., Nagy, F. & Schäfer, E. (1994). The molecular biology of photoregulated genes. In *Photomorphogenesis in Plants*, 2nd ed. Ed., R. E. Kendrick & G. H. M. Kronenberg. Kluwer, Dordrecht. pp. 559–99.

Bause, E. & Jaenicke, L. (1979). Formation of lipid-linked sugar compounds in *Volvox carteri* f. *nagariensis*. *FEBS Lett.*, **106**, 321–4.

Bause, E., Müller, T. & Jaenicke, L. (1983). Synthesis and characterization of lipid-linked mannosyl oligosaccharides in *Volvox carteri* f. *nagariensis*. *Arch. Biochem. Biophys.*, **220**, 200–7.

Beck, C. & Uhl, R. (1994). On the localization of voltage-sensitive calcium channels in the flagella of *Chlamydomonas reinhardtii*. *J. Cell Biol.*, **125**, 1119–25.

Beech, P. L., Wetherbee, R. & Pickett-Heaps, J. D. (1980). Transformation of the flagella and associated flagellar components during cell division in the coccolithophorid *Pleurochrysis carterae*. *Protoplasma*, **145**, 37–46.

Beger, H. (1927). Beiträge zur Okologie und Soziologie der luftlebigen (atmophytischen) Kieselalgen. *Deutsche Bot. Gesell.*, **45**, 385–407.

Belar, E. (1926). Der Formwechsel der Protistenkerne. *Ergeb. Fortschr. Zool.*, **6**, 1–420.

Bell, G. (1985). The origin and early evolution of germ cells as illustrated by the

Volvocales. In *The Origin and Evolution of Sex*. Ed., H. O. Halvorson & A. Monroy. Alan R. Liss, New York, pp. 221–56.

Berbee, M. L. & Taylor, J. W. (1993). Dating the evolutionary radiations of the true fungi. *Can. J. Bot.,* **71,** 1114–27.

Bergson, H. (1911). *Creative Evolution*. Translator, A. Mitchell. H. Holt, New York.

Berman, T. & Rodhe, W. (1971). Distribution and migration of *Peridinium* in Lake Kinneret. *Mitt. Int. Ver. Theor. Angew. Limnol.,* **19,** 266–76.

Bernstein, E. (1964). Physiology of an obligate photoautotroph (*Chlamydomonas moewusii*). I. Characteristics of synchronously and randomly reproducing cells, and an hypothesis to explain their population curves. *J. Protozool.,* **11,** 56–74.

Bernstein, E. (1966). Physiology of an obligate photoautotroph (*Chlamydomonas moewusii*). II. The effect of light–dark cycles on cell division. *Exp. Cell Res.,* **41,** 307–15.

Bernstein, E. (1968). Induction of synchrony in *Chlamydomonas moewusii* as a tool for the study of cell division. *Methods Cell Physiol.,* **3,** 119–45.

Bessen, M., Fay, R. B. & Witman, G. B. (1980). Calcium control of waveform in isolated flagellar axonemes of *Chlamydomonas*. *J. Cell Biol.,* **86,** 446–55.

Birchem, R. (1977). The ultrastructure of the male colonies of *Volvox carteri* f. *weismannia*. Ph.D. thesis, University of Georgia.

Birchem, R. & Kochert, G. (1979a). Development of sperm cells of *Volvox carteri* f. *weismannia* (Chlorophyceae). *Phycologia,* **18,** 409–19.

Birchem, R. & Kochert, G. (1979b). Mitosis and cytokinesis in androgonidia of *Volvox carteri* f. *weismannia*. *Protoplasma,* **100,** 1–12.

Bisalputra, T. & Stein, J. R. (1966). The development of cytoplasmic bridges in *Volvox aureus*. *Can. J. Bot.,* **44,** 1697–1702.

Blankenship, J. A. & Kindle, K. L. (1992). Expression of chimeric genes by the light-regulated *cabII*-1 promoter in *Chlamydomonas reinhardtii*: A *cabII*-1/*nit1* gene functions as a dominant selectable marker in a *nit1⁻ nit2⁻* strain. *Mol. Cell. Biol.,* **12,** 5268–79.

Blasco, D. (1978). Observations of the diel migration of marine dinoflagellates of the Baja California coast. *Mar. Biol.,* **46,** 41–7.

Blueweiss, L., Fox, H., Kudzma, V., Nakashima, D., Peters, R. & Sams, S. (1978). Relationship between body size and some life history parameters. *Oecologia,* **37,** 257–72.

Bock, F. (1926). Experimentelle Untersuchungen an kolonie-bildenen Volvocaceen. *Arch. Protistenkd.,* **56,** 321–56.

Bogdan, K. G. & Gilbert, J. J. (1984). Body size and food size in freshwater zooplankton. *Proc. Natl. Acad. Sci. USA,* **81,** 6427–31.

Bokoch, G. M. & Der, C. J. (1993). Emerging concepts in the *Ras* superfamily of GTP-binding proteins. *FASEB J.,* **7,** 750–9.

Bold, H. J. & Wynne, M. J. (1985). *Introduction to the Algae,* 2nd. ed. Prentice-Hall, Englewood Cliffs, NJ.

Bonner, J. T. (1950). Volvox: a colony of cells. *Sci. Am.,* **1950,** 52–5.

Bonner, J. T. (1952). *Morphogenesis*. Princeton University Press.

Bonner, J. T. (1959). *The Cellular Slime Molds*. Princeton University Press.

Bonner, J. T. (1965). *Size and Cycles. An Essay on the Structure of Biology*. Princeton University Press.

Bonner, J. T. (1974). *On Development. The Biology of Form.* Harvard University Press.

Bonner, J. T. (1988). *The Evolution of Complexity by Means of Natural Selection.* Princeton University Press.

Bonner, J. T. (1993). *Life Cycles. Reflections of an Evolutionary Biologist.* Princeton University Press.

Bonnet, C. (1762). *Considérations sur les corps organisés.* Marc-Michel Rey, Amsterdam.

Boscov, J. S. & Feinleib, M. E. (1979). Phototactic response of *Chlamydomonas* to flashes of light. II. Response of individual cells. *Photochem. Photobiol.,* **30,** 499–505.

Bouck, G. B. & Brown, D. L. (1973). Microtubule biogenesis and cell shape in *Ochromonas.* I. The distribution of cytoplasmic and mitotic microtubules. *J. Cell Biol.,* **56,** 340–59.

Bourrelly, P. (1966). *Les Algues d'Eau Douce. Initiation à la systématique. I. Les Algues Vertes.* N. Boubée et Cie, Paris.

Bourrelly, P. (1972). *Les Algues d'Eau Douce. Initiation à la systématique. I. Les Algues Vertes.* Rev. ed. N. Boubée, Paris.

Bowring, S., Grotzinger, J. P., Isachsen, C. E., Knoll, A. H., Pelechaty, S. M. & Kolosov, P. (1993). Calibrating rates of early Cambrian evolution. *Science,* **261,** 1293–8.

Bradley, D. M., Goldin, H. H. & Claybrook, J. R. (1974). Histone analysis in *Volvox. FEBS Lett.,* **41,** 219–22.

Bremer, K. (1985). Summary of green plant phylogeny and classification. *Cladistics,* **1,** 369–85.

Bremer, K., Humphries, C. J., Mishler, B. D. & Churchill, S. P. (1987). On cladistic relationships in green plants. *Taxon,* **36,** 339–49.

Brock, T. D., Smith, D. W. & Madigan, M. T. (1984). *Biology of Microorganisms,* 4th ed. Prentice-Hall, Englewood Cliffs, NJ.

Brooks, A. E. (1966). The sexual cycle and intercrossing in the genus *Astrephomene. J. Protozool.,* **13,** 367–75.

Brown, C. R., Doxsey, S. J., Hong-Brown, L. Q., Martin, R. L. & Welch, W. J. (1996a). Molecular chaperones and the centrosome. A role for TCP-1 in microtubule nucleation. *J. Biol. Chem.,* **271,** 824–32.

Brown, C. R., Hong-Brown, L. Q., Doxsey, S. J. & Welch, W. J. (1996). Molecular chaperones and the centrosome. A role for hsp 73 in centrosomal repair following heat shock treatment. *J. Biol. Chem.,* **271,** 833–40.

Brown, D. L., Stearns, M. E. & Macrae, T. H. (1982). Microtubule organizing centers. In *The Cytoskeleton in Plant Growth and Development.* Ed., C. W. Lloyd. Academic Press, London, pp. 55–83.

Brun, Y. V., Marczynski, G. & Shapiro, L. (1994). The expression of asymmetry during *Caulobacter* cell differentiation. *Annu. Rev. Biochem.,* **63,** 419–50.

Brunke, K. J., Anthony, J. G., Sternberg, E. J. & Weeks, D. P. (1984). Repeated consensus sequence and pseudopromoters in the four coordinately regulated tubulin genes of *Chlamydomonas reinhardi. Mol. Cell. Biol.,* **4,** 1115–24.

Brunke, K. J., Young, E. E., Buchbinder, B. U. & Weeks, D. P. (1982). Coordinate regulation of the four tubulin genes of *Chlamydomonas reinhardi. Nucl. Acids Res.,* **10,** 1295–310.

Brusca, R. C. & Brusca, G. J. (1990). *Invertebrates*. Sinauer, Sunderland, MA.

Buchanan, M. J. & Snell, W. J. (1988). Biochemical studies on lysin, a cell-wall degrading enzyme released during fertilization in *Chlamydomonas*. *Exp. Cell Res.,* **179,** 181–93.

Buchheim, M. A. & Chapman, R. L. (1991). Phylogeny of the colonial green flagellates: a study of 18S and 26S rRNA sequence data. *BioSystems,* **25,** 85–100.

Buchheim, M. A., Lemieux, C., Otis, C., Gutell, R. R., Chapman, R. & Turmel, M. (1996). Phylogeny of the Chlamydomonadales (Chlorophyceae): A comparison of ribosomal RNA sequences from the nucleus and the chloroplast. *Molec. Phylog. Evol.,* **5,** 391–402.

Buchheim, M. A., McAuley, M. A., Zimmer, E. A., Theriot, E. C. & Chapman, R. L. (1994). Multiple origins of colonial green flagellates from unicells: evidence from molecular and organismal characters. *Molec. Phylog. Evol.,* **3,** 322–43.

Buchheim, M. A., Turmel, M., Zimmer, E. A. & Chapman, R. L. (1990). Phylogeny of *Chlamydomonas* (Chlorophyta) based on cladistic analysis of nuclear 18S rRNA sequence data. *J. Phycol.,* **26,** 689–99.

Buffaloe, N. D. (1958). A comparative cytological study of four species of *Chlamydomonas*. *Bull. Torrey Bot. Club,* **85,** 157–78.

Burns, C. W. (1968). The relationship between body size of filter-feeding cladocerans and the maximum size of particle ingested. *Limnol. Oceanogr.,* **13,** 675–8.

Burr, F. A. & McCracken, M. D. (1973). Existence of a surface layer on the sheath of *Volvox*. *J. Phycol.,* **9,** 345–6.

Busk, G. (1853). Some observations on the structure and development of *Volvox globator* and its relations to other unicellular organisms. *Q. J. Microsc. Sci.,* **1,** 31–45.

Buss, L. W. (1987). *The Evolution of Individuality*. Princeton University Press.

Bütschli, O. (1883). Bemurkungen zur Gastraeatheorie. *Morph. Jahrb.,* **9,** 415–27.

Butterfield, N. J. (1994). Burgess shale-type fossils from a Lower Cambrian shallow-shelf sequence in Northwestern Canada. *Nature,* **369,** 477–9.

Butterfield, N. J., Knoll, A. H. & Swett, K. (1990). A Bangliophyte red alga from the proterozoic of Arctic Canada. *Science,* **250,** 104–7.

Callahan, A. M. & Huskey, R. J. (1980). Genetic control of sexual development in *Volvox*. *Dev. Biol.,* **80,** 419–35.

Caplen, H. S. & Blamire, J. (1980). Polyadenylated RNA of *Volvox*: Isolation and partial characterization. *Cytobios,* **29,** 115–28.

Carefoot, J. R. (1966). Sexual reproduction and intercrossing in *Volvulina steinii*. *J. Phycol.,* **2,** 150–6.

Carlgren, O. (1899). Ueber die Einwerkung des constanten galvanischen Stromes auf nieder Organismen. *Arch. Anat. Physiol.,* **4,** 49–76.

Carter, H. J. (1859). On fecundation in the two Volvoces, and their specific differences. *Ann. & Mag. Nat. Hist., ser. 3,* **3,** 1–20.

Cavalier-Smith, T. (1967). Organelle development in *Chlamydomonas reinhardii*. Ph.D. thesis, University of London.

Cavalier-Smith, T. (1974). Basal body and flagellar development during the vegetative cell cycle and the sexual cycle of *Chlamydomonas reinhardii*. *J. Cell Sci.,* **16,** 529–56.

Cavalier-Smith, T. (1981). Eukaryotic kingdoms: seven or nine? *BioSystems,* **14,** 461–81.

Cavalier-Smith, T. (1991). Intron phylogeny: a new hypothesis. *Trends Genet.,* **7,** 145–8.

Cave, M. S. & Pocock, M. A. (1951a). The aceto-carmine technic applied to the colonial Volvocales. *Stain Technol.,* **26,** 173–4.

Cave, M. S. & Pocock, M. S. (1951b). Karyological studies in the Volvocaceae. *Am. J. Bot.,* **38,** 800–11.

Cave, M. S. & Pocock, M. S. (1956). The variable chromosome number in *Astrephomene gubernaculifera. Am. J. Bot.,* **43,** 122–34.

Chapman, D. J. (1985). Geological factors and biochemical aspects of the origins of land plants. In *Geological Factors and the Evolution of Land Plants.* Ed., B. H. Tiffney. Yale University Press, pp. 23-45.

Choi, G., Przybylska, M. & Straus, D. (1996). Three abundant germ line-specific transcripts in *Volvox carteri* encode photosynthetic proteins. *Curr. Genet.,* **30,** 347–55.

Chu, G., Vollrath, D. & Davis, R. W. (1986). Separation of large DNA molecules by contour-clamped homogeneous electric fields. *Science,* **234,** 1582–85.

Cloud, P. E., Jr., Licori, G. R., Wright, L. R. & Troxel, B. W. (1969). Proterozoic eukaryotes from eastern California. *Proc. Natl. Acad. Sci. USA,* **63,** 623–30.

Coggin, S. J., Hutt, W. & Kochert, G. (1979). Sperm bundle–female somatic cell interaction in the fertilization process of *Volvox carteri* f. *weismannia. J. Phycol.,* **15,** 247–51.

Coggin, S. J. & Kochert, G. (1986). Flagellar development and regeneration in *Volvox carteri* (Chlorophyta). *J. Phycol.,* **22,** 370–81.

Cohn, F. (1875). *Die Entwickelungsgeschichte der Gattung Volvox. Festschrift dem Geheimen Medicinalrath Prof. Dr. Göppert zu seinem funfzigjährigen Doctorjubilätum.* J. U. Kern, Breslau.

Coleman, A. W. (1959). Sexual isolation in *Pandorina morum. J. Protozool.,* **6,** 249–64.

Coleman, A. W. (1962). Sexuality. In *Physiology and Biochemistry of Algae.* Ed., R. A. Lewin. Academic Press, New York, pp. 711–29.

Coleman, A. W. (1975). The long-term maintenance of fertile algal clones· experience with the genus *Pandorina* (Chlorophyceae). *J. Phycol.,* **11,** 282–6.

Coleman, A. W. (1977). Sexual and genetic isolation in the cosmopolitan algal species *Pandorina morum* (Chlorophyta). *Am. J. Bot.,* **64,** 361–8.

Coleman, A. W. (1978). Visualization of chloroplast DNA with two fluorochromes. *Exp. Cell Res.,* **114,** 95–100.

Coleman, A. W. (1979a). Feulgen microspectrophotometric studies of *Pandorina morum* and other Volvocales (Chlorophyceae). *J. Phycol.,* **15,** 216–20.

Coleman, A. W. (1979b). Sexuality in colonial green flagellates. In *Biochemistry and Physiology of Protozoa,* Vol. 1. Ed., M. Levandowsky & S. H. Hutner. Academic Press, New York, pp. 307–40.

Coleman, A. W. (1980). The biological species concept: its applicability to the taxonomy of freshwater algae. In *Proceedings of the 2nd International Symposium on Taxonomy of Algae.* Ed., T. V. Desikachary. University of Madras, pp. 22–36.

Coleman, A. W. (1982). The nuclear cell cycle in *Chlamydomonas* (Chlorophyceae). *J. Phycol.*, **18**, 192–5.

Coleman, A. W. (1983). The roles of resting spores and akinetes in chlorophyte survival. In *Survival Strategies of the Algae*. Ed., G. A. Fryxell. Cambridge University Press, pp. 1–21.

Coleman, A. W. (1996a). Are the impacts of events in the earth's history discernible in the current distributions of freshwater algae? *Hydrobiologia*, **336**, 137–42.

Coleman, A. W. (1996b). The Indian connection, crucial to reconstruction of the historical biogeography of freshwater algae: Examples among Volvocaceae (Chlorophyta). *Nova Hedwigia*, **112**, 475–80.

Coleman, A. W. & Goff, L. J. (1991). DNA analysis of eukaryotic algal species. *J. Phycol.*, **27**, 463–73.

Coleman, A. W. & Maguire, M. J. (1982a). A microspectrophotometric analysis of nuclear and chloroplast DNA in *Volvox. Dev. Biol.*, **94**, 441–50.

Coleman, A. W. & Maguire, M. J. (1982b). The nuclear cell cycle in *Chlamydomonas* (Chlorophyceae). *J. Phycol.*, **18**, 192–5.

Coleman, A. W., Suarez, A. & Goff, L. J. (1994). Molecular delineation of species and syngens in volvocacean green algae (Chlorophyta). *J. Phycol.*, **30**, 80–90.

Coleman, A.W. & Zollner, J. (1977). Cytogenetic polymorphism within the species *Pandorina morum* Bory de St. Vincent (Volvocaceae). *Arch. Protistenkd.*, **119**, 224–32.

Collins, F. S. (1909). The green algae of North America. *Tufts College Studies*, **2**, 79–480.

Collins, F. S. (1918). The green algae of North America. Second supplement. *Tufts College Studies*, **4**, 1–106.

Conway Morris, S. (1993). The fossil record and the early evolution of the Metazoa. *Nature*, **361**, 219–25.

Conway Morris, S. (1994). A paleontological perspective. *Curr. Opin. Genet. Dev.*, **4**, 802–9.

Cooke, M. C. (1882). *British Fresh-water Algae*. Williams & Norgate, London.

Coss, R. A. (1974). Mitosis in *Chlamydomonas reinhardtii*: Basal bodies and the mitotic apparatus. *J. Cell Biol.*, **63**, 325–9.

Cove, D. J. (1976a). Chlorate toxicity in *Aspergillus nidulans*: Studies of mutants altered in nitrate assimilation. *Mol. Gen. Genet.*, **146**, 147–59.

Cove, D. J. (1976b). Chlorate toxicity in *Aspergillus nidulans*: The selection and characterization of chlorate resistant mutants. *Heredity*, **36**, 191–203.

Cowell, I. G. (1994). Repression versus activation in the control of gene transcription. *Trends Biochem. Sci.*, **19**, 38–42.

Craigie, R. A. & Cavalier-Smith, T. (1982). Cell volume and the control of the *Chlamydomonas* cell cycle. *J. Cell Sci.*, **54**, 173–91.

Crawford, N. M., Smith, M., Bellisimo, D. & Davis, R. W. (1988). Sequence and nitrate regulation of the *Arabidopsis thaliana* mRNA encoding nitrate reductase, a metalloflavoprotein with three functional domains. *Proc. Natl. Acad. Sci. USA*, **85**, 5006–10.

Crayton, M. A. (1980). Presence of a sulfated polysaccharide in the extracellular matrix of *Platydorina caudata* (Volvocales, Chlorophyta). *J. Phycol.*, **16**, 80–7.

Cresnar, B., Mages, W., Müller, K., Salbaum, J. M. & Schmitt, R. (1990). Structure and expression of a single actin gene in *Volvox carteri*. *Curr. Genet.*, **18**, 337–46.

Crow, W. B. (1918). The classification of some colonial Chlamydomonads. *The New Phytologist*, **17**, 151–9.

Curry, A. M., Williams, B. D. & Rosenbaum, J. L. (1992). Sequence analysis reveals homology between two proteins of the flagellar radial spoke. *Mol. Cell. Biol.*, **12**, 3967–77.

Dangeard, P. A. (1899). Mémoire sur les Chlamydomonadinées ou l'histoire d'une cellule. *Botaniste*, **6**, 65–292.

Dangeard, P. A. (1901). Etude comparative de la zoospore et du spermatozoïde. *C. R. Hebd. Seances Acad. Sci.*, **132**, 859–61.

Darden, W. H. (1966). Sexual differentiation in *Volvox aureus*. *J. Protozool.*, **13**, 239–55.

Darden, W. H. (1968). Production of a male-inducing hormone by a parthenosporic *Volvox aureus*. *J. Protozool.*, **15**, 412–14.

Darden, W. H. (1970). Hormonal control of sexuality in *Volvox aureus*. *Ann. N. Y. Acad. Sci.*, **175**, 757–63.

Darden, W. H. (1971). A new system of male induction in *Volvox aureus* M5. *Biochem. Biophys. Res. Commun.*, **45**, 1205–11.

Darden, W. H. (1973a). Formation and assay of a *Volvox* factor–histone complex. *Microbios*, **8**, 1167–74.

Darden, W. H. (1973b). Hormonal control of sexuality in algae. In *Humoral Control of Growth and Differentiation*. Ed., J. LoBue & A. S. Gordon. Academic Press, New York, pp. 101–19.

Darden, W. H. (1980). Some properties of male-inducing pheromones from *Volvox aureus* M5. *Microbios*, **28**, 27–39.

Darden, W. H. & Sayers, E. R. (1969). Parthenospore induction in *Volvox aureus* D5. *Microbios*, **2**, 171–6.

Darden, W. H. & Sayers, E. R. (1971). The effect of selected chemical and physical agents on the male-inducing substance from *Volvox aureus*. *Microbios*, **3**, 209–14.

Dauwalder, M., Whaley, W. G. & Starr, R. C. (1980). Differentiation and secretion in *Volvox*. *J. Ultrastruct. Res.*, **70**, 318–35.

Davies, J. P. & Grossman, A. R. (1994). Sequences controlling transcription of the *Chlamydomonas reinhardtii* β2-tubulin gene after deflagellation and during the cell cycle. *Mol. Cell. Biol.*, **14**, 5165–74.

Davies, J. P., Weeks, D. & Grossman, A. R. (1992). Expression of the arylsulfatase gene from the β2-tubulin promoter in *Chlamydomonas reinhardtii*. *Nucl. Acids Res.*, **20**, 2959–65.

Davies, J. P., Yildiz, F. H. & Grossman, A. (1996). Sac1, a putative regulator that is critical for survival of *Chlamydomonas reinhardtii* during sulfur deprivation. *EMBO J.*, **15**, 2150–9.

Day, A. & Rochaix, J.-D. (1989). Characterization of transcribed dispersed repetitive DNAs in the nuclear genome of the green alga *Chlamydomonas reinhardtii*. *Curr. Genet.*, **16**, 165–76.

Deason, T. R. & Darden, W. H., Jr. (1971). The male initial and mitosis in *Volvox*. In *Contributions in Phycology*. Ed., B. C. Parker & R. M. Brown, Jr. Allen Press, Lawrence, KS, pp. 67–79.

Deason, T. R., Darden, W. H., Jr. & Ely, S. (1969). The development of the sperm packets of the M5 strain of *Volvox aureus*. *J. Ultrastruct. Res.,* **26**, 85–94.

de Guerne, J. M. (1888). Sur la dissemination des organismes d'eau douce par les Palmipèdes. *Société de Biol.,* **8**, 294–8.

De Hostis, E. L., Schilling, J. & Grossman, A. R. (1989). Structure and expression of the gene encoding the periplasmic arylsulfatase of *Chlamydomonas reinhardtii. Mol. Gen. Genet.,* **218**, 229–39.

Deininger, W., Kröger, P., Hegemann, U., Lottspeich, F. & Hegemann, P. (1995). Chlamyrhodopsin represents a new type of photoreceptor. *EMBO J.,* **14**, 5849–58.

Delsman, H. (1919). The egg cleavage of *Volvox globator* and its relation to the movement of the adult form and to cleavage types of Metazoa. *Proc. Roy. Acad. Sci. Amsterdam,* **21**, 243–51.

Denis, H. & Lacroix, J.-C. (1993). The dichotomy between germ line and somatic line, and the origin of cell mortality. *Trends Genet.,* **9**, 7–11.

deNoyelles, F., Jr. (1971). Factors affecting distribution of phytoplankton in eight replicated unfertilized and fertilized ponds. Ph.D. thesis, Cornell University.

Desnitski, A. G. (1980). Kletochnaya differentsirovka i morfognes u *Volvox* [Cell differentiation and morphogenesis in *Volvox*]. *Ontogenez.,* **11**, 339–50.

Desnitski, A. G. (1981a). Kletochnaya tsicly u *Volvox* i nekotorykh drugikh jguti-konostsev (Volvocales) [Cell cycles in *Volvox* and some other green flagellates (Volvocales)]. *Tsitologiya,* **23**, 243–53.

Desnitski, A. G. (1981b). Isucheniye razvitiya *Volvox aureus* Ehrenburg (Petergof-skaya linia P-1) [Study of the development of *Volvox aureus* Ehrenburg (Peterhof strain P-I)]. *Vestnik Leningrad St. Univ.,* **21**, 80–5.

Desnitski, A. G. (1982a). Avtoradiograficheskoye issledovaniye vklucheniya pred-shestvennikov nucleinovykh kislot v kletki *Volvox* [An autoradiographic study of incorporation of nucleic acid precursors in *Volvox* cells]. *Tsitologiya,* **24**, 172–6.

Desnitski, A. G. (1982b). K probleme regulatsii sinteza yadernoy DNK u zelenykh vodorosley [On the regulation of nuclear DNA synthesis in green algae]. *Tsitologiya,* **24**, 855–62.

Desnitski, A. G. (1982c). Osobennosti vklucheniya ^{3}H-timidina v zarodyshy *Volvox aureus* [Peculiarities of ^{3}H-thymidine incorporation in *Volvox aureus* embryos]. *Ontogenez.,* **13**, 424–6.

Desnitski, A. G. (1983a). Kletochnaya tsicly i programmirovaniye gennoy aktivnosti i kletochnoy differentsirovki [Cell cycles and the programming of gene activity and cellular differentiation]. In *Kletochnaya reproduktsia i protsessy different-sirovki [Cell Reproduction and the Processes of Differentiation]*. Ed., A. K. Don-dua. Nauka, Leningrad, pp. 216–22.

Desnitski, A. G. (1983b). Imeneniya morfologii yadra v hode vegetativnogo jiznen-nogo tsikla u *Volvox carteri* [Changes in nuclear morphology in the course of the vegetative life cycle of *Volvox carteri*]. *Vestnik Leningrad St. Univ.,* **21**, 86–7.

Desnitski, A. G. (1983c). Stanovleniye mnogokletochnosti u *Volvox* [Origin of the multicellularity of *Volvox*]. In *The Evolutionary Morphology of Invertebrates.* Ed., A. V. Ivanov. Nauka, Leningrad, pp. 68–75.

Desnitski, A. G. (1983d). Problema stanovleniya i nachalnykh etapov evolutsii mnogokletochnosti u Volvocales [The problem of establishment and of primary stages of the evolution of multicellularity in the Volvocales]. *Tsitologiya*, **25**, 635–42.

Desnitski, A. G. (1984a). Regulatsiya kletochnykh tsilov u fitoflagellyat [The regulation of cell cycles in phytoflagellates]. *Tsitologiya*, **26**, 635–42.

Desnitski, A. G. (1984b). Nekotorye osobennosti regulatsii kletochnykh deleniy u *Volvox* [Some peculiarities of cell division regulation in *Volvox*]. *Tsitologiya*, **26**, 269–74.

Desnitski, A. G. (1985a). Opredeleniye vremeni nachala drobleniya gonidiyev u *Volvox aureus* i *Volvox tertius* [Determination of the timing of the gonidial cleavage onset in *Volvox aureus* and *Volvox tertius*]. *Tsitologiya*, **27**, 227–9.

Desnitski, A. G. (1985b). Sutochniy ritm initsiatsii deleniy gonidiyev v yestestvennoy populatsii *Volvox* [Diurnal rhythm of gonidial cleavage initiation in a natural population of *Volvox*]. *Tsitologiya*, **27**, 1075–7.

Desnitski, A. G. (1985c). Vliyaniye streptomitsina na delenijya i rost kletok u trekh vidov *Volvox* [The effect of streptomycin on the divisions and growth of cells in three species of *Volvox*]. *Tsitologiya*, **27**, 921–7.

Desnitski, A. G. (1986). Vliyaniye aminopterina na kletochnye deleniya u *Volvox* [The effect of aminopterin on cell divisions in *Volvox*]. *Tsitologiya*, **28**, 545–51.

Desnitski, A. G. (1987). Vliyaniye aktinomitsina D na kletochnye deleniya u trekh vidov *Volvox* [The effect of actinomycin D on cell divisions in three species of *Volvox*]. *Tsitologiya*, **29**, 448–53.

Desnitski, A. G. (1990). Vliyaniye tsiklogeximida na kletochnye deleniya u *Volvox* [The effect of cycloheximide on cell divisions in *Volvox*]. *Bot. Zhurn.*, **75**, 181–6.

Desnitski, A. G. (1991). Mekhanizmy i evolutsionnye aspekty onotogeneza roda *Volvox* (Chlorophyta, Volvocales) [Mechanisms and evolutionary aspects of the ontogeny of the genus *Volvox* (Chlorophyta, Volvocales)]. *Bot. Zhurn.*, **76**, 657–67.

Desnitski, A. G. (1992). Cellular mechanisms of the evolution of ontogenesis in *Volvox*. *Arch. Protistenkd.*, **141**, 171–8.

Desnitski, A. G. (1993). On the origins and early evolution of multicellularity. *BioSystems*, **29**, 129–32

Desnitski, A. G. (1995a). A review on the evolution of development in *Volvox* – morphological and physiological aspects. *Eur. J. Protistol.*, **31**, 241–7.

Desnitski, A. G. (1995b). O skorosti kletochnych delenii v khode bespologo casvitiya u *Volvox globator* i *V. spermatosphaera* (Chlorophyta, Volvocales) [On the rate of cell divisions during asexual development in *Volvox globator* and *V. spermatosphaera* (Chlorophyta, Volvocales)]. *Bot. Zhurn.*, **80**, 40–2.

Desnitski, A. G. (1995c). On the origin of metazoa. *Zhurn. Obshchei Bio.*, **56**, 629–31.

Desnitski, A. G. (1996a). O geographicheskom rasprostranenii vidov roda *Volvox* (Chlorophyta, Volvocales) [The geographical distribution of the species of *Volvox* (Chlorophyta, Volvocales)]. *Bot. Zhurn.*, **81 (3)**, 28-33.

Desnitski, A. G. (1996b). Sravnitzlnyi analiz mechanizmov kletochnoy differentsiatsii v rode [Comparative analysis of cell differentiation mechanisms in the genus *Volvox* (Chlorophyta, Volvocales)]. *Bot. Zhurn.*, **81 (6)**, 1–9.

Detmers, P. A., Carboni, J. M. & Condeelis, J. (1985). Localization of actin in *Chlamydomonas* using antiactin and NBD-phallacidin. *Cell Motil.*, **5**, 415–30.

Devereux, R., Loeblich, A. R., III & Fox, G. E. (1990). Higher plant origins and the phylogeny of green algae. *J. Mol. Evol.*, **31**, 18–24.

Diener, K. R., Curry, A. M., Johnson, K. A., Williams, B. D., Lefebvre, P. A., Kindle, K. L. & Rosenbaum, J. L. (1990). Rescue of a paralyzed flagellar mutant of *Chlamydomonas* by transformation. *Proc. Natl. Acad. Sci. USA*, **87**, 5739–43.

Dietmaier, W. & Fabry, S. (1994). Analysis of introns in genes encoding small G proteins. *Curr. Genet.*, **26**, 497–505.

Dietmaier, W., Fabry, S., Huber, H. & Schmitt, R. (1995). Analysis of a family of *ypt* genes and their products from *Chlamydomonas reinhardtii*. *Gene*, **158**, 41–50.

Dobell, D. (1932). *Antony van Leeuwenhoek and His 'Little Animals': Being Some Account of the Father of Protozoology and Bacteriology and His Multifarious Discoveries in these Disciplines.* Staples, London (reprinted by Russell & Russell, New York, 1958).

Dolle, R., Pfau, J. & Nultsch, W. (1987). Role of calcium ions in motility and phototaxis of *Chlamydomonas reinhardtii*. *J. Plant Physiol.*, **126**, 467–73.

Dolzmann, R. & Dolzmann, P. (1964). Untersuchungen über die Feinstruktur und die Funktion der Plasmodesmen von *Volvox aureus* L. *Planta*, **61**, 332–46.

Domozych, D. S., Stewart, K. D. & Mattox, K. R. (1980). The comparative aspects of cell wall chemistry in the green algae (Chlorophyta). *J. Mol. Evol.*, **15**, 1–12.

Domozych, D. S., Stewart, K. D. & Mattox, K. R. (1981). *In vivo* cell wall ontogenesis in chlorophycean flagellates. 1. The cell wall, endomembrane system and interphase wall expansion. *Cytobios*, **32**, 147–65.

Donnan, L. & John, P. C. L. (1983). Cell cycle control by timer and sizer in *Chlamydomonas*. *Nature*, **304**, 630–3.

Doonan, J. H. & Grief, C. (1987). Microtubule cycle in *Chlamydomonas reinhardtii*. An immunofluorescence study. *Cell Motil. Cytoskel.*, **7**, 381–92.

Duncan, T. & Stuessy, T. F. (1984). *Cladistics: Perspectives on the Reconstruction of Evolutionary History.* Columbia University Press, New York.

Dutcher, S. K. (1986). Genetic properties of linkage group XIX in *Chlamydomonas reinhardtii*. In *Extrachromosomal Elements in Lower Eukaryotes*. Eds., R. B. Wickner, A. Hinnebusch, A. M. Lambowitz, I. C. Gunsalus & A. Hollaender. Plenum, New York, pp 303–21.

Dykstra, M. J. & Olive, L. S. (1975). *Sorodiplophrys*: an unusual sorocarp-producing protist. *Mycologia*, **67**, 873–8.

Ehler, L. L., Holmes, J. A. & Dutcher, S. K. (1995). Loss of spatial control of the mitotic spindle apparatus in a *Chlamydomonas reinhardtii* mutant strain lacking basal bodies. *Genetics*, **141**, 945–60.

Ehrenberg, C. G. (1831). *Organisation, Systematik und geographische Verbreitung der Infusionsthierchen.* Druckerei der kniglichen Akadamie der Wissenschaften, Berlin.

Ehrenberg, C. G. (1832a). Zur Erkenntniss der Organisation in der Richtung des kleinsten Raumes. *Abhandl. Kgl. Akad. Wiss. Berlin*, **1831**, 1–154.

Ehrenberg, C. G. (1832b). Ueber das Entstehen des Organischen aus einfacher sichtbarer Materie, und über die organischen Molecüle und Atomen insbesondere, als Erfahrungsgegenstände. *Ann. Physik Chemie*, **24**, 1–48.

Ehrenberg, C. G. (1838). *Die Infusionsthierchen als vollkommene Organismen.* L. Voss, Berlin.

Ehrlich, M. & Zhang, X.-Y. (1990). Naturally occurring modified nucleosides in DNA. In *Chromatography and Modification of Nucleosides*. Eds., C. W. Gehrke & K. C. T. Kuo. Elsevier, Amsterdam, pp. B327–62.

Ely, T. H. & Darden, W. H. (1972). Concentration and purification of the male-inducing substance from *Volvox aureus* M5. *Microbios*, **5**, 51–6.

Eppley, R. W., Holm-Hansen, O. & Strickland, J. D. (1968). Some observations on the vertical migrations of dinoflagellates. *J. Phycol.*, **4**, 333–40.

Errington, J. (1993). Sporulation in *Bacillus subtilis*: regulation of gene expression and control of morphogenesis. *Micobiol. Rev.*, **57**, 1–33.

Ertl, H., Hallmann, A., Wenzl, S. & Sumper, M. (1992). A novel extensin that may organize extracellular matrix biogenesis in *Volvox carteri*. *EMBO J.*, **11**, 2055–62.

Ertl, H., Mengele, R., Wenzl, S., Engel, J. & Sumper, M. (1989). The extracellular matrix of *Volvox carteri*: Molecular structure of the cellular compartment. *J. Cell Biol.*, **109**, 3493–501.

Ettl, H. (1976). Die Gattung *Chlamydomonas* Ehrenberg. *Beih. Nova Hedwigia*, **49**, 1–1122.

Ettl, H. & Schlösser, U. G. (1992). Towards a revision of the systematics of the genus *Chlamydomonas* (Chlorophyta). I. *Chlamydomonas applanata* Pringsheim. *Bot. Acta*, **105**, 323–30.

Fabry, S., Jacobsen, A., Huber, H., Palme, K. & Schmitt, R. (1993). Structure, expression, and phylogenetic relationships of a family of *ypt* genes encoding small G-proteins in the green alga *Volvox carteri*. *Curr. Genet.*, **24**, 229–40.

Fabry, S., Müller, K., Lindauer, A., Ford, C., Cornelius, T. & Schmitt, R. (1995a). The organization, structure and controlling elements of *Chlamydomonas* histone genes reveal features linking plant and animal genes. *Curr. Genet.*, **28**, 333–45.

Fabry, S., Nass, N., Huber, H., Palme, K., Jaenicke, L. & Schmitt, R. (1992). The *yptV1* gene encodes a small G-protein in the green alga *Volvox carteri*: gene structure and properties of the gene product. *Gene*, **118**, 153–62.

Fabry, S., Steigerwald, R., Bernklau, C., Dietmaier, W. & Schmitt, R. (1995b). Structure-function analysis of *Volvox* and *Chlamydomonas* small G proteins by complementation of *Saccharomyces cerevisiae* YPT/SEC mutations. *Mol. Gen. Genet.*, **247**, 265–74.

Farley, J. (1974). *The Spontaneous Generation Controversy from Descartes to Oparin*. Johns Hopkins University Press.

Feldwisch, O., Lammertz, M., Hartmann, E., Feldwisch, J., Palme, K., Jastoroff, B. & Jaenicke, L. (1995). Purification and characterization of a cAMP-binding protein of *Volvox carteri* f. *nagariensis* Iyengar. *Eur. J. Biochem.*, **228**, 480–9.

Fernández, E., Schnell, R., Ranum, L. P. W., Hussey, S. C., Silflow, C. D. & Lefebvre, P. A. (1989). Isolation and characterization of the nitrate reductase structural gene of *Chlamydomonas reinhardtii*. *Proc. Natl. Acad. Sci. USA*, **86**, 6449–53.

Ferris, P. J. & Goodenough, U. W. (1994). The mating-type locus of Chlamydomonas reinhardtii contains highly rearranged DNA sequences. *Cell*, **76**, 1135–45.

Ferris, P. J., Woessner, J. P. & Goodenough, U. W. (1996). A sex recognition glycoprotein is encoded by the *plus* mating-type gene *fus1* of *Chlamydomonas reinhardtii*. *Mol. Biol. Cell*, **7**, 1235–48.

Finkelstein, R. & Boncinelli, E. (1994). From fly head to mammalian forebrain: the story of *otd* and *Otx*. *Trends Genet.*, **10**, 310–15.

Floyd, G. L. (1978). Mitosis and cytokinesis in *Asteromonas gracilis*, a wall-less green monad. *J. Phycol.*, **14**, 440–5.

Floyd, G. L., Hoops, H. J. & Swanson, J. A. (1980). Fine structure of the zoospore of *Ulothrix belkae* with emphasis on the flagellar apparatus. *Protoplasma*, **104**, 17–31.

Foster, K. W., Saranak, J., Patel, N., Zarilli, G., Okabe, M., Kline, T. & Nakanishi, K. (1984). A rhodopsin is the functional photoreceptor for phototaxis in the unicellular eukaryote *Chlamydomonas*. *Nature*, **311**, 756–9.

Foster, K. W. & Smyth, R. D. (1980). Light antennas in phototactic algae. *Microbiol. Rev.*, **44**, 572–630.

Fott, B. (1949). *Corone*, a new genus of colonial Volvocales. *Mem. Soc. Roy. Letter. Scienc. de Boheme.*, **CL. Scienc.**, 1–9 (Not seen; cited in Starr, 1980).

Frederick, L. (1990). Phylum plasmodial slime molds. Class Myxomycota. In *Handbook of Protoctista*. Ed., L. Margulis, J. O. Corliss, M. Melkonian & D. J. Chapman. Jones & Bartlett, Boston, pp. 467–83.

Frempong, E. (1984). A seasonal sequence of diel distribution patterns for the planktonic flagellate *Ceratium hirundinella* in an eutrophic lake. *Freshwater Biol.*, **14**, 401–22.

Fritsch, F. E. (1914). Some freshwater algae from Madagascar. *Ann. Biol. Lacustre*, **7**, 40–59.

Fritsch, F. E. (1935). *The Structure and Reproduction of the Algae*. Cambridge University Press.

Fritsch, F. E. (1945). *The Structure and Reproduction of the Algae*, Vol. 1. Cambridge University Press.

Fuhs, G. W. (1969). Phosphorus content and rate of growth in the diatoms *Cyclotella nana* and *Thalassioira fluviatilis*. *J. Phycol.*, **5**, 305–21.

Fuhs, G. W. & Canelli, E. (1970). Phosphorus-33 autoradiography used to measure phosphate uptake by individual algae. *Limnol. Oceanogr.*, **15**, 962–6.

Fuhs, G. W., Demmerle, S. D., Canelli, E. & Chen, M. (1972). Characterization of phosphorus-limited algae (with reflections on the limiting-nutrient concept). In *Nutrients and Eutrophication: The Limiting Nutrient Controversy*. Ed., G. E. Likens. Allen Press, Lawrence, KS, pp. 113–32.

Fulton, A. B. (1978a). Colony development in *Pandorina morum*. I. Structure of the extracellular matrix. *Dev. Biol.*, **64**, 224–35.

Fulton, A. B. (1978b). Colony development in *Pandorina morum*. II. Colony morphogenesis and formation of the extracellular matrix. *Dev. Biol.*, **64**, 236–51.

Gaffal, K. P. (1977). The relationship between basal bodies and the motility of *Polytoma papillatum* flagella. *Experientia*, **33**, 1372–4.

Gaffal, K. P. (1988). The basal body–root complex of *Chlamydomonas reinhardtii* during mitosis. *Protoplasma*, **143**, 118–29.

Gaffal, K. P., Arnold, C.-G., Friedrichs, G. J. & Gemple, W. (1995). Morphodynamical changes of the chloroplast of *Chlamydomonas reinhardtii* during the 1st round of division. *Arch. Protistenkd.*, **145**, 10–23.

Gaffal, K. P. & el-Gammal, S. (1990). Elucidation of the enigma of the "metaphase band" of *Chlamydomonas reinhardtii*. *Protoplasma*, **156**, 139–48.

Gaffal, K. P., el-Gammal, S. & Friedrichs, G. J. (1993). Computer-aided 3D-reconstruction of the eyespot-flagellar/basal apparatus-contractile vacuoles-nucleus-associations during mitosis of *Chlamydomonas reinhardtii. Endocyto. Cell Res.,* **9,** 177–208.

Galaván, A., Cárdenas, J. & Fernández, E. (1992). Nitrate reductase regulates expression of nitrite uptake, and nitrite reductase activities in *Chlamydomonas reinhardtii. Plant Physiol.,* **98,** 422–6.

Ganf, G. G., Shiel, R. J. & Merick, C. J. (1983). Parasitism: the possible cause of the collapse of a *Volvox* population in Mount Bold Reservoir, South Australia. *Aust. J. Mar. Freshw. Res.,* **34,** 489–94.

Gerisch, G. (1959). Die Zellendifferenzierung bei *Pleodorina californica* Shaw und die Organization der Phytomonadinenkolnien. *Arch. Protistenkd.,* **104,** 292–358.

Gerschenson, L. E. & Rotello, R. J. (1992). Apoptosis: a different kind of cell death. *FASEB J.,* **6,** 2450–5.

Gilbert, S. F (1994). *Developmental Biology,* 4th ed. Sinauer, Sunderland, MA.

Gilbert, W., Marchionni, M. & McKnight, G. (1986). On the antiquity of introns. *Cell,* **46,** 151–4.

Gilles, R., Balshüsemann, D. & Jaenicke, L. (1987). Molecular signals during sexual induction of *Volvox carteri* f. *nagariensis.* In *Algal Development (Molecular and Cellular Aspects).* Eds., W. Wiessner, D. G. Robinson & R. C. Starr. Springer-Verlag, Berlin, pp. 50–7.

Gilles, R., Bittner, C., Cramer, M., Mierau, R. & Jaenicke, L. (1980). Radioimmunoassay for the sex inducer of *Volvox carteri* f. *nagariensis. FEBS Lett.,* **116,** 102–6.

Gilles, R., Bittner, C. & Jaenicke, L. (1981). Site and time of formation of the sex-inducing glycoprotein in *Volvox carteri. FEBS Lett.,* **124,** 57–61.

Gilles, R., Gilles, C. & Jaenicke, L. (1983). Sexual differentiation of the green alga *Volvox carteri.* Involvement of extracellular phosphorylated proteins. *Naturwissenschaften,* **70,** 571–2.

Gilles, R., Gilles, C. & Jaenicke, L. (1984). Pheromone-binding and matrix-mediated events in sexual induction of *Volvox carteri. Z. Naturforsch.,* **39c,** 584–92.

Gilles, R. & Jaenicke, L. (1982). Differentiation in *Volvox carteri:* study of pattern variation of reproductive cells. *Z. Naturforsch.,* **37c,** 1023–30.

Gilles, R., Moka, R., Gilles, C. & Jaenicke, L. (1985a). Cyclic AMP as an interspheroidal differentiation signal in *Volvox carteri. FEBS Lett.,* **184,** 309–12.

Gilles, R., Moka, R. & Jaenicke, L. (1985b). The extracellular matrix plays a functional role in sexual differentiation of *Volvox carteri* (abstract). *Hoppe Seyler Z. Physiol. Chem.,* **366,** 793–4.

Gilmartin, P. M., Sarokin, L., Memlink, J. & Chua, N.-H. (1990). Molecular light switches for nuclear genes. *Plant Cell,* **2,** 369–78.

Godl, K., Hallmann, A., Rappel, A. & Sumper, M. (1995). Pherophorins: a family of extracellular matrix glycoproteins from *Volvox* structurally related to the sex-inducing pheromone. *Planta,* **196,** 781–7.

Goldschmidt-Clermont, M. & Rahire, M. (1986). Sequence, evolution and differential expression of the two genes encoding variant small subunits of ribulose bisphosphate carboxylase/oxygenase in *Chlamydomonas reinhardtii. J. Mol. Biol.,* **191,** 421–32.

Goldstein, M. (1964). Speciation and mating behavior in *Eudorina. J. Protozool.,* **11,** 317–44.

Goldstein, M. (1967). Colony differentiation in *Eudorina. Can. J. Bot.,* **45,** 1591–6.

Goodenough, U. W. & Heuser, J. E. (1985). The *Chlamydomonas* cell wall and its constituent glycoproteins analyzed by the quick-freeze deep-etch technique. *J. Cell Biol.,* **101,** 1550–68.

Goodenough, U. W. & Heuser, J. E. (1988). Molecular organization of cell-wall crystals from *Chlamydomonas reinhardtii* and *Volvox carteri. J. Cell Sci.,* **90,** 717–33.

Goodenough, U. W. & St. Clair, H. S. (1975). bald-2: a mutation affecting the formation of doublet and triplet sets of microtubules in *Chlamydomonas reinhardtii. J. Cell Biol.,* **66,** 480–91.

Goodenough, U. W. & Weiss, R. L. (1975). Gametic differentiation in *Chlamydomonas reinhardtii.* III. Cell wall lysis and microfilament-associated mating structure activation in wild-type and mutant strains. *J. Cell Biol.,* **67,** 623–37.

Goroshankin, J. (1875). Gensis im Typus der palmellanartigen Algen. Versuch einer vergleichenden Morpholgie der Volvocineae. *Mitt. Kaiserl. Ges. naturf. Freunde in Moskau,* **16,** 145–59.

Gottlieb, B. & Goldstein, M. E. (1977). Colony development in *Eudorina elegans* (Chlorophyta, Volvocales). *J. Phycol.,* **13,** 358–64.

Gottlieb, B. & Goldstein, M. E. (1979). Colchicine-induced alterations in colony development in *Eudorina elegans* (Volvocales, Chlorophyta). *J. Phycol.,* **15,** 260–5.

Gould, R. R. (1975). The basal bodies of *Chlamydomonas reinhardtii.* Formation from probasal bodies, isolation, and partial characterization. *J. Cell Biol.,* **65,** 65–74.

Gould, S. J. (1977). *Ontogeny and Phylogeny.* Harvard University Press.

Gould, S. J. & Eldredge, N. (1977). Punctuated equilibria: Tempo and mode of evolution reconsidered. *Paleobiology,* **3,** 115–51.

Green, K. J. (1982). Cleavage and the formation of a cytoplasmic bridge system in *Volvox carteri.* Ph.D. thesis, Washington University.

Green, K. J. & Kirk, D. L. (1981). Cleavage patterns, cell lineages, and development of a cytoplasmic bridge system in *Volvox* embryos. *J. Cell Biol.,* **91,** 743–55.

Green, K. J. & Kirk, D. L. (1982). A revision of the cell lineages recently reported for *Volvox carteri* embryos. *J. Cell Biol.,* **94,** 741–2.

Green, K. J., Viamontes, G. I. & Kirk, D. L. (1981). Mechanism of formation, ultrastructure and function of the cytoplasmic bridge system during morphogenesis in *Volvox. J. Cell Biol.,* **91,** 756–69.

Grell, K. G. (1973). *Protozoology.* Springer-Verlag, Berlin.

Greuel, B. T. & Floyd, G. L. (1985). Development of the flagellar apparatus and flagellar orientation in the colonial green alga *Gonium pectorale* (Volvocales). *J. Phycol.,* **21,** 358–71.

Gruber, H., Goetinck, S. D., Kirk, D. L. & Schmitt, R. (1992). The nitrate reductase-encoding gene of *Volvox carteri*: map location, sequence and induction kinetics. *Gene,* **120,** 75–83.

Gruber, H., Kirzinger, S. H. & Schmitt, R. (1996). Expression of the *Volvox* gene encoding nitrate reductase: Mutation-dependent activation of cryptic splice sites and intron enhanced gene expression from a cDNA. *Plant Mol. Biol.,* **31,** 1–12.

Gruber, H. E. & Rosario, B. (1974). Variation in eyespot ultrastructure in *Chlamydomonas reinhardtii* (ac-31). *J. Cell Sci.,* **15,** 481–94.

Guerrero, M. G., Vega, J. M. & Losada, M. (1981). The assimilatory nitrate-reducing system and its regulation. *Ann. Rev. Plant Physiol.,* **32,** 169–204.

Günther, R., Bause, E. & Jaenicke, L. (1987). UDP-l-arabinose-hydroxyproline-*O*-glycosyltransferases in *Volvox carteri. FEBS Lett.*, **221**, 293–8.

Haas, E. & Sumper, M. (1991). The sexual inducer of *Volvox carteri.* Its large-scale production and secretion by *Saccharomyces cerevisiae. FEBS Lett.*, **294**, 282–84.

Haeckel, E. (1874). The Gastrea-theory, the phylogenetic classification of the animal kingdom and the homology of the germ-lamellae. *Q. J. Microsc. Sci. (new series)*, **14**, 142–65, 222–47.

Hagen, G. & Kochert, G. (1980). Protein synthesis in a new system for the study of senescence. *Exp. Cell Res.*, **127**, 451–7.

Hails, T., Jobling, M. & Day, A. (1993). Large arrays of tandemly repeated DNA sequences in the green alga *Chlamydomonas reinhardtii. Chromosoma*, **102**, 500–7.

Halder, G., Callaerts, P. & Gehring, W. (1995). Induction of ectopic eyes by targeted expression of the *eyeless* gene in *Drosophila. Science*, **267**, 1788–92.

Hallmann, A. & Sumper, M. (1994a). An inducible arylsulfatase of *Volvox carteri* with properties suitable for a reporter-gene system. Purification, characterization and molecular cloning. *Eur. J. Biochem.*, **221**, 143–50.

Hallmann, A., Rappel, A. & Sumper, M. (in press). Gene replacement by homologous recombination in the multicellular green alga *Volvox carteri. Proc. Natl. Acad. Sci. USA.*

Hallmann, A. & Sumper, M. (1994b). Reporter genes and highly regulated promoters as tools for transformation experiments in *Volvox carteri. Proc. Natl. Acad. Sci. USA*, **91**, 11562–6.

Hallmann, A. & Sumper, M. (1996). A transgenic *Volvox* expressing the *Chlorella* hexose/H⁺ symporter gene is highly useful for biochemical work. *Proc. Natl. Acad. Sci. USA.*, **93**, 669–73.

Hamada, H., Petrino, M. G. & Kakunaga, T. (1982). A novel repeated element with Z-DNA-forming potential is widely found in evolutionarily diverse eukaryotic genomes. *Proc. Natl. Acad. Sci. USA*, **79**, 6465–9.

Hamburger, C. (1905). Zur Kenntnis der *Dunaliella salina* und einer Amöbe aus Salinenwasser von Cagliari. *Arch. Protistenkd.*, **6**, 111–30.

Hand, W. B. & Haupt, W. (1971). Flagellar activity of the colony members of *Volvox aureus* during light stimulation. *J. Protozool.*, **18**, 361–4.

Happey-Wood, C. M. (1988). Ecology of freshwater planktonic green algae. In *Growth and Reproductive Strategies of Freshwater Phytoplankton.* Ed., C. D. Sandgren. Cambridge University Press, pp. 175–226.

Harold, F. M. (1966). Inorganic polyphosphates in biology: structure, metabolism and function. *Bacter. Rev.*, **30**, 772–93.

Harper, J. D. I. & John, P. C. L. (1986). Coordination of division events in the *Chlamydomonas* cell cycle. *Protoplasma*, **131**, 118–30.

Harper, J. F., Huson, K. S. & Kirk, D. L. (1987). Use of repetitive sequences to identify DNA polymorphisms linked to *regA*, a developmentally important locus in *Volvox. Genes Develop.*, **1**, 573–84.

Harper, J. F. & Mages, W. (1988). Organization and structure of *Volvox* β-tubulin genes. *Mol. Gen. Genet.*, **213**, 315–24.

Harper, R. A. (1912). The structure and development of the colony in Gonium. *Trans. Amer. Microsc. Soc.*, **31**, 65–83.

Harper, R. A. (1918). Binary fission and surface tension in the development of the colony in Volvox. *Brooklyn Bot. Gar. Mem.,* **1**, 154–66.

Harris, D. O. (1971a). Growth inhibitors produced by the green algae (Volvocaceae). *Arch. Mikrobiol.,* **76**, 47–50.

Harris, D. O. (1971b). A model system for the study of algae growth inhibitors. *Arch. Protistenkd.,* **113**, 230–4.

Harris, D. O. & James, D. E. (1974). Toxic algae. *Carolina Tips,* **37**, 13–15.

Harris, D. O. & Starr, R. C. (1969). Life history and physiology of reproduction of *Platydorina caudata* Kofoid. *Arch. Protistenkd.,* **111**, 138–55.

Harris, E. H. (1989). *The Chlamydomonas Sourcebook. A Comprehensive Guide to Biology and Laboratory Use.* Academic Press, San Diego.

Hartmann, M. (1924). Ueber die Veränderung der Koloniebildung von *Eudorina elegans* und *Gonium pectorale* unter dem Eifluss aüsserer Bedingungen. IV Mitt. der Untersuchungen über die Morphologie und Physiologie des Formwechsels der Phytomonadinen (Volvocales). *Arch. Protistenkd.,* **49**, 375–95.

Harz, H. & Hegemann, P. (1991). Rhodopsin-regulated calcium currents in *Chlamydomonas. Nature,* **351**, 489–91.

Hasle, G. R. (1950). Phototactic vertical migration in marine dinoflagellates. *Oikos,* **2**, 162–75.

Hebeler, M., Heintrich, S., Mayer, A., Leibfritz, D. & Grimme, L. H. (1992). Phosphate regulation and compartmentation in *Chlamydomonas reinhardtii* studied by in vivo ^{31}P-NMR. In *Research in Photosynthesis,* Vol. 3. Ed., N. Murata. Kluwer Academic, Dordrecht, pp. 717–20.

Heeg, J. & Rayner, N. A. (1988). Inter- and intra-specific associations as some possible predator avoidance and energy conservation strategies in planktonic rotifers. *J. Limnol. Soc. South Afr.,* **14**, 87–92.

Heimann, K., Reize, I. B. & Melkonian, M. (1989). The flagellar developmental cycle in algae: flagellar transformation in *Cyanophora paradoxa* (Glaucocystophyceae). *Protoplasma,* **148**, 106–10.

Henneguy, M. (1879a). Germination of the spore of *Volvox dioicus. Ann. Nat. Hist.,* **3**, 93.

Henneguy, M. (1879b). Sur le reproduction du *Volvox dioîque. C. R. Acad. Sci. Paris,* **83**, 287–9.

Hino, A. & Hirano, R. (1980). Relationship between body size of the rotifer *Brachionus plicatis* and the maximum size of particles ingested. *Bull. Jpn. Soc. Fish.,* **46**, 1217–22.

Hobbs, M. J. (1971). The fine structure of *Eudorina illinoisensis* (Kofoid) Pascher. *Br. Phycol. Bull.,* **6**, 81–103.

Hoek, C. van den, Stam, W. T. & Olsen, J. L. (1988). The emergence of a new chlorophytan system, and Dr. Kormann's contribution thereto. *Helgoländer Meeresunters,* **42**, 339–83.

Hofstra, J. J. (1977). Chlorate toxicity and nitrate reductase activity in tomato plants. *Plant Physiol.,* **41**, 65–9.

Holland, E.-M., Braun, F.-J., Nonnengässer, C., Harz, H. & Hegemann, P. (1996). The nature of rhodopsin-triggered photocurrents in *Chlamydomonas.* I. Kinetics and influence of divalent ions. *Biophys. J.,* **70**, 924–31.

Holmes, J. A. & Dutcher, S. K. (1989). Cellular asymmetry in *Chlamydomonas reinhardtii. J. Cell Sci.,* **94**, 273–86.

Holmes, J. A. & Dutcher, S. K. (1992). Genetic approaches to the study of cytoskeletal structure and function in *Chlamydomonas*. In *Cytoskeletons of the Algae*. Ed., D. Menzel. CRC Press, Cleveland, pp. 347–67.

Holmes, S. J. (1903). Phototaxis in *Volvox*. *Biol. Bull.,* **4,** 319–26.

Holst, O., Christoffel, V., Fründ, R., Moll, H. & Sumper, M. (1989). A phosphodiester bridge between two arabinose residues as a structural element of an extracellular glycoprotein of *Volvox carteri*. *Eur. J. Biochem.,* **181,** 345–50.

Hoops, H. J., III (1981). Ultrastructural studies of selected colonial volvocalean algae. Ph.D. thesis, Ohio State University.

Hoops, H. J. (1984). Somatic cell flagellar apparatuses in two species of *Volvox* (Chlorophyceae). *J. Phycol.,* **20,** 20–7.

Hoops, H. J. (1993). Flagellar, cellular and organismal polarity in *Volvox carteri*. *J. Cell Sci.,* **104,** 105–17.

Hoops, H. J. & Floyd, G. L. (1982a). Ultrastructure and taxonomic position of the rare volvocalean alga, *Chlorocorona bohemica*. *J. Phycol.,* **18,** 462–6.

Hoops, H. J. & Floyd, G. L. (1982b). Ultrastructure of the flagellar apparatus of *Pyrobotrys* (Chlorophyceae). *J. Phycol.,* **18,** 455–62.

Hoops, H. J. & Floyd, G. L. (1983). Ultrastructure and development of the flagellar apparatus and flagellar motion in the colonial green algae *Astrephomene gubernaculifera*. *J. Cell Sci.,* **63,** 21–41.

Hoops, H. J., Floyd, G. L. & Swanson, J. A. (1982). Ultrastructure of the biflagellate motile cells of *Ulvaria oxysperma* (Kütz.) Bliding and phylogenetic relationships among ulvaphycean algae. *Am. J. Bot.,* **69,** 150–9.

Hoops, H. J., Long, J. L. & Hilde, E. S. (1994). Flagellar apparatus structure is similar but not identical in *Volvulina steinii, Eudorina elegans,* and *Pleodorina illinoisensis* (Chlorophyta): implications for the "volvocine evolutionary lineage" (abstract). *J. Phycol. (suppl. 3)* **7,** 7.

Hoops, H. J. & Witman, G. B. (1983). Outer doublet heterogeneity reveals structural polarity related to beat direction in *Chlamydomonas* flagella. *J. Cell Biol.,* **97,** 902–8.

Hoops, H. J. & Witman, G. B. (1985). Basal bodies and associated structures are not required for normal flagellar motion or phototaxis in the green alga *Chlorogonium elongatum*. *J. Cell Biol.,* **100,** 297–309.

Hoops, H. J., Wright, R. L., Jarvik, J. W. & Witman, G. B. (1984). Flagellar waveform and rotational orientation in a *Chlamydomonas* mutant lacking normal striated fibers. *J. Cell Biol.,* **98,** 818–24.

Hori, H., Lim, B. K. & Osawa, S. (1985). Evolution of green plants as deduced from 5S rRNA sequences. *Proc. Natl. Acad. Sci. USA,* **82,** 820–3.

Horst, C. J. & Witman, G. B. (1993). *ptx1*, a nonphototactic mutant of *Chlamydomonas*, lacks control of flagellar dominance. *J. Cell Biol.,* **120,** 733–41.

Howell, S. H. & Naliboff, J. A. (1973). Conditional mutants in *Chlamydomonas reinhardtii* blocked in the vegetative cell cycle. I. An analysis of cell cycle block points. *J. Cell Biol.,* **57,** 760–72.

Howell, S. H. & Walker, L. L. (1976). Informational complexity of the nuclear and chloroplast genomes of *Chlamydomonas reinhardi*. *Biochim. Biophys. Acta,* **418,** 249–56.

Huang, B., Piperno, G., Ramanis, Z. & Luck, D. J. L. (1981). Radial spokes of *Chlamydomonas* flagella: Genetic analysis of assembly and function. *J. Cell Biol.,* **88,** 80–8.

Huang, B., Ramanis, Z., Dutcher, S. K. & Luck, D. J. L. (1982). Uniflagellar mutants of Chlamydomonas: evidence for the role of basal bodies in transmission of positional information. *Cell, 29,* 745–53.

Huang, B., Watterson, D. W., Lee, V. D. & Schibler, M. J. (1988). Purification and characterization of a basal body-associated Ca^{2+}-binding protein. *J. Cell Biol., 107,* 121–31.

Huber, H., Beyser, K. & Fabry, S. (1996). Small G proteins of two algae are localized to exocytic compartments and to flagella. *Plant Mol. Biol, 31,* 279–93.

Huber, O. & Sumper, M. (1994). Algal-CAMs: isoforms of a cell adhesion molecule in embryos of the alga *Volvox* with homology to *Drosophila* fasciclin I. *EMBO J., 13,* 4212–22.

Huber-Pestalozzi, G. (1961). Das Phytoplankton des Süsswassers. Systematik und Biologie. Teil 5: Ordnung Volvocales. In *Die Binnengewässer,* Vol. 16. Ed., A. Thienemann. Schweizerbart'sche Verlagsbuchhandlung, Stuttgart, pp. 1–744.

Huskey, R. J. (1979). Mutants affecting vegetative cell orientation in *Volvox carteri. Dev. Biol., 72,* 236–43.

Huskey, R. J. & Griffin, B. E. (1979). Genetic control of somatic cell differentiation in *Volvox.* Analysis of somatic regenerator mutants. *Dev. Biol., 72,* 226–35.

Huskey, R. J., Griffin, B. E., Cecil, P. O. & Callahan, A. M. (1979a). A preliminary genetic investigation of *Volvox carteri. Genetics, 91,* 229–44.

Huskey, R. J., Semenkovich, C. F., Griffin, B. E., Cecil, P. O., Callahan, A. M., Chace, K. V. & Kirk, D. L. (1979b). Mutants of *Volvox carteri* affecting nitrogen assimilation. *Mol. Gen. Genet., 169,* 157–61.

Hutchinson, G. E. (1961). The paradox of the plankton. *Amer. Nat., 95,* 137–45.

Hutchinson, G. E. (1967). *A treatise on Limnology. Vol. II. Introduction to Lake Biology and the Limnoplankton.* Wiley, New York.

Huth, K. (1970). Bewegung und Orientierung bei *Volvox aureus* Ehrb. I. Mechanismus der phototaktischen Reaktion. *Z. Pflanzenphysiol., 62,* 436–50.

Hutt, W. & Kochert, G. (1971). Effects of some protein and nucleic acid synthesis inhibitors on fertilization in *Volvox carteri. J. Phycol., 7,* 316–20.

Huxley, J. S. (1912). *The Individual in the Animal Kingdom.* Cambridge University Press.

Hyams, J. & Davies, D. R. (1972). The induction and characterization of cell wall mutants of *Chlamydomonas reinhardi. Mutat. Res., 14,* 381–9.

Hyams, J. S. & Borisy, G. G. (1978). Isolated flagellar apparatus of *Chlamydomonas*: characterization of forward swimming and alternation of waveforms and reversal of motion by calcium ions *in vitro. J. Cell Sci., 33,* 235–53.

Hyman, L. (1940). *The Invertebrates: Protozoa through Ctenophora.* McGraw-Hill, New York.

Hyman, L. H. (1942). The transition from the unicellular to the multicellular individual. In *Levels of Integration in Biological and Social Systems.* Ed., R. Redfield. Jacques Cattell Press, Lancaster, PA, pp. 27–42.

Ikemura, T. (1985). Codon usage and tRNA content in unicellular and multicellular organisms. *Mol. Biol. Evol., 2,* 13–34.

Ikushima, N. & Maruyama, S. (1968). Protoplasmic connections in *Volvox. J. Protozool., 15,* 136–40.

Imbault, P., Wittemer, C., Johanningmeier, U., Jacobs, J. D. & Howell, S. H. (1988). Structure of the *Chlamydomonas reinhardtii cabII*-1 gene encoding a chlorophyll a/b binding protein. *Gene, 73,* 397–407.

Ireland, G. W. & Hawkins, S. E. (1980). A method for studying the effect of inhibitors on the development of the isolated gonidia of *Volvox tertius*. *Microbios*, **28**, 185–201.

Ireland, G. W. & Hawkins, S. E. (1981). Inversion effects in *Volvox tertius*: the effects of Con A. *J. Cell Sci.*, **48**, 355–66.

Irénée-Marie, F. (1938). *Flore Desmidale de la Region de Montreal*. Laprairie, Canada (Not seen; cited in Schlichting, 1960).

Irvine, D. E. G. & John, D. M. (Eds.) (1984). *Systematics of the Green Algae*. Academic Press, London.

Iyengar, M. O. P. (1920). Observations on the Volvocaceae of Madras. *J. Ind. Bot. Soc.*, **1**, 330–6.

Iyengar, M. O. P. (1933). Contribution to our knowledge of the colonial Volvocales of South India. *J. Linn. Soc. Bot.*, **49**, 323–73.

Iyengar, M. O. P. & Desikachary, T. V. (1981). *Volvocales*. Indian Council of Agricultural Research, New Delhi.

Iyengar, M. O. P. & Ramanathan, K. R. (1951). On the structure and reproduction of *Pleodorina sphaerica* Iyengar. *Phytomorphology*, **1**, 215–24.

Jacobshagen, S. & Johnson, C. H. (1994). Circadian rhythms of gene expression in *Chlamydomonas reinhardtii*: circadian cycling of mRNA abundance of cabII, and possibly β-tubulin and cytochrome c. *Eur. J. Cell Biol.*, **64**, 142–52.

Jaenicke, L. (1979). Induction of sexuality in *Volvox carteri* f. *nagariensis*: A variation on the theme ''receptor/ligand interactions.'' In *Molecular Mechanisms of Biological Recognition*. Ed., B. Balaban. Elsevier, New York, pp. 413–18.

Jaenicke, L. (1982). *Volvox* biochemistry comes of age. *Trends Biochem. Sci.*, **7**, 61–4.

Jaenicke, L. (1991). Development: signals in the development of cryptograms. *Prog. Bot.*, **52**, 138–89.

Jaenicke, L., Feldwisch, O., Merkl, B., Cremer, A. & Haas, I. (1993). Expression of highly active sex-inducing pheromone of *Volvox carteri* f. *nagariensis* in a mammalian cell system. *FEBS Lett.*, **316**, 257–60.

Jaenicke, L. & Gilles, R. (1982). Differentiation and embryogenesis in *Volvox carteri*. In *Biochemistry of Differentiation and Morphogenesis*. Ed., L. Jaenicke. Springer-Verlag, Berlin, pp. 288–94.

Jaenicke, L. & Gilles, R. (1985). Germ-cell differentiation in *Volvox carteri*. *Differentiation*, **29**, 199–206.

Jaenicke, L., Kuhne, W., Spessart, R., Wahle, U. & Waffenschmidt, S. (1987). Cell-wall lytic enzymes (autolysins) of *Chlamydomonas reinhardtii* are (hydroxy)proline specific proteases. *Eur. J. Biochem.*, **170**, 485–92.

Jaenicke, L., van Leyen, K. & Siegmund, H.-U. (1991). Dolichyl phosphate-dependent glycosyltransferases utilize truncated cofactors. *Biol. Chem. Hoppe-Seyler*, **372**, 1021–6.

Jaenicke, L. & Waffenschmidt, S. (1979). Matrix-lysis and release of daughter spheroids in *Volvox carteri* – a proteolytic process. *FEBS Lett.*, **107**, 250–3.

Jaenicke, L. & Waffenschmidt, S. (1981). Liberation of reproductive units in *Volvox* and *Chlamydomonas*: proteolytic processes. *Ber. Deutsch. Bot. Ges.*, **94**, 375–86.

Janet, C. (1912). *Le Volvox*. Librairie Ducourtieux et Gout, Limoges.

Janet, C. (1914). *Note préliminaire sur l'oeuf du Volvox globator*. Ducourtieux et Gout, Limoges.

Janet, C. (1922). *Le Volvox. Deuxième Mémoire.* Paris.

Janet, C. (1923a). *Le Volvox. Troisième Mémoire.* Protat Frères, Macon.

Janet, C. (1923b). Sur l'ontogenese du *Volvox aureus* Ehr. *C. R. Acad. Sci. Paris,* **176,** 997–9.

Januszko, M. (1971). The effect of fertilizers on phytoplankton development in fry ponds. *Pol. Arch. Hydrobiol.,* **18,** 129–45.

Jarvik, J. W. & Suhan, J. P. (1991). The role of the flagellar transition region: inferences from the analysis of a *Chlamydomonas* mutant with defective transition region structures. *J. Cell Sci.,* **99,** 731–40.

John, P. C. L. (1987). Control points in the *Chlamydomonas* cell cycle. In *Algal Development (Molecular and Cellular Aspects).* Ed., W. Wiessner, D. G. Robinson & R. C. Starr. Springer-Verlag, Berlin, pp. 9–16.

Johnson, K. A. (1995). Keeping the beat: form meets function in the *Chlamydomonas* flagellum. *BioEssays,* **17,** 847–54.

Johnson, K. A. & Rosenbaum, J. L. (1992). Replication of basal bodies and centrioles. *Curr. Biol.,* **4,** 80–5.

Johnson, U. G. & Porter, K. R. (1968). Fine structure of cell division in *Chlamydomonas reinhardi. J. Cell Biol.,* **38,** 403–25.

Kamiya, R. & Witman, G. B. (1984). Submicromolar levels of calcium control the balance of beating between the two flagella in demembranated models of *Chlamydomonas. J. Cell Biol.,* **98,** 97–107.

Kantz, T. S., Theriot, E. C., Zimmer, E. A. & Chapman, R. L. (1990). The Pleurastrophyceae and Micromonadophyceae: a cladistic analysis of nuclear rRNA sequence data. *J. Phycol.,* **26,** 711–21.

Karn, R. C., Starr, R. C. & Hudcock, G. A. (1974). Sexual and asexual differentiation in *Volvox obversus* (Shaw) Printz, strains Wd3 and Wd7. *Arch. Protistenkd.,* **116,** 142–8.

Karsten, G. (1918). Ueber Tagesper. d. Kern-u. Zellteilung. *Zeitsch. Bot.,* **10,** 1–13.

Kaska, D. D., Myllyla, R., Gunzler, V., Gibor, A. & Kivirikko, K. I. (1988). Prolyl 4-hydoxylase from *Volvox carteri* – a low MR enzyme antigenically related to the alpha subunit of the vertebrate enzyme. *Biochem. J.,* **256,** 257–63.

Kater, J. M. (1929). Morphology and division of *Chlamydomonas* with respect to the phylogeny of the neuromotor system. *Univ. Calif. Publ. Zool.,* **133,** 125–68.

Katz, K. R. & McLean, R. J. (1979). Rhizoplast and rootlet apparatus of *Chlamydomonas moewusii. J. Cell Sci.,* **39,** 373–81.

Kay, S. A. & Miller, A. J. (1992). *Circadian Regulated cab Gene Expression in Higher Plants.* Marcel Dekker, New York.

Kazmierczak, J. (1975). Colonial Volvocales (Chlorophyta) from the upper Devonian of Poland and their paleoenvironmental significance. *Acta Paleont. Polonica,* **20,** 73–85.

Kazmierczak, J. (1976a). Devonian and modern relatives of the Precambrian Eosphaera: possible significance for the early eukaryotes. *Lethaia,* **9,** 39–50.

Kazmierczak, J. (1976b). Volvocacean nature of some Paleozoic non-radiosphaerid calcispheres and parathuramminid "Foraminifera." *Acta Paleont. Polonica,* **21,** 245–58.

Kazmierczak, J. (1979). The eukaryotic nature of *Eosphaera*-like structures from the Precambrian Gunflint Iron Formation, Canada: A comparative study. *Precambrian Res.,* **9,** 1–22.

Kazmierczak, J. (1981). The biology and evolutionary significance of Devonian volvocaceans and their Precambrian relatives. *Acta Paleont. Polonica,* **26,** 299–338.

Kelland, J. L. (1964). Inversion in Volvox. Ph.D. thesis, Princeton University.

Kelland, J. L. (1977). Inversion in *Volvox* (Chlorophyceae). *J. Phycol.,* **13,** 373–8.

Kemp, C. L., Tsao, M. S. & Thirsen, G. (1972). Ultraviolet light induced thymine dimers and repair processes in the alga *Eudorina elegans. Can. J. Microbiol.,* **18,** 1809–15.

Kemp, C. L. & Wentworth, J. W. (1971). UV radiation studies on the colonial alga *Eudorina elegans. Can. J. Microbiol.,* **17,** 1417–24.

Kikuchi, K. (1978). Cellular differentiation in *Pleodorina californica. Cytologia,* **43,** 153–160.

Kirk, D. L. (1988). The ontogeny and phylogeny of cellular differentiation in *Volvox. Trends Genet.,* **4,** 32–6.

Kirk, D. L. (1990). Genetic control of reproductive cell differentiation in *Volvox.* In *Experimental Phycology 1: Cell Walls and Surfaces, Reproduction and Photosynthesis.* Ed., W. Wiessner, D. G. Robinson & R. C. Starr. Springer-Verlag, Berlin, pp. 81–94.

Kirk, D. L. (1994). Germ cell specification in *Volvox carteri.* In *Germline Development (Ciba Symposium 184).* Eds., J. Marsh & J. Goode. Wiley, Chichester, pp. 2–30.

Kirk, D. L. (1995). Asymmetric division, cell size and germ–soma specification in *Volvox. Sem. Dev. Biol.,* **6,** 369–79.

Kirk, D. L., Baran, G. J., Harper, J. F., Huskey, R. J., Huson, K. S. & Zagris, N. (1987). Stage-specific hypermutability of the *regA* locus of Volvox, a gene regulating the germ–soma dichotomy. *Cell,* **48,** 11–24.

Kirk, D. L., Birchem, R. & King, N. (1986). The extracellular matrix of *Volvox:* a comparative study and proposed system of nomenclature. *J. Cell Sci.,* **80,** 207–31.

Kirk, D. L. & Harper, J. F. (1986). Genetic, biochemical and molecular approaches to *Volvox* development and evolution. *Int. Rev. Cytol.,* **99,** 217–93.

Kirk, D. L., Kaufman, M. R., Keeling, R. M & Stamer, K. A. (1991). Genetic and cytological control of the asymmetric divisions that pattern the *Volvox* embryo. *Development (Suppl. 1),* 67-82.

Kirk, D. L. & Kirk, M. M. (1976). Protein synthesis in *Volvox carteri. Dev. Biol.,* **50,** 413–27.

Kirk, D. L. & Kirk, M. M. (1983). Protein synthetic patterns during the asexual life cycle of *Volvox carteri. Dev. Biol.,* **96,** 493–506.

Kirk, D. L. & Kirk, M. M. (1986). Heat shock elicits production of sexual inducer in *Volvox. Science,* **231,** 51–4.

Kirk, D. L., Kirk, M. M., Stamer, K. A. & Larson, A. (1990). The genetic basis for the evolution of multicellularity and cellular differentiation in the volvocine green algae. In *The Unity of Evolutionary Biology.* Ed., E. C. Dudley. Dioscorides Press, Portland, OR, pp. 568–81.

Kirk, D. L., Viamontes, G. I., Green, K. J. & Bryant, J. L., Jr. (1982). Integrated morphogenetic behavior of cell sheets: *Volvox* as a model. In *Developmental Order: Its Origin and Regulation.* Ed., S. Subtelny & P. B. Green. Alan R. Liss, New York, pp. 247–74.

Kirk, M. M. & Kirk, D. L. (1985). Translational regulation of protein synthesis, in response to light, at a critical stage of Volvox development. *Cell,* **41,** 419–28.

Kirk, M. M., Ransick, A., McRae, S. E. & Kirk, D. L. (1993). The relationship between cell size and cell fate in *Volvox carteri. J. Cell Biol.,* **123,** 191–208.

Kirschner, O. (1883). Zur Entwickslungsgeschichte von *Volvox minor* (Stein). *Beitr. Biol. Pflanzen,* **3,** 95–103.

Klein, L. (1889a). Morphologische und biologische Studien über die Gattung *Volvox. Jahrb. Wiss. Bot.,* **20,** 133–211.

Klein, L. (1889b). Neue Beiträge zur Kenntniss der Gattung *Volvox. Ber. Dtsch. Bot. Ges.,* **7,** 42–53.

Klein, L. (1890). Vergleichende Untersuchungen über Morphologie und Biologie der Fortpflanzung bei der Gattung *Volvox. Ber. Natur. Ges. Freiburg,* **5,** 92.

Kochert, G. (1968). Differentiation of reproductive cells in *Volvox carteri. J. Protozool.,* **15,** 433–52.

Kochert, G. (1971). Ribosomal RNA synthesis in *Volvox. Arch. Biochem. Biophys.,* **147,** 318–22.

Kochert, G. (1973). Colony differentiation in green algae. In *Developmental Regulation: Aspects of Cell Differentiation.* Ed., S. J. Coward. Academic Press, New York, pp. 155–67.

Kochert, G. (1975). Developmental mechanisms in *Volvox* reproduction. In *The Developmental Biology of Reproduction.* Ed., C. I. Markert & J. Papaconstantinou. Academic Press, New York, pp. 55–90.

Kochert, G. (1978). Sexual pheromones in algae and fungi. *Ann. Rev. Plant Physiol.,* **29,** 461–86.

Kochert, G. (1981). Sexual pheromones in *Volvox* development. In *Sexual Interactions in Eukaryotic Microbes.* Ed., D. H. O'Day & P. A. Horgen. Academic Press, New York, pp. 73–93.

Kochert, G. & Crump, W. J. (1979). Reversal of sexual induction in *Volvox carteri* by ultraviolet irradiation and removal of sexual pheromone. *Gamete Res.,* **2,** 259–64.

Kochert, G. & Olson, L. W. (1970a). Ultrastructure of *Volvox carteri.* I. The asexual colony. *Arch. Mikrobiol.,* **74,** 19–30.

Kochert, G. & Olson, L. W. (1970b). Endosymbiotic bacteria in *Volvox carteri. Trans. Amer. Microsc. Soc.,* **89,** 475–8.

Kochert, G. & Sansing, N. (1971). Isolation and characterization of nucleic acids from *Volvox carteri. Biochim. Biophys. Acta,* **238,** 397–405.

Kochert, G. & Yates, I. (1970). A UV-labile morphogenetic substance in *Volvox carteri. Dev. Biol.,* **23,** 128–35.

Kochert, G. & Yates, I. (1974). Purification and partial characterization of a glycoprotein sexual inducer from *Volvox carteri. Proc. Natl. Acad. Sci., USA,* **71,** 1211–14.

Korschikoff, A. A. (1924). Zur Morphologie und Systematik der Volvocales. *Arch. Russ. Protistol.,* **3,** 153–97.

Korschikoff, A. A. (1938). On the occurrence of *Volvulina steinii* in Ukrainia. *Bull. Soc. Nat. Moscou,* **47,** 56–63.

Korschikoff, A. A. (1939). *Volvox polychlamys* sp. n. *Bull. Soc. Nat. Moscou,* **48,** 5–12.

Koufopanou, V. (1990). Evolution and development of the flagellate green algae (Chlorophyta, Volvocales). Ph.D. thesis, McGill University.

Koufopanou, V. (1994). The evolution of soma in the Volvocales. *Amer. Nat.,* **143,** 907–31.

Koufopanou, V. & Bell, G. (1991). Developmental mutants of *Volvox*: Does mutation recreate the patterns of phylogenetic diversity? *Evolution,* **45,** 1806–22.

Koufopanou, V. & Bell, G. (1993). Soma and germ: An experimental approach using *Volvox. Proc. R. Soc. Lond. Ser. B. Biol. Sci.,* **254,** 107–13.

Kropat, J., von Gromoff, E. D., Müller, F. W. & Beck, C. F. (1995). Heat shock and light activation of a *Chlamydomonas HSP70* gene involves independent regulatory pathways. *Mol. Gen. Genet.,* **48,** 727–34.

Krumlauf, R. (1994). *Hox* genes in vertebrate development. *Cell,* **78,** 191–201.

Kryutchkova, N. M. (1974). The content and size of food particles consumed by filter-feeding planktonic animals. *Hydrobiol. J.,* **10,** 89–94.

Kuhlemeir, C., Green, P. J. & Chua, N.-H. (1987). Regulation of gene expression in plants. *Ann. Rev. Plant Sci.,* **38,** 221–57.

Kulaev, I. S. (1979). *The Biochemistry of Inorganic Polyphosphates.* Wiley, New York.

Kurn, N. (1981). Altered development of the multicellular alga *Volvox carteri* caused by lectin binding. *Cell Biol. Int. Rep.,* **5,** 867–75.

Kurn, N. (1982). Inhibition of phosphate uptake by fluphenazine, a calmodulin inhibitor. Analysis of *Volvox* wild-type and fluphenazine-resistant mutant strains. *FEBS Lett.,* **144,** 68–72.

Kurn, N., Colb, M. & Shapiro, L. (1978). Spontaneous frequency of a developmental mutant in *Volvox. Dev. Biol.,* **66,** 266–9.

Kurn, N. & Duskin, D. (1980). Involvement of glycoprotein in the regulation of *Volvox* development: The effect of tunicamycin. *Israel J. Med. Sci.,* **16,** 475–6.

Kurn, N. & Duskin, D. (1982). Effects of tunicamycin on protein glycosylation and development in *Volvox carteri. Wilhelm Roux's Arch.,* **191,** 169–75.

Kurn, N. & Sela, B.-A. (1981). Altered calmodulin activity in fluphenazine-resistant mutant strains. Pleiotropic effect on development and cellular organization in *Volvox carteri. Eur. J. Biochem.,* **121,** 53–7.

Kurn, N. & Sela, B.-A. (1979). Surface glycoproteins of the multicellular alga *Volvox carteri.* Developmental regulation, exclusive Con A binding and induced redistribution. *FEBS Lett.,* **104,** 249–52.

Kuschakewitsch, S. (1923). Zur Kenntnis der Entwickslungsgeschichte von *Volvox. Bull. de l'Acad. d. Sc. de l'Oukraine,* **1,** 32–40.

Kuschakewitsch, S. (1931). Zur Kenntnis der Entwickslungsgeschichte von *Volvox* (reprinted from *Bull. de l'Acad. d. Sc. de l'Oukraine,* **1,** 32–40). *Arch. Protistenkd.,* **73,** 323–30.

Lair, N. & Ali, O. (1990). Grazing and assimilation rates of natural populations of planktonic rotifers *Keratella cochlearis, Keratella quadrata,* and *Kellicottia longispina* in a eutrophic lake (Ayadat, France). *Hydrobiologia,* **194,** 119–31.

Lam, E. & Chua, N.-H. (1990). GT-1 binding sites confer light-responsive expression in transgenic tobacco. *Science,* **248,** 471–4.

Lamport, D. L. (1974). The role of hydroxyproline-rich proteins in the extracellular matrix of plants. In *Macromolecules Regulating Growth and Development.* Ed., E. D. Hay & T. J. King. Academic Press, New York, pp. 113–30.

Lander, C. A. (1929). Oögenesis and fertilization in Volvox. *Bot. Gaz.,* **87,** 431–6.

Lang, N. J. (1963). Electron microscopy of the Volvocaceae and Astrephomenaceae. *Amer. J. Bot.,* **50**, 280–300.

Lankester, E. R. (1877). Notes on the embryology and classification of the animal kingdom: comprising a revision of speculations relative to the origin and significance of the germ-layers. *Q. J. Microsc. Sci.,* **17**, 399–454.

Larson, A., Kirk, M. M. & Kirk, D. L. (1992). Molecular phylogeny of the volvocine flagellates. *Mol. Biol. Evol.,* **9**, 85–105.

Laurens, H. & Hooker, H. D. (1918). The relative sensitivity of Volvox to spectral lights of equal radiant energy content. *Anat. Rec.,* **14**, 97–8.

Lawson, M. A. & Satir, P. (1994). Characterization of the eyespot regions of "blind" *Chlamydomonas* mutants after restoration of their photophobic responses. *J. Euk. Microbiol.,* **4**, 593–601.

Lechtreck, K.-F., McFadden, G. I. & Melkonian, M. (1989). The cytoskeleton of the naked green flagellate *Spermatozopsis similis*: isolation, whole mount electron microscopy, and preliminary biochemical and immunological characterization. *Cell Motil. Cytoskel.,* **14**, 552–61.

Lechtreck, K.-F. & Melkonian, M. (1991). An update on fibrous flagellar roots in green algae. *Protoplasma,* **164**, 38–44.

LeDizet, M. & Piperno, G. (1986). Cytoplasmic microtubules containing acetylated α-tubulin in *Chlamydomonas reinhardtii*: spatial arrangement and properties. *J. Cell Biol.,* **103**, 13–22.

Lee, K. A. & Kemp, C. L. (1975). Microspectrophotometric analysis of DNA replication in *Eudorina elegans*. *Phycologia,* **14**, 247–52.

Lee, K. A. & Kemp, C. L. (1976). Chemical estimation of DNA changes during synchronous growth of *Eudorina elegans* (Chlorophyceae). *J. Phycol.,* **12**, 85–8.

Lefebvre, P. A., Aselson, C. M. & Tam, L.-W. (1995). Control of flagellar length in *Chlamydomonas*. *Sem. Dev. Biol.,* **6**, 317–23.

Lembi, C. A. (1975). The fine structure of the flagellar apparatus of *Carteria*. *J. Phycol.,* **11**, 1–9.

Levin, P. A. & Losick, R. (1995). Generating specialized cell types by asymmetric division in *Bacillus subtilis*. *Sem. Dev. Biol.,* **6**, 335–45.

Lewin, R. A. (1954). Mutants of *Chlamydomonas moewusii* with impaired motility. *J. Gen. Microbiol.,* **11**, 459–71.

Lewin, R. A. & Lee, K. W. (1985). Autotomy of algal flagella: electron microscopic studies of *Chlamydomonas* (Chlorophyceae) and *Tetraselmis* (Prasinophyceae). *Phycologia,* **24**, 311–16.

Lewis, C. M. & Fincham, J. R. S. (1970). Genetics of nitrate reductase in *Ustilago maydis*. *Genet. Res.,* **16**, 151–63.

Li, H.-M., Washburn, T. & Chory, J. (1993). Regulation of gene expression by light. *Curr. Opin. Cell Biol.,* **5**, 455–60.

Lien, T. & Knutsen, G. (1979). Synchronous growth of *Chlamydomonas reinhardtii* (Chlorophyceae): a review of optimal conditions. *J. Phycol.,* **15**, 191–200.

Liljeström, S. & Aberg, B. (1966). Studies on the mechanisms of chlorate toxicity. *Ann. Kungl. Lantbrukshösgkolans,* **32**, 93–107.

Lindauer, A., Fraser, D., Brüderlein, M. & Schmitt, R. (1993a). Reverse transcriptase families and a *copia*-like retrotransposon, *Osser*, in the green alga *Volvox carteri*. *FEBS Lett.,* **319**, 261–6.

Lindauer, A., Müller, K. & Schmitt, R. (1993b). Two histone H1-encoding genes of the green alga *Volvox carteri* with features intermediate between plant and animal genes. *Gene,* **129,** 59–68.

Linnaeus, C. (1758). *System naturae. Regnum animale,* ed. 10. Stockholm.

Liss, M., Kirk, D. L., Beyser, K. & Fabry, S. (1997). Intron sequences provide a tool for high resolution phylogenetic analysis of volvocine algae. *Current Genet.* **31,** 214–27.

Luck, D., Piperno, G., Ramanis, Z. & Huang, B. (1977). Flagellar mutants of *Chlamydomonas*: Studies of radial spoke-defective strains by dikaryon and revertant analysis. *Proc. Natl. Acad. Sci. USA,* **74,** 3456–60.

McCracken, M. D. (1970). Differentiation in *Volvox. Carolina Tips,* **33,** 37–8.

McCracken, M. D. & Barcellona, W. J. (1976). Electron histochemistry and ultrastructural localization of carbohydrate containing substances in the sheath of *Volvox. J. Histochem. Cytochem.,* **24,** 668–73.

McCracken, M. D. & Barcellona, W. J. (1981). Ultrastructure of sheath synthesis in *Volvox rousseletii. Cytobios,* **32,** 179–87.

McCracken, M. D. & Starr, R. C. (1970). Induction and development of reproductive cells in the K-32 strains of *Volvox rousseletii. Arch. Protistenkd.,* **112,** 262–82.

McFadden, G. I., Schulze, D., Surek, B., Salisbury, J. L. & Melkonian, M. (1987). Basal body reorientation mediated by a Ca^{2+}-activated contractile protein. *J. Cell Biol.,* **105,** 903–12.

McInness, D. E. (1994). Volvox at Albert Park Lake. *Victorian Nat.,* **111,** 32–3.

McQueen, D. J. (1970). Grazing rates and food selection in *Diaptomus oregonensis* (Copepoda) from Marion Lake, British Columbia. *J. Fish. Bd. Canada,* **27,** 13–20.

Mages, H.-W., Tschochner, H. & Sumper, M. (1988). The sexual inducer of *Volvox carteri.* Primary structure deduced from cDNA sequence. *FEBS Lett.,* **234,** 407–10.

Mages, W. (1990). Molekulare Genetik von Aktin und Tubulin bei *Volvox carteri* und Transformations-Experimente mit Grünalgen. Ph.D. thesis, Universität Regensburg.

Mages, W., Cresnar, B., Harper, J. F., Brüderlein, M. & Schmitt, R. (1995). *Volvox carteri* α2- and β2-tubulin-encoding genes. regulatory signals and transcription. *Gene,* **160,** 47–54.

Mages, W., Salbaum, J. M., Harper, J. F. & Schmitt, R. (1988). Organization and structure of *Volvox* α-tubulin genes. *Mol. Gen. Genet.,* **213,** 449–58.

Mainx, F. (1929a). Ueber die Geschlechterverteilung bei *Volvox aureus. Arch. Protistenkd.,* **67,** 205–14.

Mainx, F. (1929b). Untersuchungen über den Einfluss von Aussenfatoren auf die phototaktische Stimmung. *Arch. Protistenkd.,* **68,** 105–76.

Mansour, S. L., Thomas, K. R. & Capecchi, M. R. (1988). Disruption of the proto-oncogene *int-2* in mouse embryo-derived stem cells: a general strategy for targeting mutations to non-selectable genes. *Nature,* **336,** 348–52.

Manton, I. (1952). The fine structure of plant cilia. *Symp. Soc. Exp. Biol.,* **6,** 306–19.

Manton, I. (1956). Plant cilia and associated organelles. In *Cellular Mechanisms in Differentiation and Growth.* Ed., D. Rudnick. Princeton University Press, pp. 61–71.

Manton, I. (1966). Observations on scale production in *Pyramimonas amylifera. J. Cell Sci.,* **1,** 429–38.

Marchant, H. J. (1976). Plasmodesmata in algae and fungi. In *Intercellular Communication in Plants: Studies on Plasmodesmata*. Ed., B. E. S. Gunning & A. W. Robards. Springer-Verlag, Berlin, pp. 59–80.

Marchant, H. J. (1977). Colony formation and inversion in the green alga *Eudorina elegans*. *Protoplasma*, **93**, 325–39.

Margolis-Kazan, H. & Blamire, J. (1976). The DNA of *Volvox carteri*: a biophysical and biosynthetic characterization. *Cytobios*, **15**, 201–16.

Margolis-Kazan, H. & Blamire, J. (1977). The effect of Δ^9-tetrahydrocannabinol on cytoplasmic DNA metabolism. *Biochem. Biophys. Res. Commun.*, **76**, 674–81.

Margolis-Kazan, H. & Blamire, J. (1979a). Effect of tetrahydrocannabinol and ethidium bromide on DNA metabolism and embryogenesis in *Volvox*. *Cytobios*, **26**, 75–95.

Margolis-Kazan, H. & Blamire, J. (1979b). Effect of Δ^9-tetrahydrocannabinol on cytoplasmic DNA metabolism in a somatic-cell regenerator mutant of *Volvox*. *Microbios Lett.*, **11**, 7–13.

Margulis, L. (1981). *Symbiosis in Cell Evolution*. Freeman, San Francisco.

Mast, S. O. (1907). Light reactions in lower organisms. II. *Volvox globator*. *J. Comp. Neurol. Psychol.*, **17**, 99–180.

Mast, S. O. (1917). The relation between spectral color and stimulation in the lower organisms. *J. Exp. Zool.*, **22**, 472–528.

Mast, S. O. (1919). Reversion in the sense of orientation to light in the colonial forms, *Volvox globator* and *Pandorina morum*. *J. Exp. Zool.*, **27**, 367–90.

Mast, S. O. (1926). Reactions to light in *Volvox*, with special reference to the process of orientation. *Z. Vergl. Physiol.*, **4**, 637–85.

Mast, S. O. (1927). Response to electricity in *Volvox* and the nature of galvanic stimulation. *Z. Physiol.*, **5**, 739–69.

Matagne, R. F. (1978). Fine structure of the *arg-7* cistron in *C. reinhardi*. *Mol. Gen. Genet.*, **160**, 95–9.

Matsuda, Y., Musgrave, A., van den Ende, H. & Roberts, K. (1987). Cell walls of algae in the Volvocales: their sensitivity to a cell wall lytic enzyme and labeling with an anti-cell wall glycopeptide of *Chlamydomonas reinhardtii*. *Bot. Mag. Tokyo*, **100**, 373–84.

Matsumoto-Taniura, N., Pirollet, F., Monroe, R., Gerace, L. & Westendorf, J. M. (1996). Identification of novel M phase phosphoproteins by expression cloning. *Mol. Biol. Cell*, **7**, 1455–69.

Mattox, K. & Stewart, K. (1984). Classification of the green algae: a concept based on comparative cytology. In *Systematics of the Green Algae*. Ed., D. E. G. Irvine & D. John. Cambridge University Press, pp. 29–72.

Mattox, K. R. & Stewart, K. D. (1977). Cell division in the scaly green flagellate *Heteromastix angulata* and its bearing on the origin of the Chlorophyceae. *Amer. J. Bot.*, **64**, 931–45.

Mattson, D. M. (1984). The orientation of cleavage planes in *Volvox carteri*. M.S. thesis, Washington University.

Maynard Smith, J., Burian, R., Kaufman, S., Alberch, P., Campbell, J. B. G., Lande, R., Raup, D. & Wolpert, L. (1985). Developmental constraints and evolution. *Quart. Rev. Biol.*, **60**, 265–87.

Meglitch, P. A. (1967). *Invertebrate Zoology*. Oxford University Press.

Melkonian, M. (1978). Structure and significance of cruciate flagellar root systems in green algae: comparative investigations in species of *Chlorosarcinopsis*. *Plant Syst. Evol.*, **130**, 265–92.

Melkonian, M. (1980). Ultrastructural aspects of basal body associated fibrous structures in green algae: a critical review. *BioSystems*, **12**, 85–104.

Melkonian, M. (1982). Structural and evolutionary aspects of the flagellar apparatus in green algae and land plants. *Taxon*, **31**, 255–65.

Melkonian, M. (1984a). Flagellar apparatus ultrastructure in relation to green algal classification. In *Systematics of the Green Algae*. Ed., D. E. G. Irvine & D. M. John. Academic Press, London, pp. 73–120.

Melkonian, M. (1984b). Flagellar root-mediated interactions between the flagellar apparatus and cell organelles in green algae. In *Compartments in Algal Cells and Their Interaction*. Ed., W. Weissner, D. G. Robinson & R. C. Starr. Springer-Verlag, Berlin, pp. 96–108.

Melkonian, M. (1989). Centrin-mediated motility: A novel cell motility mechanism in eukaryotic cells. *Bot. Acta*, **102**, 3–4.

Melkonian, M. (1990). Phylum Chlorophyta. In *Handbook of Protoctista*. Ed., L. Margulis, J. O. Corliss, M. Melkonian & D. J. Chapman. Jones & Bartlett, Boston, pp. 597–660.

Melkonian, M., Beech, P. L., Katsaros, C. & Schulze, D. (1992). Centrin-mediated motility in algae. In *Algal Cell Motility*. Ed., M. Melkonian. Chapman & Hall, New York, pp. 179–221.

Melkonian, M. & Preisig, H. R. (1984). Ultrastructure of the flagellar apparatus in the green flagellate *Spermatozopsis similis*. *Plant Syst. Evol.*, **146**, 145–62.

Melkonian, M., Reize, I. B. & Preisig, H. R. (1987). Maturation of a flagellum/basal body requires more than one cell cycle in algal flagellates: Studies on *Nephroselmis olivacea* (Prasinophyceae). In *Algal Development (Molecular and Cellular Aspects)*. Ed., W. Weissner, D. G. Robinson & R. C. Starr. Springer-Verlag, Berlin, pp. 102–13.

Melkonian, M. & Robenek, H. (1980). Eyespot membrane of *Chlamydomonas reinhardtii*: a freeze-fracture study. *J. Ultrastruct. Res.*, **72**, 90–102.

Melkonian, M. & Robenek, H. (1984). The eyespot apparatus of flagellated green algae: a critical review. *Prog. Phycol. Res.*, **3**, 193–268.

Melkonian, M., Schulze, D., McFadden, G. I. & Robenek, H. (1988). A polyclonal antibody (anticentrin) distinguishes between two types of fibrous flagellar roots in green algae. *Protoplasma*, **144**, 56–61.

Mengele, R. & Sumper, M. (1992). Gulose as a constituent of a glycoprotein. *FEBS Lett.*, **298**, 14–16.

Meredith, R. F. & Starr, R. C. (1975). The genetic basis of male potency in *Volvox carteri* f. *nagariensis* (Chlorophyceae). *J. Phycol.*, **11**, 265–72.

Merton, H. (1908). Ueber die Bau und die Fortpflanzung von *Pleodorina illinoisensis* Kofoid. *Z. Wiss. Zool.*, **90**, 445–77.

Messikommer, E. L. (1948). Algennachweis in Entenenkrementen. *Hydrobiologia*, **1**, 22–7.

Metzner, J. (1945a). A morphological and cytological study of a new form of Volvox-I. *Bull. Torrey Bot. Club*, **72**, 86–113.

Metzner, J. (1945b). A morphological and cytological study of a new form of Volvox-II. *Bull. Torrey Bot. Club*, **72**, 121–36.

Meyer, A. (1895). Ueber den Bau von *Volvox aureus* Ehrenb. und *Volvox globator* Ehrenb. *Bot. Centralbl.*, **63**, 225–33.

Meyer, A. (1896). Die Plasmaverbindung und die Membranen von *Volvox*. *Bot. Zeitg.*, **54**, 187–217.

Miller, C. E. & Starr, R. C. (1981). The control of sexual morphogenesis in *Volvox capensis*. *Ber. Deutsch. Bot. Ges.*, **94**, 357–72.

Miller, S. M., Schmitt, R. & Kirk, D. L. (1993). *Jordan*, an active *Volvox* transposable element similar to higher plant transposons. *Plant Cell*, **5**, 1125–38.

Mishra, N. C. & Threlkeld, S. F. H. (1968). Genetic studies in *Eudorina*. *Genet. Res. Camb.*, **11**, 21–31.

Mitchell, B. A. (1980). Evidence for polyhydroxyproline in the extracellular matrix of *Volvox*. M.S. thesis, Michigan State University.

Moestrup, Ø. (1978). On the phylogenetic validity of the flagellar apparatus in green algae and other chlorophyll a and b containing plants. *BioSystems*, **10**, 117–44.

Moestrup, Ø. (1982). Flagellar structure in algae: a review, with new observations particularly on the Chrysophyceae, Phaeophyceae (Fucophyceae), Euglenophyceae and Reckertia. *Phycologia*, **21**, 427–528.

Moka, R. (1985). Sind cyclonucleotide an der Differenzierung von *Volvox* beteiligt? Diplom thesis, Universität zu Köln (not seen; cited by Jaenicke and Gilles 1985, and Jaenicke 1991).

Moka, R. (1988). cAMP Metabolismus und sexuelle Differenzierung der diözischen Grünalgae *Volvox carteri* Ph.D. thesis, Universität zu Köln (not seen; cited by Jaenicke 1991, and Nass et al. 1994).

Morel-Laurens, N. M. L. & Feinleib, M. E. (1983). Photomovement in an "eyeless" mutant of *Chlamydomonas*. *Photochem. Photobiol.*, **37**, 189–94.

Morris, J., Kushner, S. R. & Ivarie, R. (1986). The simple repeat poly(dT-dG)•poly(dC-dA) common to eukaryotes is absent from eubacteria and archibacteria and rare in protozoans. *Mol. Biol. Evol.*, **3**, 343–55.

Moss, B. (1972). The influence of environmental factors on the distribution of fresh water algae: an experimental study. I. Introduction and the influence of calcium concentration. *J. Ecol.*, **60**, 917–32.

Moss, B. (1973a). The influence of environmental factors on the distribution of fresh water algae: an experimental study. II. The role of pH and the carbon dioxide-bicarbonate system. *J. Ecol.*, **61**, 157–77.

Moss B. (1973b). The influence of environmental factors on the distribution of fresh water algae: an experimental study. III. Effects of temperature, vitamin requirements and inorganic nitrogen compounds on growth. *J. Ecol.*, **61**, 179–92.

Moss, B. (1973c). The influence of environmental factors on the distribution of fresh water algae: an experimental study. IV. Growth of test species in natural lake waters and conclusion. *J. Ecol.*, **61**, 193–211.

Müller, A. J. & Grafe, R. (1978). Isolation and characterization of cell lines of *Nicotiana tabacum* lacking nitrate reductase. *Mol. Gen. Genet.*, **161**, 67–76.

Müller, F. W., Igloi, G. L. & Beck, C. F. (1992). Structure of a gene encoding heat-shock protein HSP70 from the unicellular alga *Chlamydomonas reinhardtii*. *Gene*, **111**, 165–173.

Müller, K., Lindauer, A., Brüderlein, M. & Schmitt, R. (1990). Organization and transcription of *Volvox* histone-encoding genes: similarities between algal and animal genes. *Gene,* **93,** 167–75.

Müller, K. & Schmitt, R. (1988). Histone genes of *Volvox carteri*: DNA sequence and organization of two H3-H4 loci. *Nucl. Acids Res.,* **16,** 4121–36.

Müller, T., Bause, E. & Jaenicke, L. (1981). Glycolipid formation in *Volvox carteri* f. *nagariensis. FEBS Lett.,* **128,** 208–12.

Müller, T., Bause, E. & Jaenicke, L. (1984). Evidence for an incomplete dolichyl-phosphate pathway of lipoglycan formation in *Volvox carteri* f. *nagariensis. Eur. J. Biochem.,* **138,** 153–9.

Nass, N., Moka, R. & Jaenicke, L. (1994). Adenylyl cyclase from the green alga *Volvox carteri* f. *nagariensis*: partial purification and characterisation. *Aust. J. Plant Physiol.,* **21,** 613–22.

Nelson, J. A. E. & Lefebvre, P. A. (1995). Targeted disruption of the *NIT8* gene in *Chlamydomonas reinhardtii. Mol. Cell. Biol.,* **15,** 5762–9.

Nelson, M. & McClelland, M. (1991). Site-specific methylation: effect on DNA modification methyltransferases and restriction endonucleases. *Nucl. Acids Res.,* **19,** 2045–71.

Nichols, G. L. & Syrett, P. J. (1978). Nitrate reductase deficient mutants of *Chlamydomonas reinhardii*. Biochemical characterization. *J. Gen. Microbiol.,* **108,** 71-77.

Nonnengässer, C., Holland, E.-M., Harz, H. & Hegemann, P. (1996). The nature of rhodopsin-triggered photocurrents in *Chlamydomonas.* II. The influence of monovalent ions. *Biophys. J.,* **70,** 932–8.

Nozaki, H. (1983). Morphology and taxonomy of two species of *Astrephomene* (Chlorophyta, Volvocales) in Japan. *J. Jpn. Bot.,* **58,** 345–52.

Nozaki, H. (1984). Newly found facets in the asexual and sexual reproduction of *Gonium pectorale* (Chlorophyta, Volvocales). *Jpn. J. Phycol.,* **32,** 130-3.

Nozaki, H. (1986a). Sexual reproduction in the colonial Volvocales (in Japanese). *Jpn. J. Phycol.,* **34,** 232-47.

Nozaki, H. (1986b). A taxonomic study of *Pyrobotrys* (Volvocales, Chlorophyta) in anaerobic pure culture. *Phycologia,* **25,** 455–68.

Nozaki, H. (1988). Morphology, sexual reproduction and taxonomy of *Volvox carteri* f. *kawasakiensis* f. nov. (Chlorophyta) from Japan. *Phycologia,* **27,** 209–20.

Nozaki, H. (1989). Morphological variation and reproduction in *Gonium viridistellatum* (Volvocales, Chlorophyta). *Phycologia,* **28,** 77–88.

Nozaki, H. (1993). Asexual and sexual reproduction in *Gonium quadratum* (Chlorophyta) with a discussion of phylogenetic relationships within the Goniaceae. *J. Phycol.,* **29,** 369–76.

Nozaki, H. (1994). Unequal flagellar formation in *Volvox* (Volvocaceae, Chlorophyta). *Phycologia,* **33,** 58–61.

Nozaki, H. & Itoh, M. (1994). Phylogenetic relationships within the colonial Volvocales (Chlorophyta) inferred from cladistic analysis based on morphological data. *J. Phycol.,* **30,** 353–65.

Nozaki, H., Itoh, M., Sano, R., Uchida, H., Watanabe, M. M. & Kuroiwa, T. (1995). Phylogenetic relationships within the colonial Volvocales (Chlorophyta) inferred from *rbc*L gene sequence data. *J. Phycol.,* **31,** 970–9.

Nozaki, H., Itoh, M., Watanabe, M. M. & Kuroiwa, T. (1996). Ultrastructure of the vegetative colonies and systematic position of *Basichlamys* (Volvocales, Chlorophyta). *Eur. J. Phycol.,* **31,** 62–72.

Nozaki, H. & Kuroiwa, T. (1991). Morphology and sexual reproduction of *Gonium multicoccum* (Volvocales, Chlorophyta) from Nepal. *Phycologia,* **30,** 381–93.

Nultsch, W. (1979). Effect of external factors on phototaxis of *Chlamydomonas reinhardtii.* III. Cations. *Arch. Microbiol.,* **123,** 93–9.

Nultsch, W. (1983). The photocontrol of movement of *Chlamydomonas. Symp. Soc. Exp. Biol.,* **36,** 521–39.

Nultsch, W., Pfau, J. & Dolle, R. (1986). Effects of calcium channel blockers on phototaxis and motility of *Chlamydomonas reinhardtii. Arch. Microbiol.,* **144,** 393–7.

Nultsch, W. & Rüffer, U. (1994). Die Orientierung freibeweglicher Organismen zum Licht, darstellt am Beispiel des Flagellatren *Chlamydomonas reinhardtii. Naturwissenschaften,* **81,** 164–74.

O'Brien, W. J. (1970). The effects of nutrient enrichment on the plankton community in eight experimental ponds. Ph.D. thesis, Michigan State University.

O'Kelley, C. J. & Floyd, G. L. (1983). The flagellar apparatus of *Enterocladia viridis* motile cells, and the taxonomic position of the resurrected family Ulvellaceae (Ulvales, Chlorphyta). *J. Phycol.,* **19,** 153–64.

O'Kelley, C. J. & Floyd, G. L. (1984). Flagellar apparatus absolute orientations and the phylogeny of the green algae. *BioSystems,* **16,** 227–51.

Olive, L. S. (1978). Sorocarp development by a newly discovered ciliate. *Science,* **202,** 530–2.

Olson, L. W. & Kochert, G. (1970). Ultrastructure of *Volvox carteri.* II. The kinetesome. *Arch. Mikrobiol.,* **74,** 31–40.

Oltmanns, F. (1892). Ueber die photometrischen Bewegungen der Pflanzen. *Flora,* **75,** 183–266.

Oltmanns, F. (1917). Ueber Phototaxis. *Z. Botan.,* **9,** 257–338.

Overton, E. (1889). Beiträg zur Kenntniss der Gattung *Volvox. Bot. Centralb.,* **39,** 65–72.

Owens, O. v. H. & Esaias, W. E. (1976). Physiological responses of phytoplankton to major environmental factors. *Ann. Rev. Plant Physiol.,* **27,** 461–83.

Pall, M. L. (1973). Sexual induction in *Volvox carteri. J. Cell Biol.,* **59,** 238–41.

Pall, M. (1974). Evidence for the glycoprotein nature of the inducer of sexuality in *Volvox. Biochem. Biophys. Res. Commun.,* **57,** 123–49.

Pall, M. L. (1975). Mutants of *Volvox* showing premature cessation of division: evidence for a relationship between cell size and reproductive cell differentiation. In *Developmental Biology: Pattern Formation, Gene Regulation.* Ed., D. McMahon & C. F. Fox. W. A. Benjamin, Menlo Park, CA, pp. 148–56.

Palmer, E. G. & Starr, R. C. (1971). Nutrition of *Pandorina morum. J. Phycol.,* **7,** 85–9.

Parke, M. & Manton, I. (1965). Preliminary observations on the fine structure of *Prasinocladus marinus. J. Mar. Biol. Assoc. U. K.,* **45,** 525–36.

Pascher, A. (1918). Flagellaten und Rhizopoden in ihren gegenseitigen Beziehungen. *Arch. Protistenkd.,* **38,** 1–88.

Pascher, A. (1927). Volvocales-Phytomonadinae. In *Die Süsswasser-Flora Deutschlands, Österreichs und der Schweiz.* Ed., A. Pascher. Gustav Fischer, Jena, pp. 1–506.

Pearsall, W. H. (1932). Phytoplankton in English lakes. 2. The composition of the phytoplankton in relation to dissolved substances. *J. Ecol.,* **20,** 241–62.

Pfeffer, W. (1888). Ueber chemotaktische Bewegungen von Bacterien, Flagellaten und Volvocineen. *Untersuch. Bot. Inst. Tübingen,* **2,** 582–662.

Pickett-Heaps, J. D. (1970). Some ultrastructural features of *Volvox*, with particular reference to the phenomenon of inversion. *Planta,* **90,** 174–90.

Pickett-Heaps, J. D. (1972). Variation in mitosis and cytokinesis in plant cells: its significance in the phylogeny and evolution of ultrastructural systems. *Cytobios,* **5,** 59–77.

Pickett-Heaps, J. D. (1975). *Green Algae. Structure, Reproduction and Evolution in Selected Genera.* Sinauer, Sunderland, MA.

Pickett-Heaps, J. D. (1976). Cell division in eukaryotic algae. *BioScience,* **26,** 445–50.

Pilarska, J. (1977). Eco-physiological studies on *Brachionus rubens* Ehrbg (Rotatoria). I. Food selectivity and feeding rate. *Pol. Arch. Hydrobiol.,* **24,** 319–28.

Pitelka, D. R. (1969). Fibrillar systems in protozoa. In *Research in Protozoology.* Ed., T.-T. Chen. Pergamon Press, Oxford, pp. 279–388.

Pitelka, D. R. (1974). Basal bodies and root structures. In *Cilia and Flagella.* Ed., M. A. Sleigh. Academic Press, New York, pp. 437–64.

Playfair, G. I. (1915). Freshwater algae of the Lismore District: with an appendix on the algal fungi and Schizomycetes. *Proc. Linn. Soc. N. S. Wales,* **40,** 310–62.

Playfair, G. J. (1918). New and rare fresh water algae. *Proc. Linn. Soc. N. S. Wales,* **43,** 497–543.

Pocock, M. A. (1933a). *Volvox* and associated algae from Kimberley. *Ann. South Afr. Mus.,* **16,** 473–521.

Pocock, M. A. (1933b). *Volvox* in South Africa. *Ann. South Afr. Mus.,* **16,** 523–625.

Pocock, M. A. (1938). *Volvox tertius* Meyer. With notes on the two other British species of *Volvox. J. Quek. Micr. Club, ser. 4,* **1,** 33–58.

Pocock, M. A. (1953). Two multicellular motile green algae, *Volvulina* Playfair and *Astrephomene*, a new genus. *Trans. Roy. Soc. S. Africa,* **34,** 103–27.

Pocock, M. A. (1955). Studies in North American Volvocales. I. The genus *Gonium. Madroño,* **13,** 49–64.

Pocock, M. A. (1960). *Haematococcus* in Southern Africa. *Trans. Roy. Soc. S. Africa,* **36,** 5–55.

Pocock, M. A. (1962). Algae from De Klip soil cultures. *Arch. Mikrobiol.,* **42,** 56–63.

Pommerville, J. C. & Kochert, G. (1981). Changes in somatic cell structure during senescence of *Volvox carteri. Eur. J. Cell Biol.,* **24,** 236–43.

Pommerville, J. C. & Kochert, G. (1982). Effects of senescence on somatic cell physiology in the green alga *Volvox carteri. Exp. Cell Res.,* **140,** 39–45.

Postlethwait, J. H., Johnson, S. L., Midson, C. N., Talbot, W. S., Gates, M., Ballinger, E. W., Africa, D., Andrews, R., Carl, T., Eisen, J. S., Horne, S., Kimmel, C. B., Hutchinson, M., Johnson, M. & Rodriguez, A. (1994). A genetic linkage map for the zebrafish. *Science,* **264,** 699–703.

Powers, J. H. (1907). New forms of Volvox. *Trans. Amer. Microsc. Soc.,* **27,** 123–49.

Powers, J. H. (1908). Further studies in *Volvox*, with descriptions of three new species. *Trans. Amer. Microsc. Soc.,* **28,** 141–75.

Prieto, R. & Fernandez, E. (1993). Toxicity of and mutagenesis by chlorate are independent of nitrate reductase activity in *Chlamydomonas reinhardtii*. *Mol. Gen. Genet.*, **237**, 429–38.

Pringsheim, E. G., (1930). Die Kultur von *Micrasterias* und *Volvox*. *Arch. Protistenkd.*, **72**, 1–48.

Pringsheim, E. G. (1958). Ueber Mixotrophie bei Flagellaten. *Planta*, **52**, 405–30.

Pringsheim, E. G. (1970). Identification and cultivation of European *Volvox* spp. *Antonie van Leeuwenhoek*, **36**, 33–43.

Pringsheim, E. G. & Pringsheim, O. (1959). Die Ernährung koloniebildender Volvocales. *Biol. Zentr.*, **78**, 937–71.

Pringsheim, N. (1870). Ueber Paarung von Schwärmsporen. *Monatsbar. Akad. Wiss. Berlin*, **1869**, 721–38.

Printz, H. (1927). Chlorophyceae. In *Die natüralichen Pflanzfamilien*, 2nd ed. Ed., A. Engler & K. Prantl. Leipzig, pp. 1–463.

Procter, V. W. (1959). Dispersal of fresh-water algae by migratory water birds. *Science*, **130**, 623–4.

Provasoli, L. & Pintner, I. J. (1959). Artificial media for freshwater algae; problems and suggestions. In *Ecology of Algae. Special Publication No. 2, Pymatuning Laboratory of Field Biology*. Ed., C. A. Tryon & R. T. Hartman. University of Pittsburgh, pp. 84–96.

Puente, P., Wei, N. & Deng, X. W. (1996). Combinatorial interplay of promoter elements constitutes the minimal determinants for light and developmental control of gene expression in *Arabidopsis*. *EMBO J.*, **15**, 3732–43.

Purton, S. & Rochaix, J.-D. (1995). Characterisation of the *ARG7* gene of *Chlamydomonas reinhardtii* and its application to nuclear transformation. *Eur. J. Phycol.*, **30**, 141–8.

Ransick, A. (1991). Reproductive cell specification during *Volvox obversus* development. *Dev. Biol.*, **143**, 185–98.

Ransick, A. (1993). Specification of reproductive cells in *Volvox*. In *Evolutionary Conservation of Developmental Mechanisms*. Ed., A. Spradling. Wiley-Liss, New York, pp. 55–70.

Rassow, J., Voos, W. & Pfanner, N. (1995). Partner proteins determine multiple functions of Hsp70. *Trends Cell Biol.*, **5**, 207–12.

Rausch, H., Larsen, N. & Schmitt, R. (1989). Phylogenetic relationships of the green alga *Volvox carteri* deduced from small-subunit ribosomal RNA comparisons. *J. Mol. Evol.*, **29**, 255–65.

Raven, P. H., Evert, R. F. & Eichorn, S. E. (1986). *Biology of Plants*, 4th ed. Worth, New York.

Rayburn, W. R. & Starr, R. C. (1974). Morphology and nutrition of *Pandorina unicocca* sp. nov. *J. Phycol.*, **10**, 42–9.

Rayns, D. G. & Godward, M. B. E. (1965). A quantitative study of mitosis in *Eudorina elegans* in culture. *J. Exp. Bot.*, **16**, 569–80.

Reichenbach, H. & Dworkin, M. (1992). The myxobacteria. In *The Prokaryotes*, Vol. 4. Ed., A. Barlows, H. G. Trüper et al. Springer-Verlag, Berlin, pp. 187–231.

Reiter, R. S., Williams, J. G. K., Feldman, K. A., Rafalski, J. A., Tingley, S. V. & Scolnick, P. A. (1992). Global and local genome mapping in *Arabidopsis thaliana* by using recombinant inbred lines and random amplified polymorphic DNAs. *Proc. Natl. Acad. Sci. USA*, **89**, 1477–81.

Reynolds, C. S. (1976). Succession and vertical distribution of phytoplankton in response to thermal stratification in a lowland mere, with special reference to nutrient availability. *J. Ecol.,* **64,** 529–41.

Reynolds, C. S. (1983). Growth-rate responses of *Volvox aureus* Ehrenb. (Chlorophyta, Volvocales) to variability in the physical environment. *Br. Phycol. J.,* **18,** 433–42.

Reynolds, C. S. (1984a). *The Ecology of Freshwater Phytoplankton.* Cambridge University Press.

Reynolds, C. S. (1984b). Phytoplankton periodicity: the interactions of form, function and environmental variability. *Freshwater Biology,* **14,** 111–42.

Rhee, G.-Y. (1972). Competition between an alga and an aquatic bacterium for phosphate. *Limnol. Oceanogr.,* **17,** 505–14.

Rhee, G.-Y. (1973). A continuous culture study of phosphate uptake, growth rate and polyphosphate in *Scenedesmus* sp. *J. Phycol.,* **9,** 495–506.

Rhee, G.-Y. (1974). Phosphate uptake under nitrate limitation by *Scenedesmus* sp. and its ecological implications. *J. Phycol.,* **10,** 470–75.

Rich, F. & Pocock, M. A. (1933). Observations on the genus *Volvox* in Africa. *Ann. South Afr. Mus.,* **16,** 427–71.

Ringo, D. L. (1967). Flagellar apparatus in *Chlamydomonas. J. Cell Biol.,* **33,** 543–71.

Roberts, K. (1974). Crystalline glycoprotein cell walls of algae; their structure, composition and assembly. *Phil. Trans. R. Soc. Lond., Biol. Sci.,* **268,** 129–46.

Roberts, K., Phillips, J., Shaw, P., Grief, C. & Smith, E. (1985). An immunological approach to the plant cell wall. In *Biochemistry of Plant Cell Walls.* Ed., C. T. Brett & J. R. Hillman. Cambridge University Press, pp. 125–54.

Roberts, R. J. & Macelis, D. (1991). Restriction enzymes and their isoschizomers. *Nucl. Acids Res.,* **19,** 2077–109.

Rosenburg, E. (ed.) (1984). *Myxobacteria: Development and Cell Interactions.* Springer-Verlag, Berlin.

Ross, I. K. (1979). *Biology of the Fungi.* McGraw-Hill, New York.

Round, F. E. (1971). The taxonomy of the Chlorophyta. II. *Br. Phycol. J.,* **6,** 235–64.

Rousselet, C. F. (1914). Remarks on two species of African *Volvox. J. Quek. Micr. Club, ser. 2,* **12,** 393–4.

Rüffer, U. & Nultsch, W. (1985). High-speed cinematographic analysis of the movement of *Chlamydomonas. Cell Motil. Cytoskel.,* **5,** 251–63.

Rüffer, U. & Nultsch, W. (1991). Flagellar photoresponses of *Chlamydomonas* cells held on micropipets. II. Change in flagellar beat patterns. *Cell Motil. Cytoskel.,* **18,** 269–78.

Runnegar, B. (1992). Evolution of the earliest animals. In *Major Events in the History of Life.* Ed., J. W. Schopf. Jones & Bartlett, Boston, pp. 65–93.

Ryder, J. A. (1889). The polar differentiation of Volvox and the specialization of possible anterior sense-organs. *Amer. Nat.,* **23,** 218–21.

Sack, L., Zeyl, C., Bell, G., Sharbel, T., Reboud, X., Bernhardt, T. & Koelwyn, H. (1994). Isolation of four new strains of *Chlamydomonas reinhardtii* (Chlorophyta) from soil samples. *J. Phycol.,* **30,** 770–3.

Sager, R. & Ishida, M. R. (1963). Chloroplast DNA in *Chlamydomonas. Proc. Natl. Acad. Sci. USA,* **50,** 725–30.

Sakeguchi, H. (1979). Effect of external ionic environment on phototaxis of *Volvox carteri. Plant Cell Physiol.,* **20,** 1643–51.

Sakeguchi, H. & Iwasa, K. (1979). Two photophobic responses in *Volvox carteri. Plant Cell Physiol.,* **20,** 909–16.

Sakeguchi, H. & Tawada, K. (1977). Temperature effect on the photoaccumulation and phobic response of *Volvox aureus. J. Protozool.,* **24,** 284–8.

Salisbury, J. L. (1988). The lost neuromotor apparatus of *Chlamydomonas* rediscovered. *J. Protozool.,* **35,** 574–7.

Salisbury, J. L. (1989a). Algal centrin: calcium-sensitive contractile organelles. In *Algae as Experimental Systems.* Ed., A. W. Coleman, L. J. Goff & J. R. Stein-Taylor. Alan R. Liss, New York, pp. 19–37.

Salisbury, J. L. (1989b). Centrin and the algal flagellar apparatus. *J. Phycol.,* **25,** 201–6.

Salisbury, J. L., Baron, A. T., Coling, D. E., Martindale, V. E. & Sanders, M. A. (1986). Calcium-modulated contractile proteins associated with the eukaryotic centrosome. *Cell Motil. Cytoskel.,* **6,** 193–7.

Salisbury, J. L., Baron, A. T. & Sanders, M. A. (1988). The centrin-based cytoskeleton of *Chlamydomonas reinhardtii.* Distribution in interphase and mitotic cells. *J. Cell Biol.,* **107,** 635–42.

Salisbury, J. L., Baron, A., Surek, B. & Melkonian, M. (1984). Striated flagellar roots: Isolation and partial characterization of a calcium-modulated contractile organelle. *J. Cell Biol.,* **99,** 962–70.

Salisbury, J. L. & Floyd, G. L. (1978). Calcium induced contraction of the rhizoplast of a quadriflagellate green alga. *Science,* **202,** 975–76.

Salisbury, J. L., Sanders, M. A. & Harpst, L. (1987). Flagellar root contraction and nuclear movement during flagellar regeneration in *Chlamydomonas reinhardtii. J. Cell Biol.,* **105,** 1799–1805.

Salonen, K., Jones, I. & Arvola, L. (1984). Hypolimnetic phosphorus retrieval by diel vertical migrations of lake phytoplankton. *Freshwater Biol.,* **14,** 431–8.

Sanders, M. A. & Salisbury, J. L. (1989). Centrin-mediated microtubule severing during flagellar excision in *Chlamydomonas reinhardtii. J. Cell Biol.,* **108,** 1751–60.

Sarma, Y. S. R. K. & Shyam, R. (1974). A new member of the colonial Volvocales (Chlorophyceae), *Pyrobotrys acuminata* sp. nov. from India. *Phycologia,* **13,** 121–4.

Scagel, R. F., Bandoni, R. J., Rouse, G. E., Schofield, W. B., Stein, J. R. & Taylor, T. M. C. (1965). *Plant Diversity: An Evolutionary Approach.* Wadsworth, Belmont, CA.

Schaechter, M. & DeLamater, E. D. (1955). Mitosis of *Chlamydomonas. Am. J. Bot.,* **42,** 417–22.

Schiebel, E. & Bornens, M. (1995). In search of a function for centrins. *Trends Cell Biol.,* **5,** 197–201.

Schiedlmeier, B. & Schmitt, R. (1994). Repetitious structure and transcription control of a polyubiquitin gene in *Volvox carteri. Curr. Genet.,* **25,** 169–77.

Schiedlmeier, B., Schmitt, R., Müller, W., Kirk, M. M., Gruber, H., Mages, W. & Kirk, D. L. (1994). Nuclear transformation of *Volvox carteri. Proc. Natl. Acad. Sci. USA,* **91,** 5080–4.

Schindler, D. W. (1971). Carbon, nitrogen and phosphorus and the eutrophication of freshwater lakes. *J. Phycol.,* **7,** 321–9.

Schindler, D. W. (1977). Evolution of phosphorus limitation in lakes. *Science*, **195**, 260–2.

Schindler, D. W., Armstrong, F. A., Holmgren, S. K. & Brundskill, G. J. (1971). Eutrophication of lake 227, Experimental Lakes Area, northwestern Ontario, by addition of phosphate and nitrate. *J. Fish Bd. Canada*, **28**, 1763–82.

Schleicher, M., Lukas, T. J. & Watterson, D. M. (1984). Isolation and characterization of calmodulin from the motile green alga *Chlamydomonas reinhardtii*. *Arch. Biochem. Biophys.*, **229**, 33–42.

Schletz, K. (1976). Phototaxis in *Volvox* – Pigments involved in the perception of light direction. *Z. Pflanzenphysiol.*, **77**, 189–211.

Schlichting, H. E. (1960). The role of waterfowl in the dispersal of algae. *Trans. Amer. Microsc. Soc.*, **79**, 160–6.

Schlipfenbacher, R., Wenzl, S., Lottspeich, F. & Sumper, M. (1986). An extremely hydroxyproline-rich glycoprotein is expressed in inverting *Volvox* embryos. *FEBS Lett.*, **209**, 57–62.

Schloss, J. A. (1990). A *Chlamydomonas* gene encodes a G protein β subunit-like polypeptide. *Mol. Gen. Genet.*, **221**, 443–52.

Schlösser, U. W. (1976). Entwickslungsstadien-und sippenspezifische Zellwand-Autolysine bei der Freisetzung von Fortpflanzungszellen in der Gattung *Chlamydomonas*. *Ber. Deutsch. Bot. Ges.*, **89**, 1–56.

Schlösser, U. W. (1984). Species-specific sporangium autolysins (cell-wall-dissolving enzymes) in the genus *Chlamydomonas*. In *Systematics of the Green Algae*. Ed., D. E. G. Irvine & D. M. John. Academic Press, London, pp. 409–18.

Schmidt, J. A. & Eckert, R. (1976). Calcium couples flagellar reversal to photostimulation in *Chlamydomonas reinhardtii*. *Nature*, **262**, 713–15.

Schmitt, R., Fabry, S. & Kirk, D. L. (1992). In search of the molecular origins of cellular differentiation in *Volvox* and its relatives. *Int. Rev. Cytol.*, **139**, 189–265.

Schmitt, R. & Kirk, D. L. (1992). Tubulins in the Volvocales: from genes to organized microtubule assemblies. In *Cytoskeletons of the Algae*. Ed., D. Menzel. CRC Press, Cleveland, pp. 369–92.

Schnell, R. A. & Lefebvre, P. A. (1993). Isolation of the Chlamydomonas regulatory gene *NIT2* by transposon tagging. *Genetics*, **134**, 737–47.

Schopf, J. W. (1974). Paleobiology of the Precambrian: the age of blue-green algae. In *Evolutionary Biology*. Ed., T. Dobzhansky, M. K. Hecht & W. C. Steere. Plenum Press, New York, pp. 1–43.

Schopf, J. W. (1992). The oldest fossils and what they mean. In *Major Events in the History of Life*. Ed., J. W. Schopf. Jones & Bartlett, Boston, pp. 29-63.

Schopf, J. W. (1993). Microfossils of the early Archean Apex chert: new evidence of the antiquity of life. *Science*, **260**, 640–6.

Schreiber, E. (1925). Zur Kentniss der Physiologie und Sexualität höherer Volvocales. *Z. Bot.*, **17**, 337–76.

Schulze, D., Robenek, H., McFadden, G. I. & Melkonian, M. (1987). Immunolocalization of a Ca^{2+}-modulated contractile protein in the flagellar apparatus of green algae: the nucleus-basal body connector. *Eur. J. Cell Biol.*, **45**, 51–61.

Segaar, P. J. (1990). The flagellar apparatus and temporary centriole-associated microtubule systems at the interphase-mitosis transition in the green alga *Gleomonas kupfferi*: An example of the spatio-temporal flexibility of microtubule organizing centers. *Acta Bot. Neerl.*, **39**, 29–42.

Segaar, P. J. & Gerritsen, A. F. (1989). Flagellar roots as vital instruments in cellular morphogenesis during multiple fission (sporulation) in the unicellular green flagellate *Brachiomonas submarina* (Chlamydomonadales, Chlorophyta). *Crypt. Bot.,* **1,** 249–74.

Segaar, P. J., Gerritsen, A. F. & De Bakker, M. A. G. (1989). The cytokinetic apparatus during sporulation in the unicellular green flagellate *Gleomonas kupfferi*: The phycoplast as a spatio-temporal differentiation of the cortical microtubule array that organizes cytokinesis. *Nova Hedwigia,* **49,** 1–23.

Segal, R. A., Huang, B., Ramanis, Z. & Luck, D. J. L. (1984). Mutant strains of *Chlamydomonas reinhardtii* that move backwards only. *J. Cell Biol.,* **98,** 2026–34.

Selmer, T., Hallmann, A., Schmidt, B., Sumper, M. & Von Figura, K. (1996). The evolutionary conservation of a novel protein modification, the conversion of cysteine to serine semialdehyde in arylsulfatase from *Volvox carteri. Eur. J. Biochem.,* **238,** 341–5.

Senft, W. H., II, Hunchberger, R. A. & Roberts, K. A. (1981). Temperature dependence of growth and phosphorus uptake in two species of *Volvox* (Volvocales, Chlorophyta). *J. Phycol.,* **17,** 323–8.

Sessoms, A. H. (1974). The isolation and characterization of morphogenetic mutants of *Volvox carteri* f. *nagariensis* Iyengar. Ph.D. thesis, University of Virginia.

Sessoms, A. H. & Huskey, R. J. (1973). Genetic control of development in *Volvox*: isolation and characterization of morphogenetic mutants. *Proc. Natl. Acad. Sci. USA,* **70,** 1335–8.

Shapiro, J. A. (1988). Bacteria as multicellular organisms. *Sci. Am.,* **256,** 82–9.

Shaw, W. R. (1894). Pleodorina, a new genus of the Volvocinae. *Bot. Gaz.,* **19,** 279–83.

Shaw, W. R. (1916). Besseyosphaera, a new genus of the Volvocaceae. *Bot. Gaz.,* **61,** 253–4.

Shaw, W. R. (1919). Campbellosphaera, a new genus of the Volvocaceae. *Philip. J. Sci.,* **15,** 439–520.

Shaw, W. R. (1922a). Copelandosphaera, a new genus of the Volvocaceae. *Philip. J. Sci.,* **21,** 207–32.

Shaw, W. R. (1922b). Janetosphaera, a new genus, and two new species of Volvox. *Philip. J. Sci.,* **20,** 477–508.

Shaw, W. R. (1922c). Merrillosphaera, a new genus of the Volvocaceae. *Philip. J. Sci.,* **21,** 87–129.

Shaw, W. R. (1923). Merrillosphaera africana at Manila. *Philip. J. Sci.,* **22,** 185–218.

Sheen, J. (1994). Feedback control of gene expression. *Photosynthesis Res.,* **39,** 427–38.

Sheen, J. Y. & Bogorad, L. (1986). Differential expression of six light-harvesting chlorophyll a/b binding proteins in maize leaf cell types. *Proc. Natl. Acad. Sci. USA,* **83,** 7811–15.

Shepherd, H. S., Ledoight, G. & Howell, S. H. (1983). Regulation of light-harvesting chlorophyll-binding protein (LHCP) mRNA accumulation during the cell cycle in Chlamydomonas reinhardi. *Cell,* **32,** 99–107.

Shoji, W., Inoue, T., Yamamoto, T. & Obinata, M. (1995). MIDA1, a protein associated with Id, regulates cell growth. *J. Biol. Chem.,* **270,** 24818–25.

Sibley, T. H., Herrgesell, P. L. & Knight, A. W. (1974). Density dependent vertical migration in the freshwater dinoflagellate *Peridinium penardii* (Lemm.) Lemm. fo. *californicum* Javorn. *J. Phycol.,* **10**, 475–7.

Siderius, M., Musgrave, A., van den Ende, H., Koerten, H., Cambier, P. & van der Meer, P. (1996). *Chlamydomonas eugametos* (Chlorophyta) stores phosphate in polyphosphate bodies together with calcium. *J. Phycol.,* **32**, 402–9.

Silflow, C. D., Chisholm, R. L., Conner, T. W. & Ranum, L. P. W. (1985). The two α-tubulin genes of *Chlamydomonas reinhardtii* code for slightly different proteins. *Mol. Cell. Biol.,* **5**, 2389–98.

Silver, P. A. & Way, J.C. (1993). Eukaryotic DnaJ homologs and the specificity of Hsp70 activity. *Cell* **74**, 5-6.

Simpson, J., Schell, J., Van Montagu, M. & Herrera-Estrella, L. (1986). The light-inducible and tissue-specific expression of a pea LHCP gene involves an upstream element combining enhancer and silencer-like properties. *Nature,* **323**, 551–3.

Skvortzow, B. W. (1957). New and rare flagellates from Manchuria, Eastern Asia. *Philip. J. Sci.,* **86**, 139–202.

Sluiman, H. (1985). A cladistic evaluation of the lower and higher green plants (Viridiplantae). *Plant Syst. Evol.,* **149**, 217–32.

Smith, A. (1776). *An Inquiry into the Nature and Cause of the Wealth of Nations* (reprinted 1952 as vol. 39 in the Great Books of the Western World series, W. Benton, Chicago).

Smith, B. G. (1907). Volvox for laboratory use. *Amer. Nat.,* **41**, 31–4.

Smith, C. L., Kloc, S. R. & Cantor, C. R. (1988). Pulsed-field gel electrophoresis and the technology of large DNA molecules. In *Genome Analysis: A Practical Approach.* Ed., K. E. Davis. IRL Press, Oxford, pp. 41–72.

Smith, G. M. (1918). The vertical distribution of *Volvox* in the plankton of Lake Monona. *Am. J. Bot.,* **5**, 178–85.

Smith, G. M. (1920). Phytoplankton of the inland lakes of Wisconsin. *Bull. Wisc. Geo. Nat. Hist. Surv.,* **57**, 1–243.

Smith, G. M. (1933). *Fresh Water Algae of the United States.* McGraw-Hill, New York.

Smith, G. M. (1944). A comparative study of the species of *Volvox. Trans. Amer. Microsc. Soc.,* **63**, 265–310.

Sogin, M. L. (1991). Early evolution and the origin of eukaryotes. *Curr. Biol.,* **1**, 457–63.

Solomonson, L. P. & Vennesland, B. (1972). Nitrate reductase and chlorate toxicity in *Chlorella vulgaris* Beijerinck. *Plant Physiol.,* **50**, 421–4.

Sommer, U. & Gliwicz, Z. M. (1986). Long-range vertical migration of *Volvox* in tropical Lake Cahora Bassa (Mozambique). *Limnol. Oceanogr.,* **31**, 650–3.

Sosa, F. M., Ortega, T. & Barea, J. L. (1978). Mutants from *Chlamydomonas reinhardii* affected in their nitrate assimilation capacity. *Plant Sci. Lett.,* **11**, 51–8.

Spallanzani, L. (1769). *Nouvelles Recherches sur les Découvertes Microscopiques et la Génération des Corps Organisés.* London.

Spemann, H. (1927). Neue Arbeiten über Organisatoren in der tierischen Entwicklung. *Naturwissenschaften,* **15**, 946–51.

Starling, D. & Randall, J. (1971). The flagella of temporary dikaryons of *Chlamydomonas reinhardii. Genet. Res.,* **18**, 107–13.

Starr, R. C. (1955). Sexuality in *Gonium sociale* (Dujardin) Warming. *J. Tennessee Acad. Sci.,* **30,** 90–3.

Starr, R. C. (1962). A new species of *Volvulina* Playfair. *Arch. Mikrobiol.,* **42,** 130–7.

Starr, R. C. (1968). Cellular differentiation in *Volvox. Proc. Natl. Acad. Sci. USA,* **59,** 1082–8.

Starr, R. C. (1969). Structure, reproduction, and differentiation in *Volvox carteri* f. *nagariensis* Iyengar, strains HK 9 and 10. *Arch. Protistenkd.,* **111,** 204–22.

Starr, R. C. (1970a). Control of differentiation in *Volvox. Dev. Biol. (Suppl.),* **4,** 59–100.

Starr, R. C. (1970b). *Volvox pocockiae,* a new species with dwarf males. *J. Phycol.,* **6,** 234–9.

Starr, R. C. (1971). Sexual reproduction in *Volvox africanus.* In *Contributions in Phycology.* Ed., B. C. Parker & R. M. Brown, Jr. Allen Press, Lawrence, KS, pp. 59–66.

Starr, R. C. (1972a). Sexual reproduction in *Volvox dissipatrix* (abstract). *Br. J. Phycol.,* **7,** 284.

Starr, R. C. (1972b). A working model for the control of differentiation during the development of the embryo of *Volvox carteri* f. *nagariensis. Soc. Bot. Fr., Memoires,* **1972,** 175–82.

Starr, R. C. (1973a). Apparatus and maintenance. In *Handbook of Phycological Methods.* Ed., J. R. Stein. Cambridge University Press, pp. 171–81.

Starr, R. C. (1973b). Special methods – dry soil samples. In *Handbook of Phycological Methods.* Ed., J. R. Stein. Cambridge University Press, pp. 161–7.

Starr, R. C. (1975). Meiosis in *Volvox carteri* f. *nagariensis. Arch. Protistenkd.,* **117,** 187–91.

Starr, R. C. (1980). Colonial chlorophytes. In *Phytoflagellates.* Ed., E. R. Cox. Elsevier, Amsterdam, pp. 147–63.

Starr, R. C. & Jaenicke, L. (1974). Purification and characterization of the hormone initiating sexual morphogenesis in *Volvox carteri* f. *nagariensis* Iyengar. *Proc. Natl. Acad. Sci. USA,* **71,** 1050–54.

Starr, R. C. & Jaenicke, L. (1988). Sexual induction in *Volvox carteri* f. *nagariensis* by aldehydes. *Sex. Plant Reprod.,* **1,** 28–31.

Starr, R. C. & Jaenicke, L. (1989). Cell differentiation in *Volvox carteri* (Chlorophyceae): the use of mutants in understanding pattern and control. In *Algae as Experimental Systems.* Ed., A. W. Coleman, L. J. Goff & J. R. Stein. Alan R. Liss, New York, pp. 135–47.

Starr, R. C., O'Neil, R. M. & Miller, C. E. (1980). L-Glutamic acid as a mediator of sexual differentiation in *Volvox capensis. Proc. Natl. Acad. Sci. USA,* **77,** 1025–28.

Starr, R. C. & Zeikus, J. A. (1993). UTEX – the culture collection of algae at the University of Texas at Austin. *J. Phycol. (Suppl.),* **29,** 1–106.

Stavis, R. L. & Hirshberg, R. (1973). Phototaxis in *Chlamydomonas reinhardtii. J. Cell Biol.,* **59,** 367–77.

Stein, J. R. (1958a). A morphological and genetic study of *Gonium pectorale. Am. J. Bot.,* **45,** 664–72.

Stein, J. R. (1958b). A morphological study of *Astrephomene gubernaculifera* and *Volvulina steinii. Am. J. Bot.,* **45,** 388–97.

Stein, J. R. (1959). The four-celled species of *Gonium. Am. J. Bot.,* **46,** 366–71.

Stein, J. R. (1965a). On cytoplasmic strands in *Gonium pectorale* (Volvocales). *J. Phycol.,* **1,** 1–5.

Stein, J. R. (1965b). Sexual populations of *Gonium pectorale* (Volvocales). *Am. J. Bot.,* **52,** 379–88.

Stein, J. R. (1966a). Effect of temperature on sexual populations of *Gonium pectorale* (Volvocales). *Am. J. Bot.,* **53,** 941–4.

Stein, J. R. (1966b). Growth and mating of *Gonium pectorale* (Volvocales) in defined media. *J. Phycol.,* **2,** 23–8.

Stewart, K. D. & Mattox, K. R. (1975). Comparative cytology, evolution and classification of the green algae with some consideration of the origins of other organisms with chlorophylls a and b. *Bot. Rev.,* **41,** 104–35.

Stewart, K. D. & Mattox, K. R. (1978). Structural evolution in the flagellated cells of green algae and land plants. *BioSystems,* **10,** 145–52.

Stewart, K. D. & Mattox, K. R. (1980). Phylogeny of phytoflagellates. In *Developments in Marine Biology,* Vol. 2. Ed., E. R. Cox. Elsevier-North Holland, Amsterdam, pp. 433–62.

Stouthamer, A. H. (1976). Biochemistry and genetics of nitrate reductase in bacteria. *Adv. Microb. Physiol.,* **14,** 315–75.

Strøm, K. M. (1925). Studies in the ecology and geographical distribution of fresh water algae and plankton. *Rev. Algol.,* **1,** 127–55.

Summers, F. M. (1938). Some aspects of normal development in the colonial ciliate *Zoothamnion alternans. Biol. Bull.,* **74,** 117–29.

Sumper, M. (1979). Control of differentiation in *Volvox carteri.* A model explaining pattern formation during embryogenesis. *FEBS Lett.,* **107,** 241–6.

Sumper, M. (1984). Pattern formation during embryogenesis of the multicellular organism *Volvox.* In *Pattern Formation. A Primer in Developmental Biology.* Ed., G. M. Malacinski & S. V. Bryant. Macmillan, New York, pp. 197-212.

Sumper, M., Berg, E., Wenzl, S. & Godl, K. (1993). How a sex pheromone might act at a concentration below 10^{-16} M. *EMBO J.,* **12,** 831–6.

Sumper, M. & Wenzl, S. (1980). Sulphation–desulphation of a membrane component proposed to be involved in control of differentiation in *Volvox carteri. FEBS Lett.,* **114,** 307–12.

Susek, R. E. & Chory, J. (1992). A tale of two genomes: Role of a chloroplast signal in coordinating nuclear and plastid genome expression. *Aust. J. Plant Physiol.,* **19,** 387–99.

Swinton, D. C. & Hanawalt, P. C. (1972). In vivo specific labeling of *Chlamydomonas* chloroplast DNA. *J. Cell. Biol.,* **54,** 592–7.

Taillon, B. E., Adler, S. A., Suhan, J. P. & Jarvik, J. W. (1992). Mutational analysis of centrin: an EF-hand protein associated with three distinct contractile fibers in the basal body apparatus of *Chlamydomonas. J. Cell Biol.,* **119,** 1613–24.

Taillon, B. E. & Jarvik, J. W. (1995). Central helix mutations in the centrosome-associated EF-hand protein centrin. *Protoplasma,* **189,** 203–15.

Takeuchi, Y., Dotson, M. & Keen, N. T. (1992). Plant transformation: a simple particle bombardment device based on flowing helium. *Plant Mol. Biol.,* **18,** 835–9.

Tam, L.-W. & Kirk, D. L. (1991a). Identification of cell-type-specific genes of *Volvox carteri* and characterization of their expression during the asexual life cycle. *Dev. Biol.*, **145**, 51–66.

Tam, L.-W. & Kirk, D. L. (1991b). The program for cellular differentiation in *Volvox carteri* as revealed by molecular analysis of development in a gonidialess/somatic regenerator mutant. *Development*, **112**, 571–80.

Tam, L.-W., Stamer, K. A. & Kirk, D. L. (1991). Early and late gene expression programs in developing somatic cells of *Volvox carteri*. *Dev. Biol.*, **145**, 67–76.

Tamm, S. (1994). Ca^{2+} channels and signaling in cilia and flagella. *Trends Cell Biol.*, **4**, 305–10.

Tautvydis, K. J. (1976). Evidence for chromosomal endoreduplication in *Eudorina californica*, a colonial alga. *Differentiation*, **5**, 35–42.

Tautvydis, K. J. (1978). Isolation and characterization of an extracellular hydroxy-proline-rich glycoprotein and a mannose-rich polysaccharide from *Eudorina californica* (Shaw). *Planta*, **140**, 213–20.

Taylor, F. J. R. (1978). Problems in the development of an explicit hypothetical phylogeny of the lower eukaryotes. *BioSystems*, **10**, 67–89.

Taylor, M. G., Floyd, G. L. & Hoops, H. J. (1985). Development of the flagellar apparatus and flagellar position in the colonial green alga *Platydorina caudata* (Chlorophyceae). *J. Phycol.*, **21**, 533–46.

Terry, O. P. (1906). Galvanotropism of *Volvox*. *Am. J. Physiol.*, **15**, 235–43.

Terzaghi, W. B. & Cashmore, A. R. (1995). Light-regulated transcription. *Ann. Rev. Plant Physiol. Plant Mol. Biol.*, **46**, 445–74.

Tetík, K. & Zadrazil, S. (1982). Characterization of DNA of the alga *Chlamydomonas geitleri* Ettl. *Biol. Plant.*, **24**, 202–10.

Tizer, M. M. (1973). Diurnal periodicity in the phytoplankton assemblage of a high mountain lake. *Limnol. Oceanogr.*, **18**, 15–30.

Toby, A. L. & Kemp, C. L. (1975). Mutant enrichment in the colonial alga, *Eudorina elegans*. *Genetics*, **81**, 243–51.

Tomsett, A. B. & Garrett, R. H. (1980). The isolation and characterization of mutants defective in nitrate assimilation in *Neurospora crassa*. *Genetics*, **93**, 649–60.

Triemer, R. E. & Brown, R. M., Jr. (1974). Cell division in *Chlamydomonas moewusii*. *J. Phycol.*, **10**, 419–33.

Tschochner, H., Lottspeich, F. & Sumper, M. (1987). The sexual inducer of *Volvox carteri*: its purification, chemical characterization and identification of its gene. *EMBO J.*, **6**, 2203–7.

Tucker, R. G. & Darden, W. H. (1972). Nucleic acid synthesis during the vegetative life cycle of *Volvox aureus* M5. *Arch. Mikrobiol.*, **84**, 87–94.

Uspenski, E. E. & Uspenskaja, W. J. (1925). Reinkultur und ungeschlechtliche Fortpflanzung des *Volvox minor* und *Volvox globator* in einer synthetischen Nährlösung. *Zeitsch. Bot.*, **17**, 273–308.

Utermöhl, H. (1924). Tiefenwanderurungen bei *Volvox*. *Schrft. Süsswasser Meereskunde*, **9**, 731–54.

Utermöhl, H. (1925). Limologische Phytoplanktonstudien. Die Besiedelung ostholsteinischer Seen mit Schwebpflanzen. *Arch. Hydrobiol. (Suppl.)*, **5**, 1–527.

Vande Berg, W. J. & Starr, R. C. (1971). Structure, reproduction and differentiation in *Volvox gigas* and *Volvox powersii*. *Arch. Protistenkd.*, **113**, 195–219.

Van Eldik, L. G., Piperno, G. & Watterson, D. M. (1980). Similarities and dissimilarities between calmodulin and a *Chlamydomonas* flagellar protein. *Proc. Natl. Acad. Sci. USA,* **77,** 4779–83.

van Leeuwenhoek, A. (1700). Concerning the worms in sheeps, livers, gnats and animalicula in the excrement of frogs. *Phil. Trans. R. Soc. Lond.,* **22,** 509–18.

Viamontes, G. I., Fochtmann, L. J. & Kirk, D. L. (1979). Morphogenesis in Volvox: Analysis of critical variables. *Cell,* **17,** 537–50.

Viamontes, G. I. & Kirk, D. L. (1977). Cell shape changes and the mechanism of inversion in *Volvox. J. Cell Biol.,* **75,** 719-730.

Waffenschmidt, S., Knittler, M. & Jaenicke, L. (1990). Characterization of a sperm lysin of *Volvox carteri. Sex. Plant Reprod.,* **3,** 1–6.

Wainwright, P. O., Hinkle, G., Sogin, M. L. & Stickel, S. K. (1993). Monophyletic origins of the Metazoa: An evolutionary link with fungi. *Science,* **260,** 340–2.

Wakarchuk, W. W., Müller, F. W. & Beck, C. F. (1992). Two GC-rich elements of *Chlamydomonas reinhardtii* with complex arrangements of directly repeated sequence motifs. *Plant Mol. Biol.,* **18,** 143–6.

Watson, M. W. (1975). Flagellar apparatus, eyespot, and behavior of *Microthamnion kuetzingianum* (Chlorophyceae) zoospores. *J. Phycol.,* **11,** 439–48.

Weinheimer, T. (1983). Cellular development in the green alga *Volvox carteri. Cytobios,* **36,** 161–73.

Weismann, A. (1892a). *Essays on Heredity and Kindred Biological Problems.* Clarendon Press, Oxford.

Weismann, A. (1892b). *The Germ-Plasm: A Theory of Heredity.* Charles Scribner's Sons, New York.

Weisshaar, B., Gilles, R., Moka, R. & Jaenicke, L. (1984). A high frequency mutation starts sexual reproduction in *Volvox. Z. Naturforsch.,* **39c,** 1159–62.

Wenzl, S. & Sumper, M. (1979). Evidence for membrane-mediated control of differentiation during embryogenesis of *Volvox carteri. FEBS Lett.,* **107,** 247–9.

Wenzl, S. & Sumper, M. (1981). Sulfation of a cell surface glycoprotein correlates with the developmental program during embryogenesis of *Volvox carteri. Proc. Natl. Acad. Sci. USA,* **78,** 3716–20.

Wenzl, S. & Sumper, M. (1982). The occurrence of differently sulphated cell surface glycoproteins correlates with defined developmental events in *Volvox. FEBS Lett.,* **143,** 311–15.

Wenzl, S. & Sumper, M. (1986a). Early event of sexual induction in *Volvox:* chemical modification of the extracellular matrix. *Dev. Biol.,* **115,** 119–28.

Wenzl, S. & Sumper, M. (1986b). A novel glycosphingolipid that may participate in embryo inversion in Volvox carteri. *Cell,* **46,** 633–9.

Wenzl, S. & Sumper, M. (1987). Pheromone-inducible glycoproteins of the extracellular matrix of *Volvox* and their possible role in sexual induction. In *Algal Development (Molecular and Cellular Aspects).* Ed., W. Wiessner, D. G. Robinson & R. C. Starr. Springer-Verlag, Berlin, pp. 58–65.

Wenzl, S., Thym, D. & Sumper, M. (1984). Development-dependent modification of the extracellular matrix by a sulphated glycoprotein in *Volvox carteri. EMBO J.,* **3,** 739–44.

West, G. S. (1910). Some new African species of *Volvox. J. Quek. Micr. Club, ser. 2,* **11,** 99–104.

West, G. S. (1918). A further contribution to our knowledge of the two African species of *Volvox. J. Quek. Micr. Club, ser. 2,* **13,** 425–8.

Wetherbee, R., Platt, S. J., Beech, P. L. & Pickett-Heaps, J. D. (1988). Flagellar transformation in the heterokont *Epipyxis pulchra* (Chrysophyceae): direct observations using image enhanced light microscopy. *Protoplasma,* **145,** 47–54.

Wetherell, D. F. & Kraus, R. W. (1975). X-ray induced mutations in *Chlamydomonas eugametos. Am. J. Bot.,* **44,** 609–19.

Wheatley, D. N. (1982). *The Centriole: A Central Enigma of Cell Biology.* Elsevier, Amsterdam.

Willadsen, P. & Sumper, M. (1982). Sulphation of a cell surface glycoprotein from *Volvox carteri.* Evidence for a membrane-bound sulfokinase working with PAPS. *FEBS Lett.,* **139,** 113–16.

Williams, B. D., Velleca, M. A., Curry, A. M. & Rosenbaum, J. L. (1989). Molecular cloning and sequence analysis of the *Chlamydomonas* gene coding for radial spoke protein 3: flagellar mutation *pf*-14 is an ochre allele. *J. Cell Biol.,* **109,** 235–45.

Williams, J. G. K., Hanafey, M. K., Rafalski, J. A. & Tingley, S. V. (1993). Genetic analysis using random amplified polymorphic DNA markers. *Methods Enzymol.,* **218,** 704–40.

Williams, J. G. K., Kubelik, A. R., Livak, K. J., Rafalski, J. A. & Tingley, S. V. (1990). DNA polymorphisms amplified by arbitrary primers are useful as genetic markers. *Nucl. Acids Res.,* **18,** 6531–5.

Williamson, W. C. (1851). On the *Volvox globator. Trans. Lit. Phil. Soc. Manchester,* ser. 2, **9,** 321–39.

Williamson, W. C. (1853). Further elucidations of the structure of *Volvox globator. Q. J. Microsc. Soc.,* **1,** 45–56.

Wingrove, J. A. & Gober, J. W. (1995). The molecular basis of asymmetric division in *Caulobacter crescentus. Sem. Dev. Biol.,* **6,** 325–33.

Wirt, N. S. (1974a). Galvanic stimulation of *Volvox globator.* I. Electroosmotic patterns associated with ion exchange properties. *J. Protozool.,* **21,** 121–5.

Wirt, N. S. (1974b). Galvanic stimulation of *Volvox globator.* II. Mechanism of galvanic response, an effect of calcium displacement. *J. Protozool.,* **21,** 126–8.

Witman, G. (1993). *Chlamydomonas* phototaxis. *Trends Cell Biol.,* **3,** 403–8.

Woessner, J. P. & Goodenough, U. W. (1994). Volvocine cell walls and their constituent glycoproteins: an evolutionary perspective. *Protoplasma,* **181,** 245–58.

Woessner, J. P., Molendijk, A. J., van Egmond, P., Klis, F. M., Goodenough, U. W. & Haring, M. A. (1994). Domain conservation in several volvocalean cell wall proteins. *Plant Mol. Biol.,* **26,** 947–60.

Wolle, F. (1887). *Fresh-water Algae of the United States.* Comenius Press, Bethlehem, PA.

Wray, J. L. (1986). The molecular genetics of higher plant nitrate assimilation. In *A Genetic Approach to Plant Biochemistry.* Ed., A. D. Blonstein & P. J. King. Springer-Verlag, Berlin, pp. 101–57.

Wright, R. L. (1985). Functional roles of basal body-associated fibrous structures in *Chlamydomonas reinhardtii:* a genetic and ultrastructural analysis. Ph.D. thesis, Carnegie-Mellon University.

Wright, R. L., Adler, S. A., Spanier, J. G. & Jarvik, J. W. (1989). Nucleus–basal body connector in *Chlamydomonas.* Evidence for a role in basal body segregation and against essential roles in mitosis or in determining cell polarity. *Cell Motil. Cytoskel.,* **14,** 516–26.

Wright, R. L., Chojnacki, B. & Jarvik, J. W. (1983). Abnormal basal body number, location and orientation in a striated fiber-defective mutant of *Chlamydomonas reinhardtii*. *J. Cell Biol.,* **96,** 1697–707.

Wright, R. L., Salisbury, J. & Jarvik, J. W. (1985). A nucleus–basal body connector in *Chlamydomonas reinhardtii* that may function in basal body localization or segregation. *J. Cell Biol.,* **101,** 1903–12.

Yates, I., Darley, M. & Kochert, G. (1975). Separation of cell types in synchronized cultures of *Volvox carteri*. *Cytobios,* **12,** 211–23.

Yates, I. & Kochert, G. (1972). Patterns of ribosomal RNA and DNA synthesis in synchronized cultures of *Volvox carteri*. *J. Phycol.,* **8,** 16.

Yates, I. & Kochert, G. (1976). Nucleic acid and protein differences in *Volvox carteri* cell types. *Cytobios,* **15,** 7–21.

Youngblom, J., Schloss, J. A. & Silflow, C. D. (1984). The two β-tubulin genes of *Chlamydomonas reinhardtii* code for identical proteins. *Mol. Cell. Biol.,* **4,** 2686–96.

Zechman, F. W., Theriot, E. C., Zimmer, E. A. & Chapman, R. L. (1990). Phylogeny of the Ulvophyceae (Chlorophyta): cladistic analysis of nuclear-encoded rRNA sequence data. *J. Phycol.,* **26,** 700–10.

Zeig, J., Hillmen, M. & Simon, M. (1978). Regulation of gene expression by site-specific inversion. *Cell,* **15,** 237–44.

Zeikus, J. A., Darden, W. H. & Brooks, D. (1980). Sexual spheroid formation in *Volvox carteri* f. *nagariensis* by phenolic extracts. *Microbios,* **28,** 173–84.

Zeikus, J. A. & Starr, R. C. (1980). The genetics and physiology of non-inducibility in *Volvox carteri* f. *nagariensis* Iyengar. *Arch. Protistenkd.,* **123,** 127–61.

Zhang, D. (1996). Regulation of nitrate assimilation in *Chlamydomonas reinhardtii*. Ph.D. thesis, University of Minnesota.

Zhang, Y. (1994). Localization of rhodopsin by immunofluorescence microscope in *Chlamydomonas reinhardtii*. *Biochem. Biophys. Res. Commun.,* **205,** 1025–35.

Zimmermann, W. (1921). Zur Entwicklungsgeschichte und Zytologie von *Volvox*. *Jahrb. Wiss. Bot.,* **60,** 256–94.

Zimmermann, W. (1923). Helgoländer Meeresalgen. I-VI. *Ber. Deutsch. Bot. Ges.,* **41,** 285–91.

Zimmermann, W. (1925). Die ungeschlechtliche Entwicklung von *Volvox*. *Naturwissenschaften,* **19,** 397–401.

Index